Big Data in Multimodal Medical Imaging

Big Data in Multimodal Medical Imaging

Edited by
Ayman El-Baz and Jasjit S. Suri

CRC Press
Taylor & Francis Group
Boca Raton London New York

CRC Press is an imprint of the
Taylor & Francis Group, an **informa** business

A CHAPMAN & HALL BOOK

CRC Press
Taylor & Francis Group
6000 Broken Sound Parkway NW, Suite 300
Boca Raton, FL 33487-2742

First issued in paperback 2021

© 2020 by Taylor & Francis Group, LLC
CRC Press is an imprint of Taylor & Francis Group, an Informa business

No claim to original U.S. Government works

ISBN 13: 978-1-03-208727-6 (pbk)
ISBN 13: 978-1-138-50453-0 (hbk

Library of Congress Cataloging-in-Publication Data

Names: El-Baz, Ayman S., editor. | Suri, Jasjit S., editor.
Title: Big data in multimodal medical imaging / edited by Ayman El-Baz and Jasjit S. Suri.
Description: Boca Raton : CRC Press, [2020] | Includes bibliographical references and index. | Summary: "There is an urgent need to develop and integrate new statistical, mathematical, visualization, and computational models with the ability to analyze Big Data in order to retrieve useful information to aid clinicians in accurately diagnosing and treating patients"-- Provided by publisher.
Identifiers: LCCN 2019025184 (print) | LCCN 2019025185 (ebook) | ISBN 9781138504530 (hardback) | ISBN 9781315146102 (ebook)
Subjects: LCSH: Diagnostic imaging--Data processing. | Big data.
Classification: LCC RC78.7.D53 B54 2020 (print) | LCC RC78.7.D53 (ebook) | DDC 616.07/54--dc23
LC record available at https://lccn.loc.gov/2019025184
LC ebook record available at https://lccn.loc.gov/2019025185

Visit the Taylor & Francis Web site at
http://www.taylorandfrancis.com

and the CRC Press Web site at
http://www.crcpress.com

Publisher's Note
The publisher has gone to great lengths to ensure the quality of this reprint but points out that some imperfections in the original copies may be apparent.

With love and affection to my mother and father,
whose loving spirit sustains me still

Ayman El-Baz

To my late loving parents, immediate family, and children

Jasjit S. Suri

Contents

Preface

This book covers the state-of-the-art approaches for big data in the field of medical image analysis and healthcare. Big data refers to the amalgamation and processing of huge datasets that are composed of different data types (e.g. clinical, genomic, imaging, pathological, etc.) and have rapidly become more massive and complex, particularly with the advent of new technologies. Broadly speaking, we are considering data of high volume that are obtained from a variety of sources and processed at high speed. Currently, big data is a very active area of research and is significantly dominating in the medical research field, where it achieved very promising results with medical imaging, content-based image retrieval, healthcare automation, diagnosis, and prediction of many diseases.

Moreover, big data analysis is now trending in many commercial applications in healthcare, which include, but are not limited to, designing computer aided diagnosis systems that could be used in helping physicians in diagnosing and selecting the optimal treatment plan for many diseases. Various studies are discussed in this book to demonstrate how powerful big data is in the medical field. This involves radiomics, multimodal image fusion, image segmentation, and registration, as well as applications to computational health information, pre-clinical dementia, cardiovascular diseases, acute renal rejection, prostate cancer, and diabetic retinopathy,

In summary, the main aim of this book is to help advance scientific research within the broad field of big data. The book focuses on major trends and challenges in this area, and it presents work aimed at identifying the new achievements and their use in biomedical analysis.

Ayman El-Baz
Jasjit S. Suri

Editors

Ayman El-Baz is a Professor, University Scholar, and Chair of the Bioengineering Department at the University of Louisville, Kentucky. Dr. El-Baz earned his B.Sc. and M.Sc. degrees in electrical engineering in 1997 and 2001, respectively. He earned his Ph.D. in electrical engineering from the University of Louisville in 2006. In 2009, Dr. El-Baz was named a Coulter Fellow for his contributions to the field of biomedical translational research. Dr. El-Baz has 17 years of hands-on experience in the fields of bio-imaging modeling and non-invasive computer-assisted diagnosis systems. He has authored or coauthored more than 500 technical articles (133 journals, 25 books, 57 book chapters, 212 refereed-conference papers, 143 abstracts, and 27 US patents and disclosures).

Jasjit S. Suri is an innovator, scientist, visionary, industrialist, and an internationally known world leader in biomedical engineering. Dr. Suri has spent over 25 years in the field of biomedical engineering/devices and its management. He received his Ph.D. from the University of Washington, Seattle, and his Business Management Sciences degree from Weatherhead, Case Western Reserve University, Cleveland, Ohio. Dr. Suri was awarded the President's Gold medal in 1980 and made Fellow of the American Institute of Medical and Biological Engineering for his outstanding contributions. In 2018, he was awarded the Marquis Life Time Achievement Award for his outstanding contributions and dedication to medical imaging and its management.

Contributors

Behnaz Abdollahi
Department of Electrical and Computer
 Engineering
University of Louisville
Louisville, Kentucky

Mohamed Abou El-Ghar
Radiology Department, Urology and
 Nephrology Center
Mansoura University
Mansoura, Egypt

Ahmed Aboelfetouh
Information Systems Department,
 Faculty of Computers and Information
Mansoura University
Mansoura, Egypt

Ismail Ben Ayed
École de technologie supérieure
Montreal, Québec

Mohammed A. Badawy
Radiology Department, Urology and
 Nephrology Center
Mansoura University
Mansoura, Egypt

Ashraf M. Bakr
Pediatric Nephrology Unit, Mansoura
 University Children's Hospital
Mansoura University
Mansoura, Egypt

Gaurav Bhatnagar
Department of Mathematics
Indian Institute of Technology Jodhpur
Jodhpur, India

Shu-Ching Chen
School of Computing & Info. Sciences
Florida International University
Miami, Florida

Lisa Duff
Biomedical Imaging Science Department,
 Leeds Institute of Cardiovascular and
 Metabolic Medicine
University of Leeds
Leeds, United Kingdom

Amy C. Dwyer
Kidney Transplantation–Kidney Disease
 Center
University of Louisville
Louisville, Kentucky

Nabila Eladawi
Bioengineering Department
University of Louisville
Louisville, Kentucky

Magdi El-Azab
Mathematics and Physical Engineering
 Department
Mansoura University
Mansoura, Egypt

Kost Elisevich
Department of Clinical Neurosciences
Spectrum Health
Grand Rapids, Michigan

Adel Elmaghraby
Computer Engineering and Computer
 Science Department
University of Louisville
Louisville, Kentucky

Mohammed Elmogy
Bioengineering Department
University of Louisville
Louisville, Kentucky

Ahmed ElTanboly
Bioengineering Department
University of Louisville
Louisville, Kentucky
and
Mathematics Department
Mansoura University
Mansoura, Egypt

Ruogu Fang
Department of Biomedical Engineering
University of Florida
Gainesville, Florida

Aaron Fenster
Robarts Research Institute
Western University
London, Ontario

Benjamin L. Franc
Department of Radiology
Stanford University
Stanford, California

Hermann B. Frieboes
Department of Electrical and Computer
    Engineering
University of Louisville
Louisville, Kentucky
and
Department of Bioengineering
University of Louisville
Louisville, Kentucky
and
James Graham Brown Cancer Center
University of Louisville
Louisville, Kentucky

Mohammed Ghazal
Electrical and Computer Engineering
    Department
Abu Dhabi University
Abu Dhabi, United Arab Emirates

Guruprasad Giridharan
Bioengineering Department
University of Louisville
Louisville, Kentucky

Stathis Hadjidemetriou
Department of Electrical Engineering
Cyprus University of Technology
Limassol, Cyprus

Hassan Hajjdiab
Electrical and Computer Engineering
    Department
Abu Dhabi University
Abu Dhabi, United Arab Emirates

Mohammad-Parsa Hosseini
Bioengineering Department
School of Engineering
Santa Clara University
Santa Clara, California

S. S. Iyengar
School of Computing & Info. Sciences
Florida International University
Miami, Florida

Robert Keynton
Bioengineering Department
University of Louisville
Louisville, Kentucky

Ashraf Khalil
Electrical and Computer Engineering
    Department
Abu Dhabi University
Abu Dhabi, United Arab Emirates

Yi Lao
Department of Radiation Oncology
Cedars Sinai Medical Center
Los Angeles, California
and
Department of Radiology
Children's Hospital Los Angeles
Los Angeles, California

Aaron Lau
Bioengineering Department
School of Engineering
Santa Clara University
Santa Clara, California

Natasha Lepore
Department of Radiology
Children's Hospital Los Angeles
Los Angeles, California
and
Department of Biomedical Engineering
University of Southern California
Los Angeles, California

Gengbo Liu
Department of Computer Engineering
    and Sciences
Florida Institute of Technology
Melbourne, Florida

Zheng Liu
School of Electrical Engineering and
    Computer Science
University of Ottawa
Ottawa, Ontario
and
National Research Council Canada
Ottawa, Ontario

Ali Mahmoud
BioImaging Laboratory
Bioengineering Department
University of Louisville
Louisville, Kentucky

Debasis Mitra
Department of Computer Engineering
    and Sciences
Florida Institute of Technology
Melbourne, Florida

Ismini Papageorgiou
Institute of Diagnostic and Interventional
    Radiology
University Hospital of Jena
Jena, Germany
and
Institute of Radiology
Südharz Hospital Nordhausen
Nordhausen, Germany

Samira Pouyanfar
School of Computing and Information
    Sciences
Florida International University
Miami, Florida

Islam Reda
Bioengineering Department
University of Louisville
Louisville, Kentucky

Alaa Riad
Information Systems Department,
    Faculty of Computers and Information
Mansoura University
Mansoura, Egypt

Ahmed Shalaby
BioImaging Laboratory
Bioengineering Department
University of Louisville
Louisville, Kentucky

Mohamed Shehata
BioImaging Laboratory
Bioengineering Department
University of Louisville
Louisville, Kentucky

Youngho Seo
Department of Radiology and Biomedical
    Imaging
University of California, San Francisco
San Francisco, California

Hamid Soltanian-Zadeh
Control and Intelligent Processing Center
    of Excellence (CIPCE)
School of Electrical and Computer
    Engineering
University of Tehran
Tehran, Iran
and
Radiology and Research Administration
Henry Ford Health System
Detroit, Michigan

Andrew Switala
Bioengineering Department
University of Louisville
Louisville, Kentucky

Fatma Taher
Computer Engineering
Zayed University
Dubai, United Arab Emirates

Jianqiao Tian
Department of Biomedical Engineering
University of Florida
Gainesville, Florida

Sinchai Tsao
Department of Radiology
Children's Hospital Los Angeles
Los Angeles, California
and
Department of Biomedical Engineering
University of Southern California
Los Angeles, California

Charalampos Tsoumpas
Biomedical Imaging Science Department
Leeds Institute of Cardiovascular and
    Metabolic Medicine
University of Leeds
Leeds, United Kingdom

Q.M. Jonathan Wu
Department of Electrical and Computer
    Engineering
University of Windsor
Windsor, Ontario

Yao Xiao
Department of Biomedical Engineering
University of Florida
Gainesville, Florida

Yimin Yang
School of Computing and Information
    Sciences
Florida International University
Miami, Florida

Jing Yuan
School of Mathematics and Statistics
Xidian University
Xi'an, China

Acknowledgments

The completion of this book could not have been possible without the participation and assistance of so many people whose names may not all be enumerated. Their contributions are sincerely appreciated and gratefully acknowledged. However, the editors would like to express their deep appreciation and indebtedness particularly to Dr. Ali H. Mahmoud, Yaser ElNakieb, Omar Dekhil, and Mohamed Ali for their endless support.

Ayman El-Baz
Jasjit S. Suri

Chapter 1

Multimodal Imaging Radiomics and Machine Learning

Gengbo Liu, Youngho Seo, Debasis Mitra, and Benjamin L. Franc

1.1 Introduction

According to Gerlinger et al. [1], single tumor biopsy samples can cause underestimation of intratumor heterogeneity. The heterogeneity in the gene, cell and tissue of the solid tumor limited the accuracy and representation of invasive detection results. In order to develop the concept of personalized medicine and attain a better understanding of tumor heterogeneity noninvasively, the concept of radiomics was proposed by Lambin et al. [2] in 2012. A suffix of "omics", such as genomics, refers to the objects of study of a certain field and its relationships to biology and medicine. As genomics attempts to relate genes to phenotype, radiomics studies relate medical images to phenotypes of

1

disease. Much of radiomics literature has focused on applications in oncologic medical imaging which presents a noninvasive, often quantitative observation of the overall shape of the tumor, the tumor development process and treatment response monitoring, providing a reliable mechanism to quantify tumor heterogeneity. Lambin et al. [2] hypothesized that microscopic-level gene or protein pattern changes can be expressed in macroscopic imaging features acquired during medical imaging. Therefore, Lambin et al. proposed that the high-throughput extraction of large amounts of image features from radiographic images would be able to capture intratumoral heterogeneity in a noninvasive way. Compared to traditional radiology practice, which is primarily based on visual interpretation and simple quantitative measurements, radiomics can dig deeper into data contained within medical images and potentially provide further objective support for clinical decisions. Due to the large number of image features generated from radiomics, many machine learning algorithms are ideal tools to explore the relationship between radiomics and tumor diagnosis, treatment and prognosis [3, 4, 5].

Four general steps of an ideal machine learning model over radiomic features have been described in the literature [3, 6] and are summarized in Figure 1.1: (a) high-quality standardized imaging data acquisition; (b) region of interest image segmentation; (c) high-throughput feature extraction from images; (d) clinical predictive model establishment. Machine learning algorithms were mainly used in the fourth step to predict the patient outcomes. In the following sections, the first three steps will be described briefly and the fourth step will be described in detail.

1.2 High-Quality Standardized Imaging Data Acquisition

The radiomics approach is based on machine learning a large amount of data. However, maintaining the same standard of image quality in diverse datasets is a problem. High-quality standardized image data acquisition is near impossible. This is owing to the fact that the parameter settings during image acquisition and reconstruction may be different between exams as well as sites, and there is no uniform standard. Even on the same equipment, contrast agent dose, image pixel size and slice thickness, filters used in the

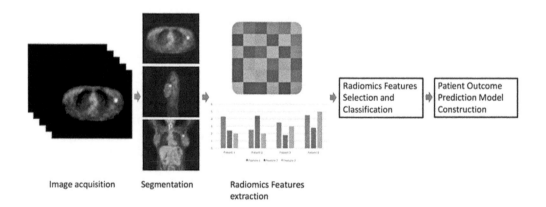

Image acquisition   Segmentation   Radiomics Features
extraction

FIGURE 1.1: Schematic diagram of workflow.

image reconstruction algorithm, etc., influence the quality of images. For this reason, many standardized image datasets are being made available for the research community. For example, the National Institutes of Health (NIH) and the National Cancer Institute (NCI) have established the Lung Image Database Consortium (LIDC) [7], The Cancer Genome Atlas (TCGA) [8] and The Cancer Imaging Archive (TCIA) [9], etc., covering medical images of the lung, brain, breast, prostate and other organs.

1.3 Segmentation of the Region of Interest from the Image

Image segmentation directly determines the accuracy of the extracted feature data. Image segmentation methods can be classified as manual, semi-automatic and fully automatic methods. The first two segmentation methods are applied widely, and the last method is still in the development stage. Manual segmentation has the advantage of high accuracy for objects with clear boundaries and is also suitable for irregular objects with blurred boundaries but has the disadvantage of being time-consuming and inefficient. This disadvantage does not meet the requirement of high-throughput data extraction from mass datasets, needed for a Big-Data radiomics approach. Semi-automatic and automatic segmentation are able to provide high-throughput data processing and have a higher reproducibility and consistency. Semi-automatic segmentation refers to the process whereby automatic segmentation is aided by manual interventions at different stages of segmentation. The semiautomatic segmentation algorithms include region-growing methods, volumetric CT-based segmentation (e.g., 3D-Slicer) [10], graph cuts-based method [11] and active contours (snake) algorithms [12]. And fully automatic segmentation refers to the process whereby segment boundaries of objects are assigned fully automatically by software. The automatic segmentation applications include white matter hyperintensities automated segmentation algorithm [13], brain tumor image analysis [14] and automatic MRI bone segmentation [15], etc.

1.4 High-Throughput Feature Extraction

Many authors have classified radiomic features in different ways [16, 17]. According to Aerts [18], radiomic features can be grouped into four groups: (1) first-order statistics; (2) shape- and size-based features; (3) second-order (long range) features; and (4) wavelet processing-based features. Here, we follow the groups classified by Aerts et al. [18].

1.4.1 First-Order Features with Image Statistics

First-order statistics features (formulas in Appendix A.1), also called image intensity features, describe information of an image related to the pixel values distribution in gray level [17]. Pixel values may be actual intensities on image or estimated.

First-order statistics features are not texture features because they ignore the spatial interaction between image pixels, whereas the texture features calculation considers the relationships between neighborhood pixels at a distance. Different first-order statistics features have been widely applied in estimating tumor heterogeneity. In a PET/CT study [19] evaluating changes in the heterogeneity of conventional parameters and first-order statistics of rectal tumors on PET/CT before and after treatment, Bundschuh et al. found that coefficient of variation (COV, a first-order feature) predicted response more accurately than conventional parameters. In a dynamic contrast-enhanced magnetic resonance imaging (DCE-MRI) study [20], first-order statistics features on images at different time points were evaluated on lung tumors in terms of temporal change, optimal time of imaging for analysis and prognostic potential. The results showed that standard deviation and entropy plateaued at 30–60 seconds after contrast administration, but skewness and kurtosis decreased very fast. Standard deviation and entropy showed high predictive ability of survival on the univariate Cox regression model, and entropy showed significant predictive ability of 2 years of progression-free survival on the multivariate Cox regression model.

1.4.2 Shape- and Size-Based Features

Shape and size of region of interest have always been used by radiologists to predict tumor phenotypes. Shape- and size-based features (formulas in Appendix A.2) are significant in some specific types of cancer. In one FDG-PET study, Naqa et al. [21] evaluated shape and texture features from PET images to predict cervix cancer and head and neck cancer patients' responses to treatment. The results showed that shape-based metrics had a higher failure risk related to the categorical prediction of treatment in head and neck cases than that from commonly used standardized uptake value (SUV) descriptive statistics. And the texture features are more relevant in the cervix cancer. Shape- and size-based features are also used to quantify the spatial changes of patients under Parkinson's disease on PET images and to classify patients between healthy controls and Parkinson's disease [22]. In another PET/CT study [23], the shape-based feature value of squamous cell lung carcinoma (SQCLC) patients who experienced recurrence was significantly higher than the rest of the patients.

1.4.3 Second-Order Features

Second-order features indicate long range interaction of pixel or voxel values in an image. These include gray level cooccurrence matrix (GLCM), gray level size zone matrix (GLSZM) and gray level run length matrix (GLRLM). As one of the important second-order features, the mathematical meaning and application of GLCM are discussed in this section.

The GLCM (formulas in Appendix A.3) is based on the second-order statistical combined conditional probability density of the image that considers the spatial relationship of pixels. Haralick et al. [24] first proposed this matrix and 14 corresponding statistics and showed those statistics reflect the information regarding the relative position of the various gray levels over the images. Although GLCM features have a good ability to distinguish different textures, GLCM computation is time consuming and computationally expensive because it needs to generate GLCM first and then extract 14 texture features. Therefore, many researchers attempt to improve the computation time in several directions. Ulaby et al. [25, 26] reported that many features in GLCM are correlated with each other and selected only four independent texture features, namely, energy, entropy, contrast and correlation. In another study [27], the results show that two parameters, energy and contrast, are considered to be the most efficient for discriminating different textures among the most relevant of 14 originally proposed parameters. In another study, to simplify the computation of GLCM, Bo et al. [28] proved that GLCM tends to a constant for different distances and angles when the distance is large enough. As one of the improvements on GLCM, gray level cooccurrence linked list (GLCLL) [29], which stores only the nonzero cooccurring probabilities, has better computational speed [30].

Many researchers have utilized GLCM features to analyze multimodal medical images. In one study, Hatt et al. [31] investigated the association between the metabolically active tumor volume (MATV) of esophageal and non-small cell lung cancer (NSCLC) through texture features generated by GLCM. In an ultrasound study [32], the authors investigated the features' differences between the normal and postradiotherapy parotid glands by extracting eight GLCM features at distance of 4 pixels and all angles in two dimensional images. The result suggested that significant differences in GLCM features were observed between the normal and postradiotherapy parotid glands. Petkovska et al. [33] reported that the GLCM features extracted from contrast-enhanced CT have a better identification ability to differentiate malignant from benign nodules than visual inspection by three experienced radiologists.

1.4.4 Wavelet-Based Features

Wavelets are small oscillatory waveforms which can be stretched into different sizes and convolved against a signal for extracting their coefficients. The coefficients are, thus, scale and translation invariant. Wavelet coefficients decompose input signal into different components, and then study each component with a resolution matched to its scale. The wavelet transform essentially splits the signal into a low pass sub-band (abbreviation notation is "L", also called approximation level) and a high pass sub-band (abbreviation notation is "H", also called detail level) at each scale. A wavelet transformed-2D image is shown in Figure 1.2 as an example. The wavelet transform-based method processes an image at multiple levels of resolution and returns a series of gray level edge maps at different resolutions. After the generation of different level of coefficients (that are 2D images themselves of varying resolutions) by wavelet transform, all the radiomic features described before, such as first-order features, shape- and size-based features and second-order features, can be extracted on the high pass and low pass sub-band images again. These could generate thousands of radiomic features [34]. Many studies have used wavelet-based features to reveal image features of medical images. Yang et al. [35, 36] used Gómes et al.'s informative classifying features and developed a feature extraction scheme using wavelet transform that further improved the diagnostic performance.

1.4.5 Feature Extraction Software

To extract the radiomic features listed above, many software packages are available. In the supplementary of Hatt et al.'s review paper on 3D-Slicer [17], many pieces of feature extraction software were listed. HeterogeneityCAD, a module in 3D-Slicer, is a radiomics extraction graphical user interface toolbox to quantify the radiomic features of a region under a segmentation mask. Pyradiomics [37] is an open-source Python script library for quantifying different radiomics. Wong et al. also provided a list of many open-source software programs for feature extraction on different modality images for head and neck cancer [38].

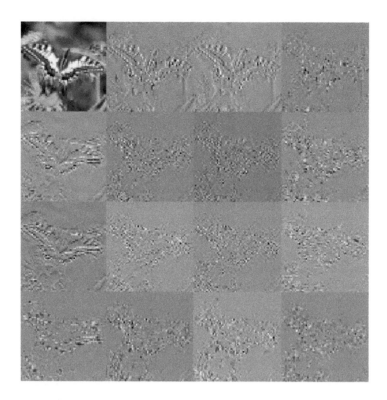

FIGURE 1.2: Wavelet transformed images, upper-left is original image and right are three different levels of wavelet coefficients. And horizontal direction goes along high-pass filtering, and vertical direction goes along low-pass filtering.

1.5 Clinical Predictive Modeling with Machine Learning

The use of a large number of radiomic features for classification and prediction under limited samples not only takes a long time to compute but may not be optimal. After a large number of high-throughput imaging features are extracted, the feature selection method is used to obtain the best-performing feature sets which are then used in turn as input to a machine learning algorithm or a statistical modeling algorithm to classify one or more target diagnostic variables. In this section, we will discuss methods for selecting features and then, for classification used in radiomics recently.

1.5.1 Feature Selection Methods

The feature selection process aims to find the most informative features among all extracted radiomic features. To guide the practical application, selected features need to have good stability and strong discrimination ability.

In much research, radiomic features are reported to be unstable within different imaging methods or at different longitudinal time points [39]. Feature stability in radiomics means feature consistency of each individual in the same cohort cluster.

Features with better stability showed higher prognostic performance, and that phenomenon may be due to reduced noise in the stable features. In one study, Aerts et al. [18] selected top features in four feature groups out of 100 most stable features which were determined by two testing datasets. According to the stability rank of test and retest CT scan dataset, they selected the four most significant radiomic features which were statistical energy, shape compactness, gray level nonuniformity in the original image and gray level nonuniformity in HLH wavelets filtered images. Their results showed that radiomic features have prognostic meaning in both lung and head and neck cancer. Leijenaar et al. [40] assessed every extracted feature between test-retest and interobserver stability of features by the intraclass correlation coefficient (ICC), which is calculated by Kruskal–Wallis one-way analysis of variance (ANOVA) method and provides an indication of feature measurements reliability. Similarity between test-retest and interobserver stability rankings of features was evaluated by Spearman's rank correlation coefficient.

In addition to being useful in improving the predictive performance of machine learning algorithms, feature selection also reduces the dimensionality of radiomics featurespace and, thus, improves computational efficiency. Feature selection methods can be classified as univariate and multivariate feature selection methods [41]. Some research investigated the role of feature selection methods on predictive models [43]. Parmar et al. [41, 42, 44] analyzed 14 feature selection methods on the predictive performance of patient survival and found that the prediction accuracy of radiomic features is influenced by the number of features, feature selection methods and pattern recognition classifiers. Among 14 feature selection algorithms, Wilcoxon test-based feature selection method (WLCX), one of the univariate methods, showed the highest performance in most of the selected features numbers.

1.5.2 Radiomic Features Classifying Outcomes

1.5.2.1 Clustering Algorithms

Clustering algorithms were used to reveal the cluster patterns of radiomic features without knowing prior patient outcome knowledge. In one recent widely cited publication, Aerts et al. [18] applied a hierarchical clustering method to reveal the clusters of patients in similar expression patterns among 440 features. By comparing with the clinical parameters, the primary tumor stage, overall stage and histology of tumors were observed to have a significant association with the cluster. In another study [42], Parmar used a consensus clustering method and showed the correlation between radiomic features and clinical parameters of two types of cancer. The term "consensus clustering algorithm" refers to finding a single (consensus) clustering which is the best fit with input dataset from the existing clusters. After that overlap, features between the feature clusters of lung and head and neck cancer cohorts were assessed by the Jaccard index matrix. In another study, a hierarchical clustering algorithm was used to identify nonredundant features, and the method is reproducible across multiple CT machines [45].

1.5.2.2 Regression Model

Linear regression is a model that provides the predictive relationship between a dependent variable and one or more independent variables. The linear regression model has been used to analyze the temporal relationship between undergoing therapy and radiomic features. For example, Yang et al. [46] compared tumor texture features of fluorodeoxyglucose (FDG) heterogeneity with tumor SUV measures on patients' PET images and classified their chemoradiotherapy responses at different time points. Temporal changes were analyzed by the linear regression model. The temporal behaviors in the early phase of treatment show that texture features are a better predictor for the outcome of therapeutic intervention than SUV measures. In another study [47], a linear regression model was also used to assess the relationship between the radiation dose and radiomic features. Li et al. [48] demonstrated significant associations between radiomic features and multigene assay recurrence scores using multiple linear regression models.

Unlike linear regression where a dependent variable is normally continuous, a logistic regression model is used when a dependent variable has only a limited number of possible or categorical values. In King et al.'s study [49], radiomic features were correlated for tumor local failure and control at 2 years. A logistic regression model was used to predict local failure by univariate and multivariate analyses of the apparent diffusion coefficients parameters, tumor stage and tumor volume. The poorly responding tumors showed a significant increase in skewness and kurtosis values than that shown by well-controlled tumors. El et al. [50] employed radiomic features to predict treatment outcomes in patients with cervical cancers and head and neck cancers. Association between different extracted radiomic features and nine patients' overall survival was investigated by a logistic regression model. The result showed that shape-based features had the highest categorical predictive capability of failure-risk, while commonly used SUV descriptive statistics had the lowest predictive ability.

The Cox regression model, referred to as the Cox proportional hazards regression model, was applied mainly for the prognosis of survival-time outcomes of tumors and other chronic diseases on one or more predictors. Cook et al. [51] measured tumor radiomic features like coarseness, contrast, busyness and complexity which were derived from a PET dataset of 53 patients with NSCLC treated with chemoradiotherapy. Cox regression was used to examine the effects of the radiomic features on the survival outcomes. Therapy responders showed lower coarseness and higher contrast and busyness than non-responders. However, none of the SUV parameters predicted treatment response accurately. Those results indicate that coarseness, busyness, and contrast are more closely associated with differentiating between responders and non-responders to chemoradiotherapy than SUV measures, and coarseness was an independent predictor of overall patient survival. In another study [52], the Cox model was trained to evaluate radiomic features' capability to predict distant metastasis.

1.5.2.3 Support Vector Machine

Support vector machine (SVM) is another machine learning algorithm for classification problems. SVM transforms images by different types of kernel functions and then finds an optimal boundary between the predictive outputs. SVM is also applied to patient outcome prediction. Some studies have used SVM for predicting patients' response to different therapies. Zhang et al. [53] studied 20 esophageal cancer patients' ^{18}F-FDG PET/CT scans both before and after trimodality therapy (chemoradiotherapy plus surgery). SVM and logistic regression (LR) models were used to predict pathologic tumor response to treatment. Compared to different models and feature groups, they found that SVM models constructed by radiomic features combined with conventional SUV measures and clinical parameters significantly improved the pathologic response prediction. Some studies [54, 55] have shown that many image features are able to differentiate tumor stages using different medical imaging modalities. In one study, Mu et al. [56] classified cervical cancer patients into early stage and advanced stage using 54 PET-based image features. One of the radiomic features, run percentage (RP), which is selected by the SVM classifier, was the most discriminative index with higher accuracy and ability to differentiate early stage and advanced stage with high accuracy. In another tumor stage study, Dong et al. [57] assessed radiomic features on PET images of 40 patients with esophageal squamous cell carcinoma. Correlations were observed between SUVmax, GLCM-entropy and GLCM-energy on the one hand and T and N stages of patients on the other. Xu et al. [58] developed a computer-aided diagnosis (CAD) system to differentiate malignant and benign bone and soft-tissue lesions on FDG PET/CT images. After selection of a subset of the most optimal radiomic features parameters, an SVM classifier was used to automatically differentiate different bone tumor tissues. The result showed that the CAD method with the combined PET and CT optimal radiomic features has a better ability in differential diagnosis than that by each of PET or CT texture analysis alone. Compared with the histological diagnosis of the lesions, the diagnostic method based on classification results of their radiomic features achieved a relatively good accuracy.

1.6 Big Data and Machine Learning

Machine Learning: The power of machine learning, both supervised and unsupervised, may be best harnessed when the training sample set is sufficiently large. Due to privacy and expense, obtaining a large number of human datasets is very challenging in medicine. However, across different laboratories and clinical facilities, data are being constantly gathered. If one can share these datasets, the use of machine learning may be as powerful as it is in many of its other application areas, such as computer vision, natural language processing, etc. This is the direction where we see that medical imaging research is making some pioneering steps [7, 8, 9]. Most recently, DeepLesion, released by NIH, contains nearly 10,600 publicly available CT scans to support the development of machine learning algorithms for medical applications [59]. Natural challenges, of course, are many, e.g., legally sharing data, standardizing data for uniform utilization by machine learning algorithms (as we have discussed in Section 1.2) and designing appropriate research questions over varieties of available data. All these challenges relate to the classical definition of Big Data, namely, volume, veracity and velocity, which pertain to the large size of data, diversity of data and rapid change of data, respectively. In the context of medical imaging these three characteristics respectively pose their own problems: (1) managing large data size over many imaging studies; (2) homogenizing data (to be accessed by individual algorithms or data processing procedures) over multiple studies acquired for different purposes and hypotheses, over different machines, possibly with multiple modalities involved; and (3) continuously increasing the size of the image database as new data are constantly added by contributors. Computer science is actively seeking solutions to the large data size problem. For example, new languages are being designed for Big Data (e.g., PlinyCompute [60]) and new architecture for deep learning as model parallelization [61]. We have discussed in this article how the radiology community is addressing the veracity of diversely acquired data. Finally, we believe these two solutions will have to adjust to the fact that the data will continuously grow and no static model for a fixed dataset will be sufficient. However, that is for the near future. In this context, the newly emerging field of topological data analysis (TDA) [62] may help in understanding the nature of a dataset. As an unsupervised machine learning model, this technique studies the topology of data points (images) in the underlying feature space. Instead of clustering the data points, TDA rather investigates more complex structures like connectivity between data. Such topological understanding may provide much better guidance to the supervised machine learning algorithms [63] or may even be embedded in the latter for a robust algorithm [64] to address the Big Data challenges.

Computing Hardware: Deep learning algorithms' miraculous success has not only dazzled all machine learning applications but also medical imaging communities. However, these algorithms are not just data hungry but also computing-resources hungry. Most of these algorithms need multiple high-performance computing nodes such as computing-focused graphics processing units (GPUs) with sufficient localized memory closer to computational unit. For example, Nvidia (Santa Clara, CA) provides a dedicated web page for medical imaging applications on their GPU platforms [65] and announced Project Clara in March 2018 [66]. Specialized hardware is being developed for deep learning algorithms by many other vendors as well.

Image Data Mart: Another sector where we see that Big Data in medical imaging needs to grow is toward Data Mart development. There are many dimensions of such image databases, e.g., imaging modalities, disease models, metadata involved with study-questions or protocols, etc. Currently, many such databases are being made available, e.g., the recent release of 100,000 chest X-ray images for 30,000 patients by NIH for lung cancer research [67]. The project will release torso images of the same type for research on abdominal diseases according to the announcement [68]. Similar other lung cancer datasets are available on the web. Eventually such large data release projects need to be consolidated, when they are too similar, or easily referred from one to another to pose higher-level research questions, e.g., for connecting genotype and phenotype of a disease. The National Cancer Informatics Program provides a step toward such an integrated view of data. It took more than 30 years for the genomics databases to provide a disciplined view to researchers for data access. We expect that experience will be harnessed to allow comprehensive access to imaging databases for scientists.

1.7 Conclusions

The machine learning models built upon unique and important radiomic features provide a new way not only to study tumor heterogeneity but also to quantitatively analyze the changes in micro-level gene or protein patterns that are hidden behind medical images. Many studies have applied radiomic features and reached promising results in precision medicine. As a new research field, radiomics still needs further exploration. With the standardization of medical imaging data, and the rapid development of various methods of image segmentation, feature extraction, feature selection and machine learning algorithms, the application of radiomic features will have far-reaching effects and lead to tremendous changes in radiology. As larger datasets are being made available to researchers and clinicians a consolidation of these datasets will take on characteristics of Big Data. Both machine learning algorithms and data organization will need to address that. This is likely to follow a similar trend in genomics. It is only natural to expect that radiomics will utilize that experience. Finally, the two fields of genomics and radiomics will strongly interact with each other to usher in a new horizon in medicine that we possibly cannot anticipate now.

References

1. Gerlinger, Marco, et al. "Intratumor heterogeneity and branched evolution revealed by multiregion sequencing." *New England Journal of Medicine* 366.10 (2012): 883–892.
2. Lambin, Philippe, et al. "Radiomics: extracting more information from medical images using advanced feature analysis." *European Journal of Cancer* 48.4 (2012): 441–446.
3. Chicklore, Sugama, et al. "Quantifying tumour heterogeneity in 18 F-FDG PET/CT imaging by texture analysis." *European Journal of Nuclear Medicine and Molecular Imaging* 40.1 (2013): 133–140.
4. Gillies, Robert J, Paul E Kinahan, and Hedvig Hricak. "Radiomics: images are more than pictures, they are data." *Radiology* 278.2 (2015): 563–577.
5. Aerts, Hugo JWL, et al. "The potential of radiomic-based phenotyping in precision medicine: a review." *JAMA Oncology* 2.12 (2016): 1636–1642.
6. Wong, Andrew J, et al. "Radiomics in head and neck cancer: from exploration to application." *Translational Cancer Research* 5.4 (2016): 371–382.
7. Armato, Samuel G, et al. "The lung image database consortium (LIDC) and image database resource initiative (IDRI): a completed reference database of lung nodules on CT scans." *Medical Physics* 38.2 (2011): 915–931.
8. Weinstein, John N, et al. "The Cancer Genome Atlas pan-cancer analysis project." *Nature Genetics* 45.10 (2013): 1113–1120.
9. Clark, Kenneth, et al. "The Cancer Imaging Archive (TCIA): maintaining and operating a public information repository." *Journal of Digital Imaging* 26.6 (2013): 1045–1057.
10. Velazquez, Emmanuel Rios, et al. "Volumetric CT-based segmentation of NSCLC using 3D-Slicer." *Scientific Reports* 3 (2013): 3529.
11. So, Ronald WK, Tommy WH Tang, and Albert CS Chung. "Non-rigid image registration of brain magnetic resonance images using graph-cuts." *Pattern Recognition* 44.10 (2011): 2450–2467.
12. Yushkevich, Paul A, et al. "User-guided 3D active contour segmentation of anatomical structures: significantly improved efficiency and reliability." *Neuroimage* 31.3 (2006): 1116–1128.
13. Samaille, Thomas, et al. "Contrast-based fully automatic segmentation of white matter hyperintensities: method and validation." *PloS one* 7.11 (2012): e48953.
14. Meier, Raphael, et al. "Clinical evaluation of a fully-automatic segmentation method for longitudinal brain tumor volumetry." *Scientific Reports* 6 (2016): 23376.
15. Dodin, Pierre, et al. "A fully automated human knee 3D MRI bone segmentation using the ray casting technique." *Medical & Biological Engineering & Computing* 49 (2011): 1413–1424.
16. Zwanenburg, Alex, et al. "Image biomarker standardisation initiative-feature definitions." arXiv preprint arXiv:1612.07003 (2016).
17. Hatt, Mathieu, et al. "Characterization of PET/CT images using texture analysis: the past, the present... any future?" *European Journal of Nuclear Medicine and Molecular Imaging* 44.1 (2017): 151–165.
18. Aerts, Hugo JWL, et al. "Decoding tumour phenotype by noninvasive imaging using a quantitative radiomics approach." *Nature Communications* 5 (2014): 4006.

19. Bundschuh, RA, J Dinges, L Neumann, M Seyfried, N Zsótér, L Papp, et al. "Textural parameters of tumor heterogeneity in 18F-FDG PET/CT for therapy response assessment and prognosis in patients with locally advanced rectal cancer." *Journal of Nuclear Medicine* 55 (2014): 891–897.

20. Yoon, SH, CM Park, SJ Park, J-H Yoon, S Hahn, and JM Goo. "Tumor heterogeneity in lung cancer: assessment with dynamic contrast-enhanced MR imaging." *Radiology* 280.3 (2016): 940–948.

21. El Naqa, I, P Grigsby, A Apte, E Kidd, E Donnelly, D Khullar, et al. "Exploring feature-based approaches in PET images for predicting cancer treatment outcomes." *Pattern Recognition* 42 (2009): 1162–1171.

22. Gonzalez, ME, K Dinelle, N Vafai, N Heffernan, J McKenzie, S Appel-Cresswell, et al. "Novel spatial analysis method for PET images using 3D moment invariants: applications to Parkinson's disease." *Neuroimage* 68 (2013): 11–21.

23. Kim, DH, JH Jung, SH Son, CY Kim, SY Jeong, SW Lee, et al. "Quantification of intratumoral metabolic macroheterogeneity on 18F-FDG PET/CT and its prognostic significance in pathologic N0 squamous cell lung carcinoma." *Clinical Nuclear Medicine* 41 (2016): e70–e75.

24. Haralick, Robert M, and Karthikeyan Shanmugam. "Textural features for image classification." *IEEE Transactions on Systems, Man, and Cybernetics* 6 (1973): 610–621.

25. Ulaby, Fawwaz T, et al. "Textural infornation in SAR images." *IEEE Transactions on Geoscience and Remote Sensing* 2 (1986): 235–245.

26. Ou, Xiang, Wei Pan, and Perry Xiao. "In vivo skin capacitive imaging analysis by using grey level co-occurrence matrix (GLCM)." *International Journal of Pharmaceutics* 460.1 (2014): 28–32.

27. Baraldi, Andrea, and Flavio Parmiggiani. "An investigation of the textural characteristics associated with gray level cooccurrence matrix statistical parameters." *IEEE Transactions on Geoscience and Remote Sensing* 33.2 (1995): 293–304.

28. Hua, BO, Ma Fu-Long, and Jiao Li-Cheng. "Research on computation of GLCM of image texture [J]." *Acta Electronica Sinica* 1.1 (2006): 155–158.

29. Duff, IS, AM Erisman, and JK Reid, 1986. *Direct Methods for Sparse Matrices*. Clarendon Press, Oxford, London, 341 pp.

30. Clausi, David A, and Yongping Zhao. "Rapid extraction of image texture by co-occurrence using a hybrid data structure." *Computers and Geosciences* 28.6 (2002): 763–774.

31. Hatt, M, M Majdoub, M Vallières, F Tixier, CC Le Rest, D Groheux, et al. "18F-FDG PET uptake characterization through texture analysis: investigating the complementary nature of heterogeneity and functional tumor volume in a multi-cancer site patient cohort. *Journal of Nuclear Medicine* 56 (2015): 38–44.

32. Yang, Xiaofeng, et al. "Ultrasound GLCM texture analysis of radiation-induced parotid-gland injury in head-and-neck cancer radiotherapy: an in vivo study of late toxicity." *Medical Physics* 39.9 (2012): 5732–5739.

33. Petkovska, Iva, et al. "Pulmonary nodule characterization: a comparison of conventional with quantitative and visual semi-quantitative analyses using contrast enhancement maps." *European Journal of Radiology* 59.2 (2006): 244–252.

34. Masoumi, Majid, and A Ben Hamza. "Shape classification using spectral graph wavelets." *Applied Intelligence* 47 (2017): 1–14.

35. Gómez, Walter, WCA Pereira, and Antonio Fernando C Infantosi. "Analysis of co-occurrence texture statistics as a function of gray-level quantization for classifying breast ultrasound." *IEEE Transactions on Medical Imaging* 31.10 (2012): 1889–1899.

36. Yang, Min-Chun, et al. "Robust texture analysis using multi-resolution gray-scale invariant features for breast sonographic tumor diagnosis." *IEEE Transactions on Medical Imaging* 32.12 (2013): 2262–2273.

37. Van Griethuysen, Joost JM, et al. "Computational radiomics system to decode the radiographic phenotype." *Cancer Research* 77.21 (2017): e104–e107.

38. Wong, Andrew J, et al. "Radiomics in head and neck cancer: from exploration to application." *Translational Cancer Research* 5.4 (2016): 371–382.

39. Yip, Stephen SF, and HJWL Aerts. "Applications and limitations of radiomics." *Physics in Medicine & Biology* 61.13 (2016): R150–R166.
40. Leijenaar, Ralph TH, et al. "Stability of FDG-PET Radiomic features: an integrated analysis of test-retest and inter-observer variability." *Acta Oncologica* 52.7 (2013): 1391–1397.
41. Parmar, Chintan, et al. "Machine learning methods for quantitative radiomic biomarkers." *Scientific Reports* 5 (2015): 13087.
42. Parmar, Chintan, et al. "Radiomic machine-learning classifiers for prognostic biomarkers of head and neck cancer." *Frontiers in Oncology* 5 (2015): 272.
43. Hawkins, SH, et al. "Predicting outcomes of nonsmall cell lung cancer using CT image features. *IEEE Access* 2 (2014): 1418–1426.
44. Wu, Weimiao, et al. "Exploratory study to identify radiomics classifiers for lung cancer histology." *Frontiers in Oncology* 6 (2016): 71.
45. Hunter, Luke A, et al. "High quality machine-robust image features: identification in nonsmall cell lung cancer computed tomography images." *Medical Physics* 40.12 (2013): 121916.
46. Yang, Fei, et al. "Temporal analysis of intratumoral metabolic heterogeneity characterized by textural features in cervical cancer." *European Journal of Nuclear Medicine and Molecular Imaging* 40.5 (2013): 716–727.
47. Cunliffe, Alexandra, et al. "Lung texture in serial thoracic computed tomography scans: correlation of radiomics-based features with radiation therapy dose and radiation pneumonitis development." *International Journal of Radiation Oncology* Biology* Physics* 91.5 (2015): 1048–1056.
48. Li, Hui, et al. "MR imaging radiomics signatures for predicting the risk of breast cancer recurrence as given by research versions of MammaPrint, Oncotype DX, and PAM50 gene assays." *Radiology* 281.2 (2016): 382–391.
49. King, Ann D, et al. "Head and neck squamous cell carcinoma: diagnostic performance of diffusion-weighted MR imaging for the prediction of treatment response." *Radiology* 266.2 (2013): 531–538.
50. El Naqa, Issam, et. al. "Exploring feature-based approaches in PET images for predicting cancer treatment outcomes." *Pattern Recognition* 42.6 (2009): 1162–1171.
51. Cook, Gary JR, et al. "Are pretreatment 18F-FDG PET tumor textural features in nonsmall cell lung cancer associated with response and survival after chemoradiotherapy?" *Journal of Nuclear Medicine* 54.1 (2013): 19–26.
52. Coroller, Thibaud P, et al. "CT-based radiomic signature predicts distant metastasis in lung adenocarcinoma." *Radiotherapy and Oncology* 114.3 (2015): 345–350.
53. Zhang, Hao, et al. "Modeling pathologic response of esophageal cancer to chemoradiation therapy using spatial-temporal 18 F-FDG PET features, clinical parameters, and demographics." *International Journal of Radiation Oncology* Biology* Physics* 88.1 (2014): 195–203.
54. Xiuhua, Guo, Sun Tao, and Liang Zhigang. "Prediction models for malignant pulmonary nodules based-on texture features of CT image." In Homma, Noriyasu, ed., *Theory and Applications of CT Imaging and Analysis*. InTech, 2011.
55. Xian, Guang-ming. "An identification method of malignant and benign liver tumors from ultrasonography based on GLCM texture features and fuzzy SVM." *Expert Systems with Applications* 37.10 (2010): 6737–6741.
56. Mu, Wei, et al. "Staging of cervical cancer based on tumor heterogeneity characterized by texture features on 18F-FDG PET images." *Physics in Medicine and Biology* 60.13 (2015): 5123.
57. Dong, Xinzhe, et al. "Three-dimensional positron emission tomography image texture analysis of esophageal squamous cell carcinoma: relationship between tumor 18F-fluorodeoxyglucose uptake heterogeneity, maximum standardized uptake value, and tumor stage." *Nuclear Medicine Communications* 34.1 (2013): 40–46.
58. Xu, Rui, et al. "Texture analysis on 18F-FDG PET/CT images to differentiate malignant and benign bone and soft-tissue lesions." *Annals of Nuclear Medicine* 28.9 (2014): 926–935.
59. Yan, Ke, et al. "Deep Lesion: automated mining of large-scale lesion annotations and universal lesion detection with deep learning." *Journal of Medical Imaging* 5.3 (2018): 036501.

60. Zou, Jia, et al. "PlinyCompute: a platform for high-performance, distributed, data-intensive tool development." *Proceedings of the 2018 International Conference on Management of Data*. ACM, 2018.
61. Dettmers, Tim. "8-bit approximations for parallelism in deep learning." arXiv preprint arXiv:1511.04561 (2015).
62. Carlsson, G. "Topology and data." *AMS Bulletin* 46.2 (2009): 255–308.
63. Basri, Ronen, and David Jacobs. "Efficient representation of low-dimensional manifolds using deep networks." arXiv preprint arXiv: 1602.04723 (2016).
64. Hofer, Christoph, et al. "Deep learning with topological signatures." *Advances in Neural Information Processing Systems* 30 (2017).
65. http://www.nvidia.com/object/medical_imaging.html
66. https://blogs.nvidia.com/blog/2018/03/28/ai-healthcare-gtc/
67. Wang, Xiaosong, et al. "Chestx-ray8: hospital-scale chest x-ray database and benchmarks on weakly-supervised classification and localization of common thorax diseases." *2017 IEEE Conference on Computer Vision and Pattern Recognition (CVPR)*. IEEE, 2017.
68. https://www.nih.gov/news-events/news-releases/nih-clinical-center-provides-one-largest-publicly-available-chest-x-ray-datasets-scientific-community.

Appendix

A.   First-Order Statistics Features

First-order statistics calculation formulas and most of their measurement meanings are described as follows, where matrix X represents a three-dimensional image matrix with N voxels, and p represents the first-order histogram with N_l discrete intensity levels, and $P(i)$ is the normalized percentage of the first-order histogram and equals to $p(i)/\Sigma p(i)$.

1: Energy

$$\textbf{Energy} = \sum_{i=1}^{N} X(i)^2$$

2: Entropy

$$\textbf{Entropy} = \sum_{i=1}^{N_l} P(i) \log_i P(i)$$

3: Uniformity

$$\textbf{Uniformity} = \sum_{i=1}^{N_l} P(i)^2$$

4: Maximum

$$\textbf{Maximum} = \max X(i)$$

5: Mean

$$\textbf{Mean} = \frac{1}{N} \sum_{i=1}^{N} X(i)$$

6: Median

$$\textbf{Median} = \text{median} X(i)$$

7: Minimum

$$\textbf{Minimum} = \min X(i)$$

8: Range

$$\textbf{Range} = \min X(i) - \max X(i)$$

9: Root Mean Square

$$\textbf{Root Mean Square} = \sqrt{\frac{1}{N} \sum_{i=1}^{N} X(i)^2}$$

10: Skewness

$$\textbf{Skewness} = \frac{\frac{1}{N} \sum_{i=1}^{N} \left(X(i) - \bar{X} \right)^3}{\left(\frac{1}{N} \sum_{i=1}^{N} \left(X(i) - \bar{X} \right)^2 \right)\left(\frac{3}{2} \right)}$$

11: Kurtosis

$$\textbf{Kurtosis} = \frac{\frac{1}{N} \sum_{i=1}^{N} \left(X(i) - \bar{X} \right)^4}{\left(\frac{1}{N} \sum_{i=1}^{N} \left(X(i) - \bar{X} \right)^2 \right)^2}$$

12: Mean Absolute Deviation

$$\textbf{Mean Absolute Deviation} = \frac{1}{N} \sum_{i=1}^{N} \left| X(i) - \bar{X} \right|$$

13: Standard Deviation

$$\textbf{Standard Deviation} = \sqrt{\frac{1}{N} \sum_{i=1}^{N} \left(X(i) - \bar{X} \right)^2}$$

14: Variance

$$\textbf{Variance} = \frac{1}{N} \sum_{i=1}^{N} \left(X(i) - \bar{X} \right)^2$$

B. Shape- and Size-Based Features

Shape and size features provide the descriptions of shape and size about the region of interest (e.g., a tumor on a medical image). The shape- and size-based features are described as follows, where V and A represent the volume and the surface area of the region of interest. Volume is equal to the number of pixels in the interest region multiplied by the size of single voxels.

1: Compactness

$$\textbf{Compactness 1} = \frac{V}{\sqrt{\pi}\sqrt{A^3}}$$

$$\textbf{Compactness 2} = \frac{36\pi V^2}{A^3}$$

2: Maximum 3D Diameter

$$\textbf{Maximum 3D Diameter} = \text{maxDistance}\left(\forall v_i, \forall v_j\right)$$

3: Spherical Disproportion

$$\textbf{Spherical Disproportion} = \frac{A}{4\pi R^2}$$

4: Sphericity

$$\textbf{Sphericity} = \frac{\sqrt[3]{\pi}\sqrt[3]{36V^2}}{A}$$

5: Area

$$\textbf{Area} = \sum_{i=1}^{N} \frac{1}{2}\left|a_i b_i \times a_i c_i\right|$$

6: Ratio

$$\textbf{Ratio} = \frac{A}{V}$$

C. GLCM Features

The image features generated from GLCM are described as follows.

1: Auto correlation

$$\textbf{Auto correlation} = \sum_{i=1}^{N_g}\sum_{j=1}^{N_g} ijP(i,j)$$

2: Cluster Prominence (CP)

$$\mu_x(i) = \frac{\sum_{j=1}^{N_g} P(i,j)}{N_g}$$

$$\mu_y(j) = \frac{\sum_{i=1}^{N_g} P(i,j)}{N_g}$$

$$\mathbf{CP} = \sum_{i=1}^{N_g} \sum_{j=1}^{N_g} \left[i + j - \mu_x(i) - \mu_y(j) \right]^4 P(i,j)$$

3: Cluster Shade (CS)

$$\mathbf{CS} = \sum_{i=1}^{N_g} \sum_{j=1}^{N_g} \left[i + j - \mu_x(i) - \mu_y(j) \right]^3 P(i,j)$$

4: Cluster Tendency (CT)

$$\mathbf{CT} = \sum_{i=1}^{N_g} \sum_{j=1}^{N_g} \left[i + j - \mu_x(i) - \mu_y(j) \right]^2 P(i,j)$$

5: Contrast

$$\mathbf{Contrast} = \sum_{i=1}^{N_g} \sum_{j=1}^{N_g} \left| i - j \right|^2 P(i,j)$$

6: Correlation

$$\mathbf{Correlation} = \frac{\sum_{i=1}^{N_g} \sum_{j=1}^{N_g} ij P(i,j) - \mu_i(i) \mu_j(j)}{\sigma_x(i) \sigma_y(j)}$$

7: Difference Entropy (DE)

$$\mathbf{p_{x-y}(k)} = \sum_{i=1}^{N_g} \sum_{j=1}^{N_g} P(i,j), \left| i - j \right| = k, k = 0,1\ldots,N_g - 1$$

$$\mathbf{Difference\ Entropy} = -\sum_{i=0}^{N_g-1} P_{x-y}(i) \log_2 \left[P_{x-y}(i) \right]$$

8: Dissimilarity

$$\textbf{Dissimilarity} = \sum_{i=1}^{N_g} \sum_{j=1}^{N_g} |i-j|^2 P(i,j)$$

9: Energy

$$\textbf{Energy} = \sum_{i=1}^{N_g} \sum_{j=1}^{N_g} \left[P(i,j) \right]^2$$

10: Entropy

$$\textbf{Entropy} = -\sum_{i=1}^{N_g} \sum_{j=1}^{N_g} P(i,j) \log_2 \left[P(i,j) \right]$$

11: Homogeneity

$$\textbf{Homogeneity 1} = \sum_{i=1}^{N_g} \sum_{j=1}^{N_g} \frac{P(i,j)}{1+|i-j|}$$

$$\textbf{Homogeneity 2} = \sum_{i=1}^{N_g} \sum_{j=1}^{N_g} \frac{P(i,j)}{1+|i-j|^2}$$

12: Information Measure of Correlation (IMC)

$$\mathbf{p_x(i)} = \sum_{j=1}^{N_g} P(i,j)$$

$$\mathbf{p_y(j)} = \sum_{i=1}^{N_g} P(i,j)$$

$$\textbf{HX} = -\sum_{i=1}^{N_g} p_x(i) \log_2 p_x(i)$$

$$\textbf{HY} = -\sum_{i=1}^{N_g} p_y(i) \log_2 p_y(i)$$

$$\textbf{HXY 1} = -\sum_{i=1}^{N_g} \sum_{j=1}^{N_g} P(i,j) \log \left(p_x(i) \, p_y(j) \right)$$

$$\text{HXY 2} = -\sum_{i=1}^{N_g}\sum_{j=1}^{N_g} p_x(i)\,p_y(j)\log\big(p_x(i)\,p_y(j)\big)$$

$$\text{IMC 1} = \frac{HXY - HXY1}{\max(HX, HY)}$$

$$\text{IMC 2} = \sqrt{1 - e^{-2(HXY2 - HXY)}}$$

13: Inverse Difference Moment Normalized (IDMN)

$$\text{IDMN} = \sum_{i=1}^{N_g}\sum_{j=1}^{N_g}\frac{P(i,j)}{1+\left(\dfrac{i-j}{N_g}\right)^2}$$

14: Inverse Difference Normalized (IDN)

$$\text{IDN} = \sum_{i=1}^{N_g}\sum_{j=1}^{N_g}\frac{P(i,j)}{1+\dfrac{|i-j|}{N_g}}$$

15: Inverse Variance

$$\text{Inverse Variance} = \sum_{i=1}^{N_g}\sum_{j=1}^{N_g}\frac{P(i,j)}{(i-j)^2}$$

16: Maximum Probability

$$\text{Maximum Probability} = \max P(i,j)$$

17: Sum Entropy

$$\mathbf{p_{x+y}(k)} = \sum_{i=1}^{N_g}\sum_{j=1}^{N_g} P(i,j), i+j = k, k = 2,3\ldots,2N_g$$

$$\text{Sum Entropy} = -\sum_{i=2}^{2N_g} P_{x+y}(i)\log_2\big[P_{x+y}(i)\big]$$

18: Sum Average

$$\text{Sum Average} = \sum_{i=1}^{2N_g}\big[iP_{x+y}(i)\big]$$

19: Sum Variance

$$\text{Sum Variance} = \sum_{i=2}^{2N_g}(i-SE)^2 P_{x+y}(i)$$

20: Variance

$$\text{Variance} = \sum_{i=1}^{N_g} \sum_{j=1}^{N_g} (i - \mu)^2 P(i,j)$$

D. GLRLM Features

1: Short Run Emphasis (SRE)

$$\text{SRE} = \frac{\sum_{i=1}^{N_g} \sum_{j=1}^{N_r} \left[\dfrac{p(i,j \mid \theta)}{j^2} \right]}{\sum_{i=1}^{N_g} \sum_{j=1}^{N_g} p(i,j \mid \theta)}$$

2: Long Run Emphasis (LRE)

$$\text{LRE} = \frac{\sum_{i=1}^{N_g} \sum_{j=1}^{N_r} j^2 p(i,j \mid \theta)}{\sum_{i=1}^{N_g} \sum_{j=1}^{N_r} p(i,j \mid \theta)}$$

3: Gray Level Nonuniformity (GLN)

$$\text{GLN} = \frac{\sum_{i=1}^{N_g} \left[\sum_{j=1}^{N_r} p(i,j \mid \theta) \right]^2}{\sum_{i=1}^{N_g} \sum_{j=1}^{N_g} p(i,j \mid \theta)}$$

4: Run Length Nonuniformity (RLN)

$$\text{RLN} = \frac{\sum_{j=1}^{N_r} \left[\sum_{i=1}^{N_g} p(i,j \mid \theta) \right]^2}{\sum_{i=1}^{N_g} \sum_{j=1}^{N_g} p(i,j \mid \theta)}$$

5: Run Percentage (RP)

$$\text{RP} = \sum_{i=1}^{N_g} \sum_{j=1}^{N_r} \frac{p(i,j \mid \theta)}{N_p}$$

6: Low Gray Level Run Emphasis (LGLRE)

$$\text{LGLRE} = \frac{\sum_{i=1}^{N_g} \sum_{j=1}^{N_r} \dfrac{p(i,j \mid \theta)}{i^2}}{\sum_{i=1}^{N_g} \sum_{j=1}^{N_r} p(i,j \mid \theta)}$$

7: High Gray Level Run Emphasis (HGLRE)

$$\mathbf{HGLRE} = \frac{\sum_{i=1}^{N_g} \sum_{j=1}^{N_r} i^2 p(i,j \mid \theta)}{\sum_{i=1}^{N_g} \sum_{j=1}^{N_r} p(i,j \mid \theta)}$$

8: Short Run Low Gray Level Emphasis (SRLGLE)

$$\mathbf{SRLGLE} = \frac{\sum_{i=1}^{N_g} \sum_{j=1}^{N_r} \frac{p(i,j \mid \theta)}{i^2 j^2}}{\sum_{i=1}^{N_g} \sum_{j=1}^{N_r} p(i,j \mid \theta)}$$

9: Short Run High Gray Level Emphasis (SRHGLE)

$$\mathbf{SRHGLE} = \frac{\sum_{i=1}^{N_g} \sum_{j=1}^{N_r} \frac{p(i,j \mid \theta) i^2}{j^2}}{\sum_{i=1}^{N_g} \sum_{j=1}^{N_r} p(i,j \mid \theta)}$$

10: Long Run Low Gray Level Emphasis (LRLGLE)

$$\mathbf{LRLGLE} = \frac{\sum_{i=1}^{N_g} \sum_{j=1}^{N_r} \frac{p(i,j \mid \theta) j^2}{i^2}}{\sum_{i=1}^{N_g} \sum_{j=1}^{N_r} p(i,j \mid \theta)}$$

11: Long Run High Gray Level Emphasis (LRHGLE)

$$\mathbf{LRHGLE} = \frac{\sum_{i=1}^{N_g} \sum_{j=1}^{N_r} p(i,j \mid \theta) i^2 j^2}{\sum_{i=1}^{N_g} \sum_{j=1}^{N_r} p(i,j \mid \theta)}$$

E. Neighborhood Gray Tone Difference Matrix (NGTDM) Features

1: Coarseness

$$\mathbf{Coarseness} = \left[\sum_{i=1}^{N_g} p_i \times s_i \right]^{-1}$$

2: Contrast

$$\mathbf{Contrast} = \left[\frac{1}{N_g \times (N_g - 1)} \sum_{i=1}^{N_g} \sum_{j=1}^{N_g} p_i \times p_j \times (i - j)^2 \right] \left[\frac{1}{n} \sum_{i=1}^{N_g} s_i \right]$$

3: Busyness

$$\text{Busyness} = \frac{\sum_{i=1}^{N_g} p_i \times s_i}{\sum_{i=1}^{N_g} \sum_{j=1}^{N_g} \left| i \times p_i - j \times p_j \right|}$$

4: Complexity

$$\text{Complexity} = \sum_{i=1}^{N_g} \sum_{j=1}^{N_g} \frac{\left| i - j \right| \times \left(p_i s_i + p_j s_j \right)}{n \times \left(p_i + p_j \right)}$$

Chapter 2

Multimodal Medical Image Fusion in NSCT Domain

Gaurav Bhatnagar, Zheng Liu, and Q.M. Jonathan Wu

2.1   Introduction

In the recent years, medical imaging has attracted increasing attention due to its critical role in healthcare. However, different types of imaging modalities such as X-ray, computed tomography (CT), magnetic resonance imaging (MRI), magnetic resonance angiography (MRA), and other modalities provide limited information, which may be common among different modalities or unique to an individual modality. For example, X-ray and CT can provide images of dense structures like bones and implants with less distortion, but they cannot detect physiological changes [1]. Similarly, normal and pathological soft tissue can be better visualized by MRI image whereas PET can be used to provide better information on blood flow and flood activity with low spatial resolution. As a result, the anatomical and functional medical images need to be combined for a compendious view. For this purpose, multimodal image fusion has been identified as a promising solution which aims to integrate information from multiple modality images to obtain a more complete and accurate description of the same object [2]. Multimodal medical image fusion not only helps in diagnosing diseases, but it also reduces the storage cost by reducing storage to a single fused image instead of multiple-source images.

In general, the multimodal fusion can be categorized into three categories according to merging stage [3]: (1) pixel level fusion, (2) feature level fusion, and (3) decision level fusion. Among these, the pixel level fusion is the simplest and most straight-forward process, which directly combines the pixel data of original source images. In contrast, the latter two strategies are complicated ones as they involve the feature extraction, co-registration and classification of the features to obtain the final fused image. A detailed block diagram of all three strategies is depicted in Figure 2.1. Though these strategies work on different philosophies, the motivation is similar: to synthesize a fused image that is more informative for human visual perception and computer processing. The fusion of medical images usually belongs to the pixel level fusion due to the advantages of containing the original measured quantities, easy implementation and computational efficiency [3, 4]. Hence, in this chapter, all of the efforts are concentrated on pixel level fusion, and the terms image fusion or fusion are used for pixel level fusion.

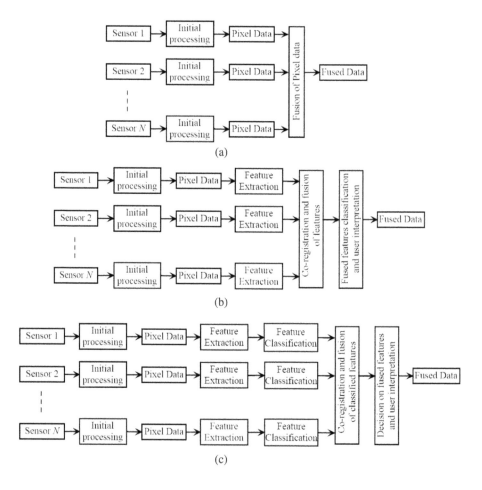

FIGURE 2.1: Block diagram of (a) pixel level, (b) feature level and (c) decision level fusion.

So far, extensive work has been done on pixel level image fusion techniques [4–51] with various techniques dedicated to multimodal medical image fusion [14, 15, 19, 20, 26–28, 32, 33, 36, 42, 48, 51]. These techniques have been broadly classified into four categories namely: (1) spatial domain techniques, (2) transform domain techniques, (3) sparse representation techniques, and (4) hybrid techniques. The spatial domain techniques are the simplest and the earliest techniques based on the modification of the pixel intensities. These techniques generally introduce lower contrast in the fused image and therefore are not suitable for the medical image fusion. On the other hand, transform domain techniques first project the source images onto some orthogonal basis to bifurcate the complementary information contained in them. The complementary information is then fused to obtain the final image. The well-known orthogonal basis includes principal component analysis (PCA) [7], independent component analysis (ICA), contrast pyramid (CP) [8], gradient pyramid (GP) [9] filtering and multiscale transforms [10–33]. Among these, PCA, ICA, CP and GP based techniques are not suitable for medical image fusion since the image features are sensitive to the human visual system exists in different scales [26]. In contrast, the development in the theory of multiscale transforms recognized and identified these transforms as an ideal tool for image fusion. These transforms include wavelet transform [10–16], framelet transform [17], dual-tree complex wavelet transform [18], rippet transform [19], curvelet transform [20, 21], contourlet transform [22–30], and shearlet transform [31–33]. However, it is argued that wavelet transform is good at isolated discontinuities but not good at edges and textured regions. Further, it captures limited directional information along vertical, horizontal and diagonal directions [20]. These issues are rectified in newly introduced multiscale decompositions, viz., contourlet, shearlet and their non-subsampled versions. Contourlet is a "true" 2-D sparse representation for two-dimensional signals like images where sparse expansion is expressed by contour segments. As a result, it can capture two-dimensional geometrical structures in visual information much more effectively than traditional multiscale methods [23]. The contourlet transform and its variants are efficient for fusion problems, though it will be more time consuming and has limited directionality [30]. Recently, the shearlet also has been successfully applied for image fusion [32]. Both the transforms are equally efficient for the multimodal image fusion, however, shearlet transform has the advantage over contourlet as it has more computationally efficient implementation and encompassing directionality [34].

In the third category, fusion has been done in the sparse representation of the source images [35–42]. The sparse representation of an image is motivated by simulating sparse coding mechanism of human vision system. In this representation an image is generally described by a sparse linear combination of patches selected from the overcomplete dictionary [43]. The sparse representation is successfully applied to image fusion, but it is lacking in its own fusion strategies which is its only drawback [44]. The fusion strategies are generally adopted from spatial/transform domain techniques such as window-based activity level measurement, choose-max and weighted average-based coefficients combined. It is evident that different categories have their specific strengths and weaknesses. For instance, the transform domain techniques can extract structures located at different scales but are not able to represent the low-frequency component sparsely. In contrast, sparse representation-based techniques enable a substantive overview which is exquisitely close to the source images, but the size of the dictionary is generally very large. Consequently, hybrid techniques try to combine the strengths of different categories into a single and structured way [45–49]. However, there is no unified process to combine different strategies of different categories. Therefore, the hybridization is application specific and may not be used in the applications it is not intended for. A comprehensive review of the multimodal image fusion techniques and their performance assessment methodologies can be found in [50].

In this chapter, a novel fusion framework is proposed for multimodal medical images based on non-subsampled contourlet transform. The core idea is to perform NSCT on the source images followed by the fusion of low- and high-frequency coefficients. The phase congruency and directive contourlet contrast features are unified as the fusion rules for low- and high-frequency coefficients. The phase congruency provides a contrast- and brightness-invariant representation of low-frequency coefficients whereas directive contrast efficiently determines the frequency coefficients from the clear parts in the high-frequency coefficients. The combinations of these two can preserve more details in source images and further improve the quality of the fused image. The efficiency of the proposed framework is carried out by the extensive fusion experiments on different multimodal CT/MRI, PET/MRI, SPECT/MRI, and X-ray/bone scan datasets. Further visual and quantitative analysis shows that the proposed framework provides a better fusion outcome when compared to conventional image fusion techniques.

The rest of the chapter is organized in the following manner: The brief description of used methodologies, which include $la\beta$-color space and phase congruency, is given in Section 2.2. A detailed introduction of non-subsampled contourlet transform is given in Section 2.3. In Section 2.4, the proposed multimodal medical image fusion framework is explained thoroughly followed by the experimental results and discussions in Section 2.5. Finally, the concluding remarks are described in Section 2.6.

2.2   Preliminaries

This section provides the description of concepts on which the proposed framework is based. These concepts include $la\beta$-color space and phase congruency and are described as follows.

2.2.1 $la\beta$-Color Space

Color space, also known as the color model, is an abstract mathematical model which simply describes the range of colors as tuples of numbers. A color space is essential to understand the color-perceiving capabilities of a specific device. In other words, it represents what a camera can see, a monitor can display or a printer can print. There are a variety of color spaces, such as RGB, CMY, HSV, HIS, and $la\beta$. Among these, $la\beta$ has the advantage as it is designed to approximate human vision and its l component closely matches human perception of lightness [52]. The RGB-to-$la\beta$ color space conversion can be summarized as follows. First, the RGB color space is converted to LMS cone space as

$$\begin{bmatrix} L \\ M \\ S \end{bmatrix} = \begin{bmatrix} 0.3811 & 0.5783 & 0.0402 \\ 0.1967 & 0.7244 & 0.0782 \\ 0.0241 & 0.1288 & 0.8444 \end{bmatrix} \begin{bmatrix} R \\ G \\ B \end{bmatrix} \qquad (2.1)$$

The data in LMS cone space show a great deal of skew, and this can be eliminated by converting LMS cone space channels to logarithmic color space, i.e.,

$$\Gamma = \lg L, \Omega = \lg M, \Psi = \lg S \tag{2.2}$$

The logarithmic color space is further transformed in three orthogonal color space ($la\beta$) as

$$\begin{bmatrix} l \\ a \\ \beta \end{bmatrix} = \begin{bmatrix} \dfrac{1}{\sqrt{3}} & 0 & 0 \\ 0 & \dfrac{1}{\sqrt{6}} & 0 \\ 0 & 0 & \dfrac{1}{\sqrt{2}} \end{bmatrix} \begin{bmatrix} 1 & 1 & 1 \\ 1 & 1 & -2 \\ 1 & -1 & 0 \end{bmatrix} \begin{bmatrix} \Gamma \\ \Omega \\ \Psi \end{bmatrix} \tag{2.3}$$

In $la\beta$ color space, l represents an achromatic channel whereas a and β are chromatic yellow–blue and red–green channels, and these channels are symmetrical and compact. A comparative illustration of RGB and $la\beta$ color spaces is depicted in Figure 2.2. The inversion, $la\beta$ to RGB space, is done by the following inverse operations.

$$\begin{bmatrix} \Gamma \\ \Omega \\ \Psi \end{bmatrix} = \begin{bmatrix} 1 & 1 & 1 \\ 1 & 1 & -1 \\ 1 & -2 & 0 \end{bmatrix} \begin{bmatrix} \dfrac{1}{\sqrt{3}} & 0 & 0 \\ 0 & \dfrac{1}{\sqrt{6}} & 0 \\ 0 & 0 & \dfrac{1}{\sqrt{2}} \end{bmatrix} \begin{bmatrix} l \\ a \\ \beta \end{bmatrix} \tag{2.4}$$

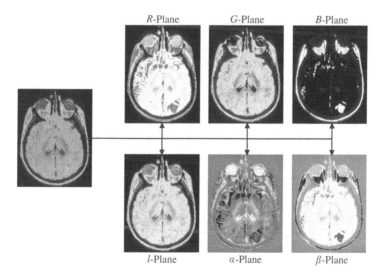

FIGURE 2.2: Comparative illustration of RGB and $la\beta$ color spaces.

and

$$\begin{bmatrix} R \\ G \\ B \end{bmatrix} = \begin{bmatrix} 4.4679 & -3.5873 & 0.1193 \\ -1.2186 & 2.3809 & -0.1624 \\ 0.0497 & -0.2439 & 1.2045 \end{bmatrix} \begin{bmatrix} 10^{\Gamma} \\ 10^{\Omega} \\ 10^{\Psi} \end{bmatrix} \tag{2.5}$$

2.2.2 Phase Congruency

Phase congruency is a measure of feature perception in the images which provides an illumination and contrast invariant feature extraction method [53, 54]. This approach is based on the Local Energy Model, which postulates that significant features can be found at points in an image where the Fourier components are maximally in phase. Furthermore, the angle at which phase congruency occurs signifies the feature type. The phase congruency approach to feature perception has been used for feature detection. First, logarithmic Gabor filter banks at different discrete orientations are applied to the image, and the local amplitude and phase at a point (x, y) are obtained. The phase congruency, $P_{x,y}^{o}$, is then calculated for each orientation o as follows.

$$P_{x,y}^{o} = \frac{\sum_{n} W_{x,y}^{o} \left\lfloor A_{x,y}^{o,n} \left(\cos\left(\phi_{x,y}^{o,n} - \tilde{\phi}_{x,y}^{o}\right) - \left|\sin\left(\phi_{x,y}^{o,n} - \tilde{\phi}_{x,y}^{o}\right)\right| \right) - T \right\rfloor}{\sum_{n} A_{x,y}^{o,n} + \varepsilon} \tag{2.6}$$

where $W_{x,y}^{o}$ is the weight factor based on the frequency spread, $A_{x,y}^{o,n}$ and $\phi_{x,y}^{o,n}$ are the respective amplitude and phase for the scale n, $\tilde{\phi}_{x,y}^{o}$ is the weighted mean phase, T is a noise threshold constant and ε is a small constant to avoid divisions by zero. The symbols $\lfloor \; \rfloor$ denote that the enclosed quantity is equal to itself when its value is positive and zero otherwise. Only energy values that exceed T, the estimated noise influence, are counted in the result. The appropriate noise threshold, T, is readily determined from the statistics of the filter responses to the image. An illustration of the phase congruency can be seen in Figure 2.3. For more details on phase congruency measure and its implementation see [54].

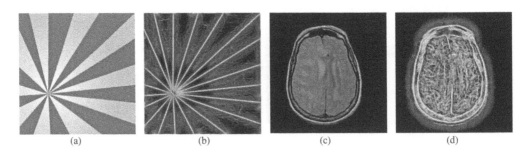

(a)   (b)   (c)   (d)

FIGURE 2.3: An illustration of phase congruency: (a,b) Rays and (c,d) CT images.

2.3 Non-Subsampled Contourlet Transform (NSCT)

NSCT, based on the theory of contourlet transform (CT), is a kind of multi-scale and multi-direction computation framework for the discrete images [55]. NSCT can be described mainly in three phases: (1) construction of non-subsampled pyramid structure that ensures the multiscale property, (2) construction of non-subsampled directional filter bank that gives directionality, and (3) the synchronization of non-subsampled pyramid structure and non-subsampled DFB to define the NSCT. The detailed description of these phases can be summarized as follows.

2.3.1 Construction of Non-Subsampled Pyramid (NSP)

This phase essentially introduces the multiscale property by using a two-channel non-subsampled filter bank. Similar to non-subsampled wavelet transform computed with à trous algorithm, one low-frequency and one high-frequency image can be produced at each NSP decomposition level. The subsequent NSP decomposition stages are carried out to decompose the low-frequency component available iteratively to capture the singularities in the image. As a result, NSP can result in $k+1$ sub-images, which consist of one low- and k high-frequency images having the same size as the source image where k denotes the number of decomposition levels. For the jth level, the ideal support of the low-pass filter is the region $[-\pi/2^{j}, \pi/2^{j}]^{2}$. Consequently, the ideal support of the high-pass filter is the region $[-\pi/2^{j-1}, \pi/2^{j-1}]^{2} / [-\pi/2^{j}, \pi/2^{j}]^{2}$. The up-sampling of first level filters is then performed to get the filters for subsequent stages. As a result, the multiscale property is achieved without additional filters. An illustration of NSP with three levels is given in Figure 2.4.

2.3.2 Construction of Non-Subsampled Directional Filter Bank (NS-DFB)

The NS-DFB consists of two-channel non-subsampled filter banks which are constructed by combining the directional fan filter banks and resampling operations. This combination resulted in a tree-structured filter bank that divides the frequency plane into directional wedges. The NS-DFB is then constructed by decimating the

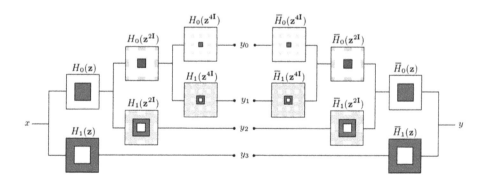

FIGURE 2.4: Three-stage non-subsampled pyramid decomposition.

down-samplers and up-samplers in the filter bank. Therefore, NS-DFB allows the direction decomposition with l stages in high-frequency images from NSP at each scale and produces 2^l directional sub-images with the same size as the source image. As a result, the NS-DFB offers the NSCT with the multi-direction property and provides more precise directional information. A four-channel NS-DFB constructed with two-channel fan filter banks is illustrated in Figure 2.5.

2.3.3 Synchronization of NSP and NS-DFB

The NSP and NS-DFB are combined to construct the NSCT. The combining procedure is depicted in Figure 2.6. While constructing the NSCT, one has to be careful in applying the directional filters to the coarser scales. This is due to the tree-structure of the NS-DFB as the directional response at the lower and upper frequency components generally suffer from aliasing. The aliasing can create problems in the upper scales of

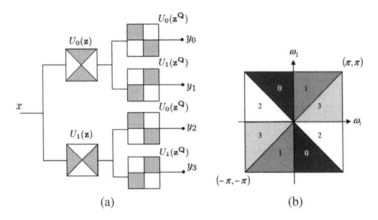

(a)   (b)

FIGURE 2.5: (a) Four-channel non-subsampled directional filter bank. (b) Corresponding frequency distribution.

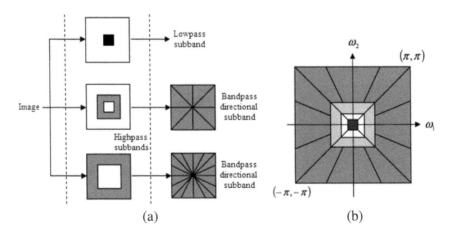

(a)   (b)

FIGURE 2.6: Non-subsampled contourlet transform: (a) Synchronization of NSP and NS-DFB, (b) idealized frequency partitioning obtained with the synchronization.

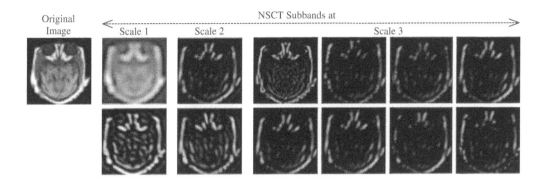

FIGURE 2.7: Three level NSCT of an MRI image.

the NSP. This situation can be avoided by judiciously up-sampling the NS-DFB filters. This can be done using an à trous filtering algorithm proposed in [9], which essentially ensures that each filter in the NS-DFB tree has the same complexity as that of the building-block fan NSFB. Therefore, the use of up-sampled filters does not increase computational complexity. A visual illustration of the NSCT can be seen in Figure 2.7.

2.4 Proposed Multimodal Medical Image Fusion Framework

In this section, we have discussed some of the motivating factors in the design of our approach to multimodal medical image fusion. The proposed framework is based on the directive contrast and phase congruency in the NSCT domain, which takes a pair of source images denoted by A and B to generate a composite image F. The basic condition in the proposed framework is that all the source images must be registered in order to align the corresponding pixels. The block diagram of the proposed framework is depicted in Figure 2.8, but before describing it, the definition of directive contrast is first described, which is as follows.

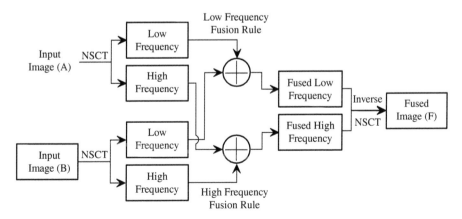

FIGURE 2.8: Block diagram of proposed multimodal medical image fusion framework.

2.4.1 Directive Contrast in NSCT Domain

The contrast feature measures the difference of the intensity value at some pixel from the neighboring pixels. The human visual system is highly sensitive to the intensity contrast rather than the intensity value itself. Generally, the same intensity value looks like a different intensity value depending on intensity values of neighboring pixels. Therefore, local contrast is developed and is defined as [11]

$$C = \frac{L - L_B}{L_B} = \frac{L_H}{L_B} \tag{2.7}$$

where L is the local luminance and L_B is the luminance of the local background. Generally, L_B is regarded as local low-frequency, and hence, $L - L_B = L_H$ is treated as local high-frequency. This definition is further extended as directive contrast for multimodal image fusion [12]. These contrast extensions take high-frequency as the pixel value in the multiresolution domain. However, considering a single pixel is insufficient to determine whether the pixels are from clear parts or not. Therefore, the directive contrast is integrated with the sum-modified-Laplacian [56] to get more accurate salient features.

In general, the larger absolute values of high-frequency coefficients correspond to the sharper brightness in the image and lead to the salient features such as edges, lines, region boundaries, and so on. However, these are very sensitive to the noise, and therefore, the noise will be taken as the useful information and the actual information in the fused images will be misinterpreted. Hence, a proper way to select high-frequency coefficients is necessary to ensure better information interpretation. Hence, the sum-modified-Laplacian is integrated with the directive contrast in NSCT domain to produce accurate salient features. Mathematically, the directive contrast in NSCT domain is given by

$$D_{l,\theta}(i,j) = \begin{cases} \dfrac{SML_{l,\theta}(i,j)}{I_l(i,j)}, & \text{if } I_l(i,j) \neq 0 \\ SML_{l,\theta}(i,j), & \text{if } I_l(i,j) = 0 \end{cases} \tag{2.8}$$

where $SML_{l,\theta}$ is the sum-modified-Laplacian of the NSCT frequency bands at scale l and orientation θ. On the other hand, $I_l(i, j)$ is the low-frequency sub-band at the coarsest level (l). The sum-modified-Laplacian is defined by following equation

$$SML_{l,\theta}(i,j) = \sum_{x=i-m}^{i+m} \sum_{y=j-n}^{j+n} \nabla^2_{l,\theta} I(x,y) \tag{2.9}$$

where

$$\nabla^2_{l,\theta} I(i,j) = \left| 2I_{l,\theta}(i,j) - I_{l,\theta}(i-\text{step},j) - I_{l,\theta}(i+\text{step},j) \right|$$
$$+ \left| 2I_{l,\theta}(i,j) - I_{l,\theta}(i,j-\text{step}) - I_{l,\theta}(i,j+\text{step}) \right| \tag{2.10}$$

In order to accommodate for possible variations in the size of texture elements, a variable spacing (step) between the pixels is used to compute SML and is always equal to 1 [56]. Further, the relationship between the contrast sensitivity threshold and background intensity is non-linear, which makes the human visual system highly sensitive to contrast variation [57]. Hence, the above integration must be improved to provide better details by exploiting the visibility of low-frequency coefficients in the above-mentioned definition. Hence, the directive contrast in NSCT domain is given as

$$
D_{l,\theta}(i,j) = \begin{cases} \left(\dfrac{1}{I_l(i,j)}\right)^a \dfrac{SML_{l,\theta}(i,j)}{I_l(i,j)}, & \text{if } I_l(i,j) \neq 0 \\ SML_{l,\theta}(i,j), & \text{if } I_l(i,j) = 0 \end{cases} \tag{2.11}
$$

where a is a visual constant representing the slope of the best-fitted lines through high-contrast data, which is determined by physiological vision experiments, and it ranges from 0.6 to 0.7 [57]. The proposed definition of directive contrast, defined by Equation 2.11, not only extracts more useful features from high-frequency coefficients but also effectively deflects noise to be transferred from high-frequency coefficients to fused coefficients.

2.4.2 Proposed Fusion Framework

In this subsection, the proposed fusion framework will be discussed in detail. Considering two perfectly registered source images A and B the proposed image fusion approach consists of the following steps:

1. Perform ℓ-level NSCT on the source images to obtain one low-frequency and a series of high-frequency sub-images at each level and direction θ, i.e.,

$$
A : \left\{ \mathcal{C}_\ell^A, \mathcal{C}_{l,\theta}^A \right\} \text{ and } B : \left\{ \mathcal{C}_\ell^B, \mathcal{C}_{l,\theta}^B \right\} \tag{2.12}
$$

where $\mathcal{C}_\ell^*$ are the low-frequency sub-images and $\mathcal{C}_{l,\theta}^*$ represent the high-frequency sub-images at level $l \in [1, \ell]$ in the orientation θ.

2. *Fusion of low-frequency sub-images*: The coefficients in the low-frequency sub-images represent the approximation component of the source images. The simplest way is to use the conventional averaging methods to produce the composite bands. However, this cannot give the fused low-frequency component of high quality for medical image because it leads to reduced contrast in the fused images. Therefore, a new criterion is proposed here based on the phase congruency. The complete process is described as follows.

    • First, the features are extracted from low-frequency sub-images using the phase congruency extractor (Equation 2.6), denoted by $P_{c_\ell^A}$ and $P_{c_\ell^B}$ respectively.

    • Fuse the low-frequency sub-images as

$$
\mathcal{C}_\ell^F(x,y) = \begin{cases} \mathcal{C}_\ell^A(x,y), & \text{if } P_{\mathcal{C}_\ell^A}(x,y) > P_{\mathcal{C}_\ell^B}(x,y) \\ \mathcal{C}_\ell^B(x,y), & \text{if } P_{\mathcal{C}_\ell^A}(x,y) < P_{\mathcal{C}_\ell^B}(x,y) \\ \dfrac{\displaystyle\sum_{k \in A,B} \mathcal{C}_\ell^k(x,y)}{2}, & \text{if } P_{\mathcal{C}_\ell^A}(x,y) = P_{\mathcal{C}_\ell^B}(x,y) \end{cases} \tag{2.13}
$$

3. *Fusion of high-frequency sub-images*: The coefficients in the high-frequency sub-images usually include details component of the source image. It is noteworthy that the noise is also related to high-frequencies and may cause miscalculation of the sharpness value and therefore affect the fusion performance. Therefore, a new criterion is proposed here based on directive contrast. The whole process is described as follows.

   - First, the directive contrast for NSCT high-frequency sub-images at each scale and orientation using Equations 2.8–2.10), denoted by $D_{\mathcal{C}_{l,\theta}^A}$ and $D_{\mathcal{C}_{l,\theta}^B}$ at each level $l \in [1, \ell]$ in the direction θ.
   - Fuse the high-frequency sub-images as

$$
\mathcal{C}_{l,\theta}^F(x,y) = \begin{cases} \mathcal{C}_{l,\theta}^A(x,y), & \text{if } D_{\mathcal{C}_{l,\theta}^A}(x,y) \geq D_{\mathcal{C}_{l,\theta}^B}(x,y) \\ \mathcal{C}_{l,\theta}^A(x,y), & \text{if } D_{\mathcal{C}_{l,\theta}^A}(x,y) < D_{\mathcal{C}_{l,\theta}^B}(x,y) \end{cases} \tag{2.14}
$$

4. Perform ℓ-level inverse NSCT on the fused low-frequency ($\mathcal{C}_\ell^F$) and high-frequency ($\mathcal{C}_{l,\theta}^F$) sub-images, to get the fused image ($\mathcal{F}$).

2.4.3   Extension to Multispectral Image Fusion

The IHS transform is a widely used multispectral image fusion method in the research community. It works in a simple way to convert a multispectral image from RGB to IHS color space. Fusion is then performed by fusing the I-component and source panchromatic image followed by the inverse IHS conversion to get the fused image. The IHS-based process can preserve the same spatial resolution as the source panchromatic image but seriously distorts the spectral (color) information in the source multispectral image. Therefore, the IHS model is not suitable for multimodal medical image fusion because a little distortion can lead to wrong diagnosis. This drawback can be rectified by incorporating different operations or a different color space such that undesirable cross-channel artifacts will not occur. Such a color space is *laβ*-space which is based on the fact that human perception research assumes that natural image processing is ideally done by the human visual system. The RGB-to-*laβ* color space conversion can be seen in Section 2.1.

The proposed fusion algorithm can easily be extended for the multispectral images by utilizing proposed fusion rules in *laβ* color space (see Figure 2.9). The core idea is to transform multispectral image from the RGB color space to the *laβ* color space using the process given above. Now, the panchromatic image and the achromatic channel (l) of the multispectral image are fused using the proposed fusion algorithm followed by the inverse *laβ*-to-RGB conversion to get the final fused image.

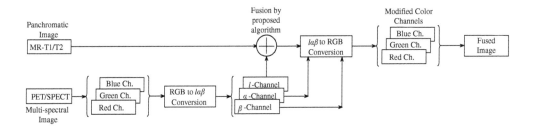

FIGURE 2.9: Block diagram for the multispectral image fusion: Synchronization of proposed fusion algorithm in $la\beta$ color space.

2.5 Results and Discussions

Some general requirements for fusion algorithm are: (1) it should be able to extract complementary features from input images, (2) it must not introduce artifacts or inconsistencies according to the human visual system and (3) it should be robust and reliable. Generally, these can be evaluated subjectively or objectively. The former relies on human visual characteristics and the specialized knowledge of the observer; hence it is vague, time-consuming and poor-repeatable but is typically accurate if performed correctly. The other one is relatively formal and easily realized by the computer algorithms, which generally evaluate the similarity between the fused and source images. However, selecting a proper consistent criterion with the subjective assessment of the image quality is rigorous. Hence, there is a need to create an evaluation system. Therefore, first an evaluation index system is established to evaluate the proposed fusion algorithm. These indices are determined according to the statistical parameters.

2.5.1 Evaluation Index System

1) *Normalized mutual information*: Mutual information (MI) is a quantitative measure of the mutual dependence of two variables. It usually shows measurement of the information shared by two images. Mathematically, MI between two discrete random variables U and V is defined as:

$$MI(U,V) = \sum_{u \in U} \sum_{v \in V} p(u,v) \log_2 \frac{p(u,v)}{p(u)\,p(v)} \qquad (2.15)$$

where $p(u, v)$ is the joint probability distribution function of U and V whereas $p(u)$ and $p(v)$ are the marginal probability distribution functions of U and V respectively. Based on the above definition, the quality of the fused image with respect to input images A and B can be expressed as

$$Q_{MI} = 2\left[\frac{MI(A,F)}{H(A)+H(F)} + \frac{MI(B,F)}{H(B)+H(F)} \right] \qquad (2.16)$$

where $H(A)$, $H(B)$ and $H(F)$ are the marginal entropies of images A, B and F respectively.

2) *Structural similarity-based metric:* Structural similarity (SSIM) is designed by modeling any image distortion as the combination of loss of correlation, radiometric and contrast distortion. Mathematically, SSIM between two variables U and V is defined as:

$$SSIM(U,V) = \frac{\sigma_{UV}}{\sigma_U \, \sigma_V} \frac{2\mu_U \, \mu_V}{\mu_U^2 + \mu_V^2} \frac{2\sigma_U \, \sigma_V}{\sigma_U^2 + \sigma_V^2} \tag{2.17}$$

where μ_U, μ_V are mean intensity and σ_U, σ_V, σ_{UV} are the variances and covariance respectively. Based on the definition of SSIM, a new way to use SSIM for the image fusion assessment is proposed in [58] and is defined as

$$Q_S = \begin{cases} \lambda(w)SSIM(A,F \mid w) + (1 - \lambda(w))SSIM(B,F \mid w), \\ \qquad\qquad \text{if } \ SSIM(A,B \mid w) \geq 0.75 \\ \max\left[SSIM(A,F \mid w), SSIM(B,F \mid w)\right], \\ \qquad\qquad \text{if } \ SSIM(A,B \mid w) < 0.75 \end{cases} \tag{2.18}$$

where w is a sliding window of size, which moves pixel by pixel from the top-left to the bottom-right corner and $\lambda(w)$ is the local weight obtained from the local image salience. See [58] for the detailed implementation of the aforementioned metric.

3) *Edge-based similarity measure*: The edge-based similarity measure gives the similarity between the edges transferred in the fusion process. Mathematically, $Q^{AB/F}$ is defined as

$$Q^{AB/F} = \frac{\displaystyle\sum_{i=1}^{M} \sum_{j=1}^{N} \left[Q_{i,j}^{AF} \, w_{i,j}^{x} + Q_{i,j}^{BF} \, w_{i,j}^{y} \right]}{\displaystyle\sum_{i=1}^{M} \sum_{j=1}^{N} \left[w_{i,j}^{x} + w_{i,j}^{y} \right]} \tag{2.19}$$

where A, B, and F represent the input and fused images respectively. The definitions of Q^{AF} and Q^{BF} are the same and given as

$$Q_{i,j}^{AF} = Q_{g,i,j}^{AF} \, Q_{a,i,j}^{AF}, \quad Q_{i,j}^{BF} = Q_{g,i,j}^{BF} \, Q_{a,i,j}^{BF} \tag{2.20}$$

where Q_g^{*F} and Q_a^{*F} are the edge strength and orientation preservation values at location (i, j) respectively for images A and B. The dynamic range for $Q^{AB/F}$ is $[0,1]$, and it should be as close to 1 as possible for better fusion.

2.5.2   Experiments on CT/MRI Image Fusion

To evaluate the performance of the proposed image fusion approach, four different datasets of the human brain are considered (see Figure 2.10). These images are characterized in two different groups: (1) CT–MRI and (2) MR-T1–MR-T2. The images

MR-CT Data Set   MR_T1-MR_T2 Data Set

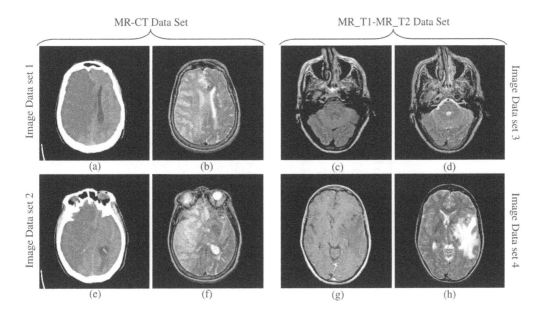

FIGURE 2.10: Multimodal medical image data sets: (a,e) CT image, (b,f) MRI image, (c,g) MR-T1 image, (d,h) MR-T2 image.

in Figure 2.10(a,e) and 10(b,f) are CT and MRI images whereas Figure 2.10(c,g) and 10(d,h) are T1-weighted MR image (MR-T1) and T2-weighted MR image (MR-T2). The corresponding pixels of the two input images have been perfectly co-aligned. All images have the same size of 256×256 pixels, with 256-level gray scale. The proposed medical fusion technique is applied to these image sets.

It can be seen that due to various imaging principles and environments, the source images with different modalities contain complementary information. For all these image groups, the results of the proposed fusion framework are compared with the traditional PCA (MS rule), contrast pyramid [8], gradient pyramid [9], wavelet [14], and contourlet [26] based methods.

The comparison of statistical parameters for fused images according to different fusion algorithms is shown in Table 2.1 and visually in Figure 2.11. From the figure and table, it is clear that the proposed algorithms not only preserve spectral information but also improve the spatial detail information in comparison to the existing algorithms (highlighted by red arrows), which can also be justified by the obtained maximum values of evaluation indices (see Table 2.1). The PCA algorithm gives baseline results. For all experimental images, PCA-based methods give poor results relative to other algorithms. This was expected because this method has no scale selectivity; therefore it cannot capture prominent information localized in different scales. This limitation is rectified in pyramid- and multiresolution-based algorithms but at the cost of quality, i.e. the contrast of the fused image is reduced which is greater in pyramid-based algorithms and comparatively less in multiresolution-based algorithms. Among multiresolution-based algorithms, the proposed algorithm based on NSCT performs better. The main reason behind the better performance is the use of the directive contrast which essentially takes advantage of both contrast and visibility. Further, the shift-invariance property of NSCT produces a clearer and more natural fused image

TABLE 2.1: Evaluation Indices for Fused Medical Images

Images	Indices	PCA	Contrast [8]	Gradient [9]	Wavelet [14]	Contourlet [26]	Proposed
Image Dataset 1	Q_{MI}	**1.5645**	1.0372	0.9417	0.8812	1.0380	1.0813
	Q_S	0.8415	0.8059	0.7495	0.7551	0.7968	**0.8726**
	$Q^{AB/F}$	0.5226	0.6863	0.7055	0.6669	0.7424	**0.7560**
Image Dataset 2	Q_{MI}	0.9436	0.9412	0.8466	0.8340	0.9463	**0.9681**
	Q_S	0.7357	0.7545	0.6253	0.7247	0.7663	**0.7705**
	$Q^{AB/F}$	0.5545	0.5922	0.7235	0.7038	0.7776	**0.7825**
Image Dataset 3	Q_{MI}	1.0259	1.1723	0.8688	1.0286	1.1405	**1.1865**
	Q_S	0.9054	0.9068	0.8083	0.9043	0.9259	**0.9527**
	$Q^{AB/F}$	0.6730	0.5909	0.6655	0.6403	0.6916	**0.6991**
Image Dataset 4		**1.4094**	1.0672	0.9556	0.9535	1.0684	1.0695
		0.7945	0.8389	0.7815	0.7497	0.8112	**0.8117**
		0.4408	0.4301	0.5636	0.5326	0.6780	**0.6783**

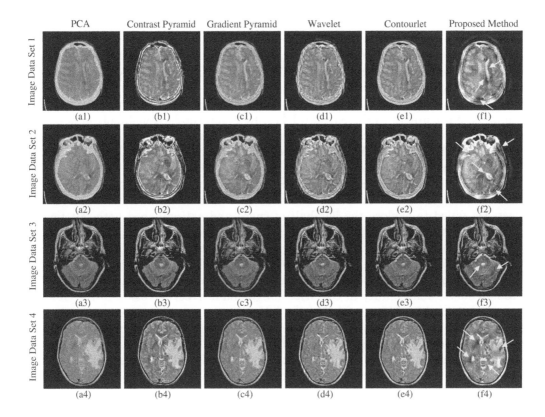

FIGURE 2.11: The multimodal medical image fusion results of different fusion algorithms: Fused images from: (a,f,k,p) PCA-based technique, (b,g,l,q) gradient pyramid-based technique, (c,h,m,r) wavelet-based technique, (d,i,n,s) contourlet-based technique, (e,j,o,t) proposed technique.

than other multiresolution-based fused results. This is also justified by the fact that shift-invariant decomposition overcomes pseudo-Gibbs phenomena successfully and improves the quality of the fused image around edges.

2.5.3 Clinical Examples of PET/MRI and SPECT/MRI Image Fusion

Despite the great success of the MRI-CT fusion, its role in neuroscience is considered to be limited compared with the potential of PET-MRI and SPECT/MRI fusion. PET can provide functional eloquent brain areas such as motor or speech regions by using specific activation tasks. On the other hand, single-photon emission computed tomography (SPECT) images reveal the metabolic change that has significant clinical value [59]. Therefore, in the modern era PET/MRI and SPECT/MRI fusion are analyzed over MRI-CT fusion for better diagnosis in different diseases. In order to demonstrate the practical value of the proposed scheme in medical imaging, three clinical cases are considered where PET/MRI and SPECT/MRI medical modalities are used. These include the case of Alzheimer's, subacute stroke and brain tumor respectively. The images have been downloaded from the Harvard University site (www.med.harvard.edu/AANLIB/home.html).

The first case is of a 70-year-old man who began experiencing difficulty with memory about nine months prior to imaging. He had a history of atrial fibrillation and was taking warfarin. He had become lost on several occasions and had difficulty orienting himself in unfamiliar circumstances. This man is affected by the disease, namely Alzheimer's (highlighted by red arrows). Figure 2.12(a–b) shows the MRI and PET images of the person. The MRI image showed a globally widened hemispheric sulci, which is more prominent in the parietal lobes. Regional cerebral metabolism is markedly abnormal, with hypometabolism in the anterior temporal and posterior parietal regions. These changes are bilateral, but the right hemisphere is slightly more affected than the left, and the posterior cingulate is relatively spared.

Figure 2.13 shows the subacute stroke case of a 65-year-old man who suddenly experienced tingling in the left hand and arm and on examination had a syndrome of left neglect: He failed to explore the left half of space and extinguished both left tactile and left visual stimuli when presented on both sides simultaneously. The MRI study revealed that the frontal pole in the old infract is replaced with the high signal of cerebrospinal fluid left after liquefaction necrosis (highlighted by red arrow). The beginning of new symptoms corresponds to the right parietal infarction with hyperperfusion. There is a subtle abnormality in the MRI image and a hyperperfusion in the SPECT image (highlighted by red arrow). Figure 2.14 shows the recurrent tumor case of a 51-year-old woman who sought medical attention because of gradually increasing right hemiparesis (weakness) and hemianopia (visual loss). At craniotomy, left parietal anaplastic astrocytoma was found. A right frontal lesion was biopsied. The evolution of high tumor thallium uptake, indicating astrocytoma recurrence, is revealed by the SPECT

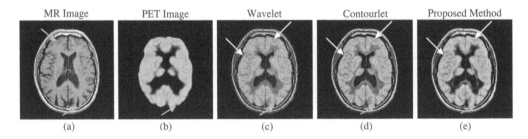

FIGURE 2.12:	Brain images of the man affected with Alzheimer's: (a) MRI image, (b) PET image; fused images by (c) wavelet, (d) contourlet, (e) proposed method.

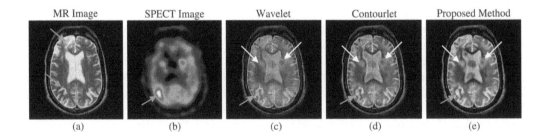

FIGURE 2.13:	Brain images of the man affected with subacute stroke: (a) MRI image, (b) SPECT image; fused images by (c) wavelet, (d) contourlet, (e) proposed method.

MR Image	SPECT Image	Wavelet	Contourlet	Proposed Method

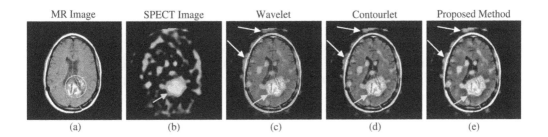

(a)	(b)	(c)	(d)	(e)

FIGURE 2.14: Brain images of the man with recurrent tumor: (a) MRI image, (b) SPECT image; fused images by (c) wavelet, (d) contourlet, (e) proposed method.

study, which is highlighted by a red arrow in the SPECT image whereas a large region of mixed signal on the MRI image gives the signs of the possibility of active tumor (encircled).

Here, the results are compared with the best two algorithms obtained with the earlier analysis, i.e., Guihong et al. [14] and Yang et al. [26]. From Figures 2.12–2.14, it can be observed that all the fusion algorithms have fairly good spatial information, but the spectral distortions are somewhat high in the existing algorithms, i.e., spectral information is lost in the case of existing algorithms which is greater in the case of [14] and comparatively lesser in [26]. The color information is also distorted in the existing algorithms (shown with the white arrows). On the contrary, the color information is least distorted and the spatial details are as clear as the original MRI image, and the spectral features are also natural. Therefore, the proposed method not only preserves the crucial features existing in both original images but also improves the color information when compared to existing methods.

2.5.4 Clinical Examples of X-ray and Bone Scintigraphy Image Fusion

The efficiency of the proposed scheme is further assessed using two more clinical cases where X-ray and bone scintigraphy medical imaging modalities are used. The damage in the bones such as fracture or dislocated joint or foreign object location can primarily be determined using an X-ray. In contrast, certain bone abnormalities such as bone inflammation (bone pain due to a fracture), light fractures, bone damage detection (due to certain infections) and cancer of the bone can be identified by bone scintigraphy [60]. These abnormalities may not be visible in X-rays. The bone scintigraphy result is generally perceived as hot and cold spots. Hot spots usually indicate the affected area in the bones, and the perfect location can be obtained by combining X-rays with bone scintigraphy.

The first case is of a 55-year-old policeman with about one-week onset of right forefoot pain. The X-ray was performed for the possibility of fracture, and no fracture was found (see Figure 2.15(a–b)). A bone scintigraphy was then performed, which demonstrated increased vascularity with moderate periarticular activity in the right first and second MTP joints and at the right second/third tars-metatarsal joints and proximal metatarsal articulations. There was normal activity within the metatarsal shafts as demonstrated by the fused image, however, indicating inflammatory arthropathy/synovitis. The second case is of a 72-year-old female with a medical history of osteoarthritis in the left hand and a past history of melanoma. A three-phase bone scintigraphy was performed on the wrists and left hand wherein the delayed bone scintigraphy showed bilateral increased tracer activity at the bases of the thumbs (see Figure 2.16(a–b)). After fusion with the X-ray, increased focal activity is observed in the carpometacarpal joint of the thumb, both the trapezium and trapezoid and the adjacent junction with the scaphoid.

Here, the results of the proposed technique are again compared with the best two algorithms obtained from earlier analysis, i.e., Guihong et al. [14] and Yang et al. [26]. Similar observations, which were observed from the previous sub-section, are discovered from Figures 2.15–2.16. In other words, it can be observed that all the fusion algorithms have fairly good spatial information, but the spectral information is lost in the case of existing algorithms which is greater in the case of [14] and comparatively lesser

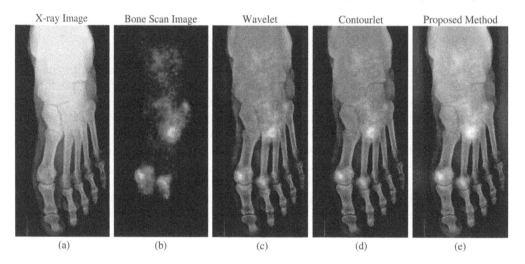

X-ray Image	Bone Scan Image	Wavelet	Contourlet	Proposed Method
(a)	(b)	(c)	(d)	(e)

FIGURE 2.15: Forefoot images of the man affected with inflammatory arthropathy: (a) X-ray image, (b) bone scintigraphy image; fused images by (c) wavelet, (d) contourlet, (e) proposed method.

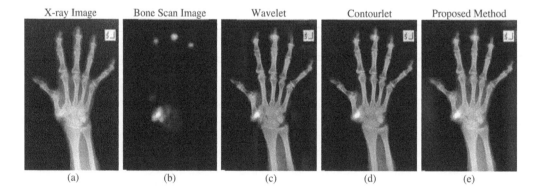

X-ray Image	Bone Scan Image	Wavelet	Contourlet	Proposed Method
(a)	(b)	(c)	(d)	(e)

FIGURE 2.16: Hand images of the women affected with osteoarthritis: (a) X-ray image, (b) bone scintigraphy image; fused images by (c) wavelet, (d) contourlet, (e) proposed method.

in [26]. It is also clear that the additional clinical information that was not evident from either X-ray or bone scintigraphy was interpreted in isolation which was useful to improve diagnostic utility and/or interpretive certainty.

2.6 Conclusions

In this chapter, a novel image fusion framework is proposed for multimodal medical images, which is based on non-subsampled contourlet transform, phase congruency and directive contrast. After the source images are decomposed by NSCT transform, two different rules are used by which more information can be preserved in the fused image. A phase congruency-based rule is proposed and used to fuse low-frequency components. In contrast, directive contrast-based rule is proposed and used to fuse high-frequency components. In the experiments, several groups of CT/MRI and MR-T1/MR-T2 images are fused using conventional fusion algorithms and the proposed framework. The objective and subjective comparisons demonstrate that the proposed framework not only preserves the image details but also meliorates the image visually compared to the existing fusion methods.

References

1. F. Maes, D. Vandermeulen and P. Suetens, "Medical image registration using mutual information," *Proceedings of IEEE*, vol. 91, no. 10, 2003, pp. 1699–1721.
2. R.S. Blum and Z. Liu, *Multi-Sensor Image Fusion and Its Applications*, CRC Press, 2005.
3. H.B. Mitchell, *Image Fusion: Theories, Techniques and Applications*, Springer-Verlag, 2010.
4. S. Li, X. Kang, L. Fang, J. Hu and H. Yin, "Pixel-level image fusion: A survey of the state of the art," *Information Fusion*, vol. 33, 2017, pp. 100–112.
5. G. Bhatnagar, Q.M. Jonathan Wu and B. Raman, "Real time human visual system based framework for image fusion," in *Proceedings of International Conference on Signal and Image Processing*, Trois-Rivieres, Quebec, Canada, 2010, pp. 71–78.
6. A. Cardinali and G.P. Nason, "A statistical multiscale approach to image segmentation and fusion," in *Proceedings of International Conference on Information Fusion*, Philadelphia, PA, USA, 2005, pp. 475–482.
7. P.S. Chavez and A.Y. Kwarteng, "Extracting spectral contrast in Landsat thematic mapper image data using selective principal component analysis," *Photogrammetric Engineering and Remote Sensing*, vol. 55, 1989, pp. 339–348.
8. A. Toet, L.V. Ruyven and J. Velaton, "Merging thermal and visual images by a contrast pyramid," *Optical Engineering*, vol. 28, no. 7, 1989, pp. 789–792.
9. V.S. Petrovic and C.S. Xydeas, "Gradient-based multiresolution image fusion," *IEEE Transactions on Image Processing*, vol. 13, no. 2, 2004, pp. 228–237.
10. H. Li, B.S. Manjunath and S.K. Mitra, "Multisensor image fusion using the wavelet transform," *Graph Models Image Processing*, vol. 57, no. 3, 1995, pp. 235–245.
11. A. Toet, "Hierarchical image fusion," *Machine Vision and Applications*, vol. 3, no. 1, 1990, pp. 1–11.
12. G. Bhatnagar and B. Raman, "A new image fusion technique based on directive contrast," *Electronic Letter on Computer Vision and Image Analysis*, vol. 8, no. 2, 2009, pp. 18–38.
13. S. Li, B. Yang and J. Hu, "Performance comparison of different multiresolution transforms for image fusion," *Information Fusion*, vol. 12, no. 2, 2011, pp. 74–84.

14. Q. Guihong, Z. Dali and Y. Pingfan, "Medical image fusion by wavelet transform modulus maxima," *Optics Express*, vol. 9, 2001, pp. 184–190.

15. Y. Yang, D.S. Park, S. Huang and N. Rao, "Medical image fusion via an effective wavelet-based approach," *EURASIP Journal on Advances in Signal Processing*, vol. 2010, 2010, pp. 44-1–44-13.

16. R. Redondo, F. Sroubek, S. Fischer and G. Cristobal, "Multifocus image fusion using the log-Gabor transform and a multisize windows technique," *Information Fusion*, vol. 10, no. 2, 2009, pp. 163–171.

17. G. Bhatnagar and Q.M.J. Wu, "An image fusion framework based on human visual system in framelet domain," *International Journal of Wavelets, Multiresolution and Information Processing*, vol. 10, no. 1, 2012, pp. 12500021–30.

18. J.J. Lewis, R.J.O. Callaghan, S.G. Nikolov, D.R. Bull and N. Canagarajah, "Pixel- and region-based image fusion with complex wavelets," *Information Fusion*, vol. 8, no. 2, 2007, pp. 119–130.

19. S. Das, M. Chowdhury and M.K. Kundu, "Medical image fusion based on ripplet transform type-I," *Progress In Electromagnetics Research B*, vol. 30, 2011, pp. 355–370.

20. F.E. Ali, I.M. El-Dokany, A.A. Saad and F.E. Abd El-Samie, "Curvelet fusion of MR and CT images," *Progress In Electromagnetics Research C*, vol. 3, 2008, pp. 215–224.

21. F. Nencini, A. Garzelli, S. Baronti and L. Alparone, "Remote sensing image fusion using the curvelet transform," *Information Fusion*, vol. 8, no. 2, 2007, pp. 143–156.

22. X. Qu, J. Yan, H. Xiao and Z. Zhu, "Image fusion algorithm based on spatial frequency-motivated pulse coupled neural networks in nonsubsampled contourlet transform domain," *Acta Automatica Sinica*, vol. 34, no. 12, 2008, pp. 1508–1514.

23. Q. Zhang and B.L. Guo, "Multifocus image fusion using the nonsubsampled contourlet transform," *Signal Processing*, vol. 89, no. 7, 2009, pp. 1334–1346.

24. S. Yang, M. Wang, Y. Lu, W. Qi and L. Jiao, "Fusion of multiparametric SAR images based on SW-nonsubsampled contourlet and PCNN," *Signal Processing*, vol. 89, no. 12, 2009, pp. 2596–2608.

25. Y. Chai, H. Li and X. Zhang, "Multifocus image fusion based on features contrast of multiscale products in nonsubsampled contourlet transform domain," *Optik - International Journal for Light and Electron Optics*, vol. 123, no. 7, 2012, pp. 569–581.

26. L. Yang, B.L. Guo and W. Ni, "Multimodality medical image fusion based on multiscale geometric analysis of contourlet transform," *Neurocomputing*, vol. 72, 2008, pp. 203–211.

27. T. Li and Y. Wang, "Biological image fusion using a NSCT based variable-weight method," *Information Fusion*, vol. 12, no. 2, 2011, pp. 85–92.

28. G. Bhatnagar, Q.M. Jonathan Wu and Z. Liu, "Directive contrast based multimodal medical image fusion in NSCT domain," *IEEE Transactions on Multimedia*, vol. 15, no. 5, 2013, pp. 1014–1024.

29. S. Yang, M. Wang, L. Jiao, R. Wu and Z. Wang, "Image fusion based on a new contourlet packet," *Information Fusion*, vol. 11, no. 2, 2010, pp. 78–84.

30. K.P. Upla, M.V. Joshi and P.P. Gajjar, "An edge preserving multiresolution fusion: Use of contourlet transform and MRF prior," *IEEE Transactions on Geoscience and Remote Sensing*, vol. 53, no. 6, 2015, pp. 3210–3220.

31. Q. Miao, C. Shi, P. Xu, M. Yang and Y. Shi, "A novel algorithm of image fusion using shearlets," *Optics Communications*, vol. 284, no. 6, 2011, pp. 1540–1547.

32. L. Wang, B. Li and L. Tian, "Multi-modal medical image fusion using the inter-scale and intra-scale dependencies between image shift-invariant shearlet coefficients," *Information Fusion*, vol. 19, no. 1, 2014, pp. 20–28 .

33. L. Wang, B. Li and L. Tian, "EGGDD: An explicit dependency model for multi-modal medical image fusion in shift-invariant shearlet transform domain," *Information Fusion*, vol. 19, no. 1, 2014, pp. 29–37.

34. G. Easley, D. Labate and W.-Q. Lim, "Sparse directional image representations using the discrete shearlet transform," *Applied and Computational Harmonic Analysis*, vol. 25, no. 1, 2008, pp. 25–46.

35. B. Yang and S. Li, "Multifocus image fusion and restoration with sparse representation," *IEEE Transactions on Instrumentation and Measurement*, vol. 59, no. 4, 2010, pp. 884–892.
36. S. Li, H. Yin and L. Fang, "Group-sparse representation with dictionary learning for medical image denoising and fusion," *IEEE Transactions on Biomedical Engineering*, vol. 59, no. 12, 2012, pp. 3450–3459.
37. C. Chen, Y. Li, W. Liu and J. Huang, "Image fusion with local spectral consistency and dynamic gradient sparsity," in *Proceedings of IEEE Conference on Computer Vision and Pattern Recognition*, Columbus, OH, 2014, pp. 2760–2765.
38. B. Yang and S. Li, "Pixel-level image fusion with simultaneous orthogonal matching pursuit," *Information Fusion*, vol. 13, no. 1, 2012, pp. 10–19.
39. H. Yin and S. Li, "Multimodal image fusion with joint sparsity model," *Optical Engineering*, vol. 50, no. 6, 2011, pp. 067007.1–067007.10.
40. H. Yin, S. Li and L. Fang, "Simultaneous image fusion and super-resolution using sparse representation," *Information Fusion*, vol. 14, 2013, pp. 229–240.
41. M. Nejati, S. Samavi and S. Shirani, "Multi-focus image fusion using dictionary-based sparse representation," *Information Fusion*, vol. 25, 2015, pp. 72–84.
42. M. Kim, D.K. Han and H. Ko, "Joint patch clustering-based dictionary learning for multimodal image fusion," *Information Fusion*, vol. 27, no. 1, 2016, pp. 198–214.
43. B.A. Olshausen and J.F. David, "Emergence of simple-cell receptive field properties by learning a sparse code for natural images," *Nature*, vol. 381, no. 6583, 1996, pp. 607–609.
44. Q. Zhang and M. Levine, "Robust multi-focus image fusion using multi-task sparse representation and spatial context," *IEEE Transactions on Image Processing*, vol. 25, no. 5 2016, pp. 2045–2058.
45. Y. Liu, S. Liu and Z. Wang, "A general framework for image fusion based on multiscale transform and sparse representation," *Information Fusion*, vol. 24, no. 1, 2015, pp. 147–164.
46. Y. Jiang and M. Wang, "Image fusion with morphological component analysis," *Information Fusion*, vol. 18, no. 1, 2014, pp. 107–118.
47. J. Wang, J. Peng, X. Feng, G. He, J. Wu and K. Yan, "Image fusion with nonsubsampled contourlet transform and sparse representation," *Journal of Electronic Imaging*, vol. 22, no. 4, 2013, pp. 043019–043019.
48. S. Daneshvar and H. Ghassemian, "MRI and PET image fusion by combining IHS and retina-inspired models," *Information Fusion*, vol. 11, no. 2, 2010, pp. 114–123.
49. Y. Zhang and G. Hong, "An IHS and wavelet integrated approach to improve pan-sharpening visual quality of natural colour IKONOS and QuickBird images," *Information Fusion*, vol. 6, no. 3, 2005, pp. 225–234.
50. Z. Liu, E. Blasch, G. Bhatnagar, V. John, W. Wu and R.S. Blum, "Fusing synergistic information from multi-sensor images: An overview from implementation to performance assessment," *Information Fusion*, vol. 42, 2018, pp. 127–145.
51. N. Boussion, M. Hatt, F. Lamare, C.C.L. Rest and D. Visvikis, "Contrast enhancement in emission tomography by way of synergistic PET/CT image combination," *Computer Methods and Programs in Biomedicine*, vol. 90, no. 3, 2008, pp. 191–201.
52. D.L. Ruderman, T.W. Cronin and C.C. Chiao, "Statistics of cone responses to natural images: Implications for visual coding," *Journal of the Optical Society of America-A*, vol. 15, no. 8, 1998, pp. 2036–2045.
53. P. Kovesi, "Image features from phase congruency," *Videre: A Journal of Computer Vision Research*, vol. 1, no. 3, 1999, pp. 2–26.
54. P. Kovesi, "Phase congruency: A low-level image invariant," *Psychological Research Psychologische Forschung*, vol. 64, no. 2, 2000, pp. 136–148.
55. A.L. da Cunha, J. Zhou and M.N. Do, "The nonsubsampled contourlet transform: Theory, design, and applications," *IEEE Transactions on Image Processing*, vol. 15, no. 10, 2006, pp. 3089–3101.
56. W. Huang and Z. Jing, "Evaluation of focus measures in multi-focus image fusion," *Pattern Recognition Letters*, vol. 28, no. 4, 2007, pp. 493–500.

57. A.B. Watson, "Efficiency of a model human image code," *Journal of the Optical Society of America-A*, vol. 4, no. 12, 1987, pp. 2401–2417.
58. C. Yang, J. Zhang, X. Wang and X. Liu, "A novel similarity based quality metric for image fusion," *Information Fusion*, vol. 9, 2008, pp. 156–160.
59. M.D. Devous, "Single-photon emission computed tomography in neurotherapeutics," *NeuroRx*, vol. 2, no. 2, 2005, pp. 237–249.
60. G. Currie, R. Pearce and J. Wheat, "Planar fusion: A pictorial review," *The Internet Journal of Radiology*, vol. 11, no. 1, 2009, pp. 1–5.

Chapter 3

Computer Aided Diagnosis in Pre-Clinical Dementia: From Single-Modal Metrics to Multi-Modal Fused Methodologies

Yi Lao, Sinchai Tsao, and Natasha Lepore

3.1   Introduction

As our population ages, the prevalence of dementia has been markedly increasing. Alzheimer's disease (AD), which affects 10% of people over 65 and 50% of people over 85, is the most common type of dementia [1]. The average cost of care in the final five years of life for a person with dementia is above $287,000, which is 64% higher than those who died of heart disease and 66% higher than those with cancer [2]. The global costs of dementia were an estimated total of $604 billion in 2010, and this number will increase by 85% by 2030 based on a conservative estimation [3]. In the past decades, encouraging advances have been made in medical interventions to improve the quality of life of AD patients. However, no disease-modifying therapies of AD have been developed to date [4, 5]. Emerging evidence has shown that cardiovascular disease (CVD) and preclinical cardiovascular risk factors are linked to the etiology of dementia, including AD [6–13]. Specifically, some findings suggest a direct influence of vascular diseases in accelerating amyloid β accumulation [11, 14]. The entanglement of cardiovascular and

47

neural factors is further evidenced by the recently hypothesized connection between the locus coeruleus (LC) in the pons of the brainstem and AD. In this scenario, AD is mediated by the integrated modulatory function of LC on the heart rate, attention, memory and cognitive functions [15]. While an effective treatment for AD is still out of reach, there are established therapeutic strategies for CVD, and its risk factors are also clinically modifiable [16–18]. Therefore, disentangling the effects of CVD and its risk factors on the development of AD has implications for symptom management and may potentially alter clinical outcomes for pre-dementia patients.

Neuronal disturbances, notably the accumulation of amyloid plaques and neurofibrillary tangles, are thought to begin years before the onset of clinical symptoms [19, 20]. There is growing evidence showing the vulnerability of brain cortical and subcortical structures in early AD and mild cognitive impairment (MCI) – a precursor to AD and other types of dementia [21–24]. Therefore, differentiating the effects of different CVD profiles on the anatomy of the brain in MCI would provide important insights into the effects of preventable CVD factors on the initial course of AD. Nevertheless, efforts aimed at differentiating vascular diseases from MCI report inconsistent results [25–27]. In particular, Hayden et al. identified a set of memory and executive tests in prodromal vascular dementia (VaD) that are distinguishable from prodromal AD [25]. Nordlund et al. confirmed the differences in executive function between MCI subjects with and without vascular disease and also reported differences in speed, attention, and visuospatial functions in these two groups [26]. However, no differences between vascular and non-vascular types of MCI were found by other studies [27]. These discrepancies may partially be caused by different inclusion criteria for vascular disease (i.e. with or without stroke), coupled to analysis techniques that do not have the required detection sensitivity. This highlights the need for sensitive and reliable algorithms that can help in decoupling the vascular component of preclinical dementia and thus aid in early diagnosis and therapeutic design.

Being the largest white matter (WM) structure in the brain, the corpus callosum (CC) plays a fundamental role in integrating and communicating multiple brain domains and is involved in perceptual, cognitive, memory, learning, and volitional functions [28]. Given the extensive and broad projections between the CC and the cortex, cortical focal or diffuse alterations may have a secondary trans-synaptical effect on CC anatomy. Thus, the investigation on CC may yield a better coverage of underlying risk to functional domains than that of any other single subcortical structure. Enlarge cortical sulci and ventricles are two typical brain hallmarks seen in patients suffering from AD and other types of dementia. Situated right above the lateral ventricles and bridging the two hemispheres, the CC is one of the putative areas affected by AD. In fact, CC alterations are readily seen in early AD patients [29–32] and, to a lesser extent, in MCI patients [33, 34]. Due to the high demand for blood supply from several main arterial systems [35], the CC is also susceptible to being affected in vascular diseases. Specifically, the impact of vascular factors such as hypertension and hypoperfusion on CC was reported in both human [36–38] and animals studies [39, 40]. Moreover, increased stroke risk is particularly correlated with WM abnormalities in the genu of the CC [36]. Therefore, anatomical alterations of the CC may serve as potential discriminators of the concurrent but possibly different effects of vascular and neurodegenerative components.

To assess CC anatomical alterations, a technique based on single imaging modalities is a typical choice. Structural MRI, particularly T1-weighted MRI scans, has been effective in deciphering brain parenchyma loss [22, 29–32 41–43]. On brain T1-MRI scans, previous studies have attempted to investigate CC impairments in early AD using CC mid-sagittal cross-sectional area [31], partitioned CC subregions [42, 43], or CC circularity [32]. Being able to characterize WM microstructure, diffusion tensor imaging (DTI) has been shown as a promising tool in discriminating MCI from normal controls [44–51]. Specifically, reduced WM integrity, reflected by changes in diffusivity metrics from DTI, has been associated with MCI in several WM regions using comparisons from manually placed region of interests [44–46, 49], automatically parcellated functional regions [51] or estimated WM skeletons [47, 48, 50]. However, these findings involve coarsely defined or non-specific regions that span the whole brain and may not be useful in localizing the underlying alterations of specific white matter tracts. Further, parenchyma and diffuse injuries often occur concomitantly in WM structures such as the CC, while above studies regarded each aspect on its own [31, 32, 41–43, 49, 51] or by comparing them side-by-side [33, 48]. Therefore, innovative ways to integrate morphological and microstructural subcortical data are needed to provide a more complete picture of CC changes brought on by brain injury and to develop more specific neural phenotypic imaging biomarkers.

In this chapter, we will describe an innovative, truly combined analysis of structural and diffusion MRI, and we will evaluate its application in prodromal dementia. Specifically, we will group the enrolled MCI subjects into high and low vascular risk profiles (will be referred to as MCI-l and MCI-h groups in the following sections) and then conduct pairwise statistical analyses on the fused morphological and diffusion properties among these two MCI subgroups as well as on aging controls without cognitive impairment. The aims are two-fold: (1) to test whether the vascular component affects distinct regional alterations that may help us to distinguish different MCI subtypes; (2) to further evaluate the feasibility and sensitivity of using the presented T1 and DTI fusion method to analyze subcortical alterations.

3.2   A Multi-Modal MRI Fused Analysis

3.2.1   Study Population

Fifty-eight subjects aged 66 to 89 were enrolled and grouped based on their clinical dementia rating (CDR) and vascular risk profile. Subjects' vascular risks were evaluated using Framingham cardiovascular risk profile (FCRP) – an estimate of the ten-year cardiovascular risk of an individual [52] – and their previous histories of myocardial infarction. Specifically, patients were grouped into 15 MCI subjects with low vascular risk (76.40 ± 7.65 years, CDR=0.5, low FCRP scores), 18 MCI subjects with high vascular risk (78.39 ± 5.69 years, CDR=0.5, high FCRP scores or had previous clinical diagnosis of myocardial infarction) and 25 healthy controls (76.68 ± 6.40 years). Subjects with confounding neurological conditions, such as stroke, were excluded.

Brain T1 and DT-MR scans of all the subjects were obtained using a 3T MRI scanner (SIEMENS Trio TIM, Siemens Healthcare, Erlangen, Germany). DTI data were acquired using an echo-planar imaging (EPI) sequence, with a voxel size of $2 \times 2 \times 2$ mm^3, resolution of $128 \times 128 \times 60$, b-value of 1000 s/mm^2, and 60 gradient directions. Anatomical data were acquired using a 3D magnetization-prepared rapid gradient-echo (MPRAGE) sequence, with a voxel size of $1 \times 1 \times 1$ mm^3, resolution of $256 \times 256 \times 192$, TE=2.98 ms, TR=2500 ms, and TI=1100 ms. A brief summary of the T1 and DTI sequences can be found in Table 3.1.

TABLE 3.1:　Brief Summary of Used MRI Sequences

	T1 Weighted (MPRAGE)	Diffusion Weighted (EPI)
ST (mm)	1	2
VS	$1\times1\times1$	$2\times2\times2$
Resolution	$256\times256\times192$	$128\times128\times64$
TR/TE (ms)	2500/2.98	9000/101
TI (ms)	1100	–
NA	1	1
NGD	–	60
b-value (s/mm^2)	–	1000

FOV=field of view, ST=slice thickness, VS=voxel size, TR/TE=time of repetition/time of echo, TI=time of inversion, FA=flip angle, NA=number of averages, NGD=number of gradient directions.

Each subject, in addition to being imaged via T1-MRI and DTI, was evaluated on the Mini-Mental State Exam (MMSE) as a marker of cognitive function, as well as a standardized battery of neuropsychological tests, consisting of verbal memory summary score (MEMSC) for verbal memory, non-verbal memory summary score (NVMEMSC) for non-verbal memory, executive function summary score (EXECSC) to measure executive function, and global cognition summary score (GLOBSC) to assess global cognition. These measures have previously been described in the literature and are commonly used in neuropsychological assessments [53].

3.2.2　Image Preprocessing

All the T1 data were preprocessed and linearly registered to the same template space – selected randomly from one of the controls that was previously transformed to MNI space [54]. On the linearly registered T1 images, each subject's corpus callosum (CC) was manually traced on the mid-sagittal plane, and the lateral boundaries were determined where the CC starts to radiate into and merge with cerebral white matter. Subsequently, 3D surface representations and conformal mesh grids of the CC were constructed using a conformal mapping program [55]. One-to-one correspondence between vertices was obtained through constrained harmonic-based registration [55].

All the DTI data were first preprocessed, which included brain masking, eddy current correction, echo-planar imaging distortion correction, and tensor estimation. To truly integrate DTI and T1 information, we transformed the DT images from each subject to their corresponding T1 space, using linear registration between the b0-weighted and T1 images. The linear transformation matrices saved from the T1 registration were then applied on the linearly aligned b0 images, to transform the diffusion information to the space of the T1 template. After each of the linear registrations, the diffusion tensors were resampled using the b0 transformation matrices and rotated according to the underlying anatomy. These steps were achieved using MedINRIA [56].

3.2.3　Calculation of Shape Descriptors

The main shape descriptors used here were derived from the surface-based registration in Section 2.2. After the constrained harmonic registration, a Jacobian of the transformation (J) was calculated for each vertex on the surface. Its determinant (det J), which represents the extent of surface warping in the corresponding location during registration, is commonly used in statistical shape analysis to capture changes within surfaces. To further incorporate directional changes in surfaces, the deformation tensor ($\sqrt{JJ^T}$), which represents a 2D ellipse with axes showing the extent and direction of changes between the two surfaces at that location, is introduced in multivariate statistical analysis. Since deformation tensors do not form a vector space where Euclidean space defined statistical formulae are not directly applied, projections of deformation tensors on the log-Euclidean space ($\log\sqrt{JJ^T}$) are often chosen to facilitate computations [57–60]. In this work, both det J and $\log\sqrt{JJ^T}$ were selected in the following statistical analyses as shape descriptors.

3.2.4 Integration of Diffusion Parameters

In order to establish correspondences between DTI and T1 information, surface-based sampling was performed, and a simple illustration can be found in Figure 3.1. To project diffusion indices of each of the CCs onto its surface, we first calculated midlines of all the 3D CCs and then collected diffusion parameters to each surface vertex along its corresponding radius to the midline, specifically using:

$$\left\| \frac{(\overrightarrow{X-M}) \times (\overrightarrow{P-M})}{\|\overrightarrow{X-M}\|} \right\| \leq R. \tag{3.1}$$

and

$$(\overrightarrow{X-P}) \cdot (\overrightarrow{P-M}) \geq 0. \tag{3.2}$$

Here X, M, P are the (x, y, z) coordinates of a vertex in the surface, the corresponding point of the vertex in the midline and a voxel within the 3D representations, respectively,

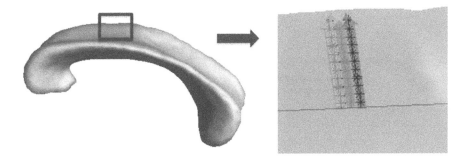

FIGURE 3.1: Surface of CC and the illustration of sampled voxels. The red line on the right side of the figure is the midline of the CC, blue stars represent voxels within CC, and pink, yellow, and green crosses represent surface vertices. In the direction perpendicular to the midline and pointing to each vertex, voxels within pink, yellow, and green areas are projected to the vertices with the corresponding colors. Mean index of projected voxels is assigned to the vertex for later statistics.

while R represents a pre-defined distance between P to the line of $\overline{X-M}$. The sampling process can be more intuitively visualized in Figure 3.1. The first equation is used to assign the voxels close enough ($\leq R$) to the corresponding radius of the vertex, while the second equation constrains the sampling to the voxels within the space between the midpoint to the vertex (not in the prolongation direction). According to our previous pilot study, we chose $R=0.6$ mm³ to make sure each of the vertices has some voxels assigned and to minimize overlap with neighboring vertices [61].

3.2.5 Group Comparisons

Our statistical analyses were conducted using either the morphometry information, the diffusion information or a combination of both. A brief illustration of the statistical pipeline can be found in Figure 3.2. Specifically, vertex-wise univariate student t-tests or multivariate Hotelling's T^2 tests were performed based on the following variables:

1. Morphometry information: univariate det J and multivariate ($s1$, $s2$, $s3$) from the logged deformation tensors, which are independent elements of the matrix (see Figure 3.3, first and second rows).
2. Diffusion information: univariate measurement (mean fractional anisotropy (FA)) along the radius of the CC for each vertex and multivariate λ_1 and λ_2 (see Figure 3.3,

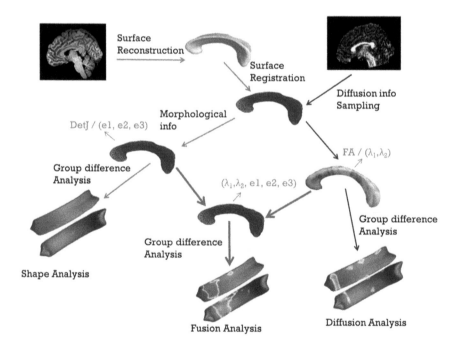

FIGURE 3.2: Illustration of the proposed statistical pipeline. Statistical analyses were conducted using anatomical features from three aspects: the morphometry information only (indicated using blue arrows), the diffusion information only (indicated using red arrows), and a fusion of both (indicated using purple arrows). The corresponding parameters used in each of the analyses were marked in green.

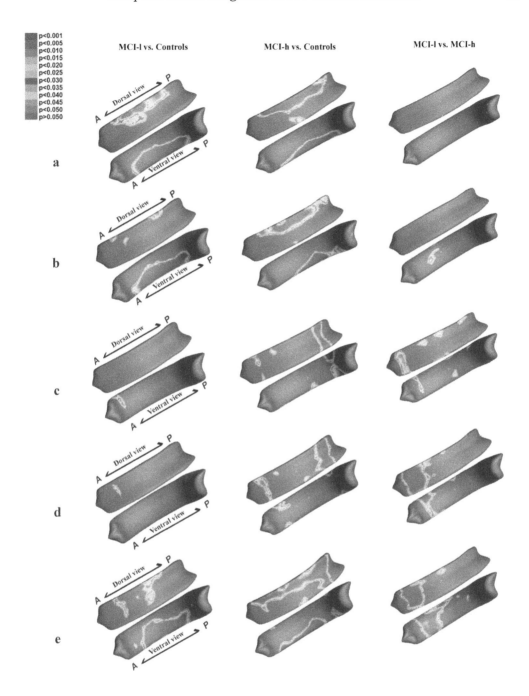

FIGURE 3.3: Group analysis of MCI-l vs. controls (first column), MCI-h vs. controls (second column), and MCI-l vs. MCI-h (third column) using five different measures: (a) det J; (b) ($s1$, $s2$, $s3$); (c) mean FA; (d) (λ_1, λ_2); (e) (λ_1, λ_2, $s1$, $s2$, $s3$). Vertex-wise corresponding p-values are color-coded according to the color bar in the upper left corner. P-maps are smoothed using heat kernel algorithm [69]. In addition, whole structure-wise corrected p-values are presented in Table 3.2.

third and fourth rows). Note: we did not include λ_3. Being small, this value is susceptible to noise and may reduce detection power. While this is fine for additive measures such as FA as the effect will be negligible (as the value is small), it is a much bigger issue when analyzing a multivariate vector of statistics, where each eigenvalue is treated as an independent measure.

3. A fusion of morphometry ($s1$, $s2$, $s3$) and diffusion indices (λ_1 and λ_2) (see Figure 3.3, fifth row).

Here, one of the primary purposes is to determine a method that can sensitively detect underlying anatomical differences between our MCI groups. Our general criteria for measurements selection were: firstly, to use the most representative or generally used measurements in both shape and diffusion analysis for comparison and secondly, to compare these to what we hypothesized would be a joint structural and diffusion measure with enough sensitivity to detect subtle underlying differences between groups, based on ours and others' prior studies. Other commonly used univariate diffusion metrics, such as mean diffusivity (MD) and radial diffusivity (RD), showed significant but less powerful results than FA. Similarly, other combinations of morphometry and diffusion, such as (λ_1, λ_2, det J), showed less significant results than (λ_1, λ_2, $s1$, $s2$, $s3$) (see Table 3.2). Therefore, we mainly investigated the above five most representative univariate or multivariate measurements in our following analysis.

Given the fact that our subjects were from a relatively large age range (66–89 years), we used linear regression to factor out the effect of age. For each feature value separately, we have:

$$F = \beta_0 + \beta_1 \times \text{age} + \beta_2 \times \text{group} + \text{error}. \tag{3.3}$$

Where F is one of the features we previously obtained, β_0, β_1, β_2 are the corresponding correlation coefficients. Groups are coded as dummy variable: 0 for controls, 1 for MCI-l group, and 2 for MCI-h group. All the following statistics were performed on linearly regressed features.

For each of the tests, two types of permutations were performed: a vertex-based one to avoid the normal distribution assumption and one over the whole segmented image to correct for multiple comparisons, as described in [58, 62]. Permutation-based corrections are independent of the distributions of the statistics, which are commonly non-parametric in voxel- or vertex-wise analyses. Furthermore, permutation-based multiple comparison corrections are less stringent than conventional sequential correction methods [63], as they do not assume independence of neighboring voxels or vertices, and they are widely accepted in brain image analyses [58, 60, 64–66]. In each of the permutation tests, 10,000 permutations were employed.

3.2.6 Correlation Analysis

The imaging measurements were further evaluated with respect to overall mental status demonstrated by Mini Mental State Examination scores, as well as four domain specific neuropsychological measurements: MEMSC, NVMEMSC, EXECSC, and GLOBSC. Four subjects with missing neuropsychological test data were excluded, leaving a total of 54 subjects for the correlation analysis. After controlling for age, Pearson's correlation analyses were conducted to determine specific contributions of regional CC to cognitive performance, in terms of shape (represented by det J) or WM integrity

TABLE 3.2: Structure-Wise Corrected *p*-Values for Different Measurements Are Displayed

2*Measurements	2*Aspects	2*MCI-l vs. ctls	2*MCI-h vs. ctls	2*MCI-l vs. MCI-h
det J	Shape	LightCyan0.0309	LightCyan0.0292	0.9901
$(s1, s2, s3)$	Shape	LightCyan0.0425	LightCyan0.0415	0.1550
FA	Diffusion	0.4810	Grey0.0815	Grey0.0813
RD	Diffusion	0.4123	Grey0.0872	0.3769
MD	Diffusion	0.3903	Grey0.1018	0.6219
(λ_1, λ_2)	Diffusion	0.3440	Grey0.0731	Grey0.0673
$(\lambda_1, \lambda_2, \det J)$	Fusion	LightCyan0.0434	LightCyan0.0154	Grey0.0697
$(\lambda_1, \lambda_2, s1, s2, s3)$	Fusion	LightCyan0.0173	LightCyan0.0153	LightCyan0.0107

All the *p*-values were corrected using a permutation based analysis with 10,000 permutations. Significance is set to $p < 0.05$, and is highlighted in light cyan. *P*-values implying trends are highlighted in light grey.

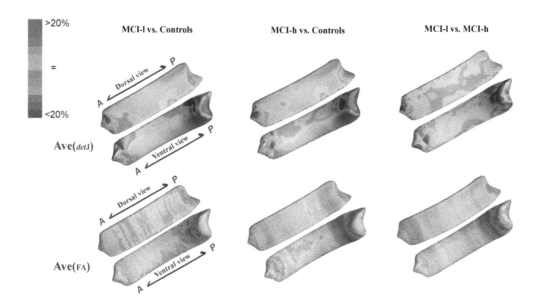

FIGURE 3.4: Average map of det J and mean FA between groups are color-coded according to the color bar in the upper left corner. When these results are compared with Figure 3.3, we can see the main direction of change: nearly all the significance areas fell in the Controls > MCI-l, Controls > MCI-h, as well MCI-l > MCI-h areas.

(represented by mean FA). To evaluate the association of neuropsychological performances with the combined feature of shape and diffusion properties of CC (represented by $(\lambda_1, \lambda_2, s1, s2, s3)$, we continued to perform a distance correlation – a generalization to the classical bivariate measurements of dependence [67]. Similar to the permutation corrections employed in group-wise comparisons, 10,000 permutations were applied in each of the correlation tests, as described in [62, 68].

3.3 Results

Figure 3.3 shows vertex-wise group differences among three groups based on five different measures: det J, $(s1, s2, s3)$, mean FA, $(\lambda_1, \lambda_2,)$, and fused $(\lambda_1, \lambda_2, s1, s2, s3)$. The corresponding structure-wise corrected p-values are displayed in Table 3.2. The final statistical results on the CC surface in Figure 3.3 are smoothed using heat kernel algorithm as described in [69].

The MCI-l group showed alterations spanning the mid-body and the posterior surface of the CC as compared to controls, with significant structure-wise differences detected by det J, $(s1, s2, s3)$, and $(\lambda_1, \lambda_2, s1, s2, s3)$ measurements; the MCI-h group presented broad areas of alterations mainly located in the dorsal anterior, mid-body, and splenium of the CC compared to controls, with significant structure-wise differences detected by det J, $(s1, s2, s3)$, and $(\lambda_1, \lambda_2, s1, s2, s3)$ measurements, as well as trends detected by mean FA and (λ_1, λ_2) measurements. For the MCI-h vs. MCI-l, the main clusters were located in the genu of the CC, and the fusion measurements reached structure-wise significance, while mean FA and (λ_1, λ_2) showed trends. It is important to note that up-sampling the relatively low-resolution DTI data resulted in same or similar diffusion indices appearing in surrounding vertices on the CC surface, thus causing the band-like areas shown in the significance map (see Figure 3.3).

To intuitively understand the direction of alterations, we also mapped the average maps of vertex-wise det J and FA between three groups, as shown in Figure 3.6. Comparing Figure 3.6 with Figure 3.3, we can see that nearly all the significance areas fell in the Controls > MCI-l, Controls > MCI-h, as well as MCI-l > MCI-h areas. These findings point to shrinkage and reduced WM integrity in the CCs of MCI-l and MCI-h patients as compared to those of normal controls.

3.3.1   Correlation Analysis Results

The vertex-wise significant p-map and correlation coefficients r-map from Pearson's correlation analyses between neuroanatomical measurements (det J and mean FA) and five neuro-cognitive indices (MMSE, MEMSC, NVMEMSC, EXECSC, GLOBSC) are displayed in Figures 3.5 and 3.6. The corresponding structural-wise corrected p-values are shown in Table 3.3.

For surface shape measurements, represented by det J, significant regional correlations are seen in clusters mainly located in anterior and posterior CC; two of the five correlation tests (EXECSC and GLOBSC) hit structure-wise significances according to Table 3.3. As to WM integrity, represented by mean FA along radial direction, four of the five correlation tests (MMSE, MEMSC, EXECSC and GLOBSC) showed anatomically meaningful correlations with WM integrity in the dorsal anterior CC. According to Table 3.3, EXECSC showed a structure-wise correlation with mean FA that represented a trend ($p = 0.0857$), and two of the five measurements (MEMSC and GLOBSC) reached structure-wise significance. As to the combined shape and diffusion features $(\lambda_1, \lambda_2, s1, s2, s3)$, areas of significant correlations are mainly consistent with those detected by shape and WM integrity separately, while nonverbal memory scores (NVMEMSC) showed anatomically meaningful correlations in the posterior CC, which were not fully captured by shape or diffusion feature-based bivariate correlations.

3.4   Discussion and Conclusion

In this chapter, we introduced a T1 and DTI fused 3D corpus callosum analysis in MCI subjects with high and low cardiovascular risk profiles. Using the presented method, the MCI-h group showed widespread atrophy and reduced WM integrity spanning nearly the whole CC as compared to controls, with the largest cluster located on the posterior end. In the group analysis between the MCI-l group and the controls, similar alterations were mainly shown in the middle to the posterior regions. When comparing the MCI-h and the MCI-l groups, our fusion method detected significant disparities in the dorsal anterior CC. These findings together indicate a consistent influence of MCI on the midbody to the posterior end of the CC and, importantly, a distinct effect of cardiovascular profile on the genu. Moreover, these same regions presented significant correlations with neurophysiological battery tests including MEMSC, EXECSC, and

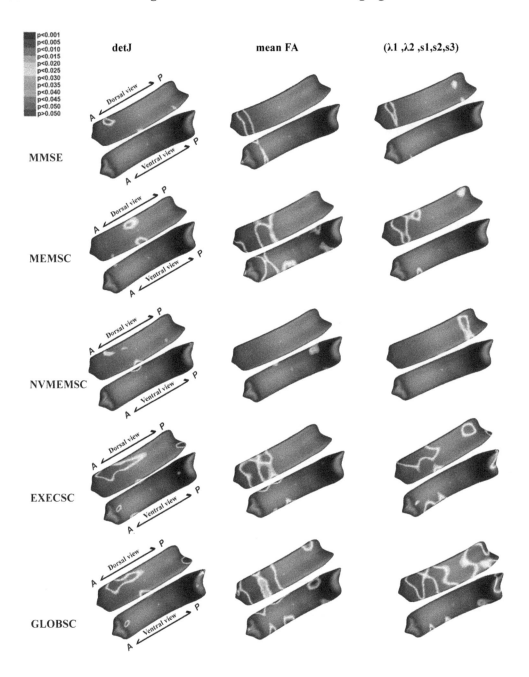

FIGURE 3.5: Vertex-wise significance results of correlation analyses between det J as well as mean FA vs. 5 neuropsychological scores. P-maps are smoothed using heat kernel algorithm [69].

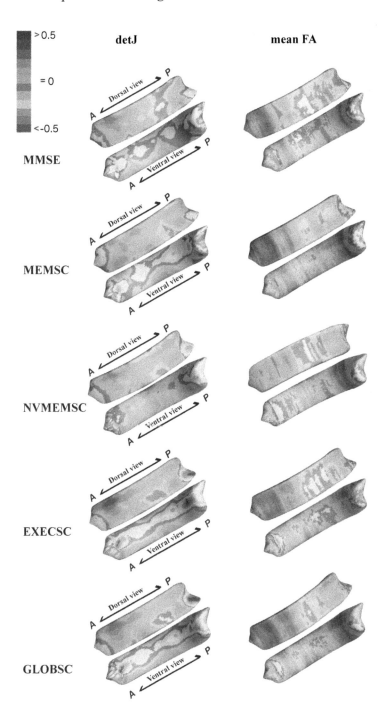

FIGURE 3.6: Vertex-wise correlation coefficient maps have been generated based on det *J* (left column) and mean FA (right column), respectively. Comparing this figure with Figure 3.5, we can see the direction of the correlation analyses: nearly all the significant regions represent positive correlations.

TABLE 3.3: Structure-Wise Corrected p-Values for Different Measurements Are Displayed

Measurements	det J	mean FA	$(\lambda_1, \lambda_2, s1, s2, s3)$
MMSE	0.5290	0.1639	0.7184
MEMSC	0.3322	LightCyan0.0147	0.2886
NVMEMSC	0.2902	0.2736	0.3857
EXECSC	LightCyan0.0411	Grey0.0857	LightCyan0.0066
GLOBSC	LightCyan0.0131	LightCyan 0.0208	LightCyan0.0048

All the p-values were corrected using a permutation based analysis with 10,000 permutations. Significance is set to $p < 0.05$, and is highlighted in light cyan. P-values implying trends are highlighted in light grey.

GLOBSC. Our findings provide important anatomical supports to the co-existence of MCI-subtypes and may yield new insights on the distinct role of cardiovascular components in the etiology of dementia. The T1 and DTI fusion analysis presented in this study yields higher statistical detection power and may provide a new direction in analyzing subcortical WM structures.

3.4.1 Significance of the Study

In the past decades, MCI has drawn increasing attention as a way to study the early evolution of AD and as a potential target for early interventions. However, not all MCI patients will convert to AD as it is not a homogenous state and may also precede other types of dementia, such as vascular dementia (VaD). One of the main difficulties in accurately predicting the MCI–AD conversion is due to the co-morbidity and shared etiology with other types of disease [70, 71]. In particular, CVD – precursors of VaD – are also important risk factors of AD [6, 7]. Epidemiological studies have shown that cardiovascular risk factors such as hypertension, high cholesterol and diabetes are highly associated with cognitive decline and AD [7, 72, 73]. Nevertheless, no established mechanisms clarify how CVD participates in the development of AD and whether there is a dissociable impact of CVD and cardiovascular risk factors (CRF).

A considerable number of researchers suggested a selective cognitive decline pattern associated with vascular pathology, and efforts have been made to differentiate vascular disease from AD or MCI using cognitive performance. Ingles et al. investigated the neuropsychological performance in elderly subjects five years before diagnosis and reported a selectively low abstract reasoning performance in subjects who evolved toward vascular cognitive impairment compared to those who converted to AD or remained normal [74]. Marra et al. reported executive functioning problems in the vascular form of MCI subjects, whereas those with the degenerative form of MCI were mainly impaired in episodic memory tasks [75]. Similar to these, greater impairments in executive function have been reported in MCI with a vascular component [25, 26, 76, 77]. However, there is no consensus on the executive dysfunction predominance of vascular pathology. A neurophysiological study aiming to discriminate cerebrovascular disease from AD observed a slightly severe but non-significant executive dysfunction other than memory failure in autopsy-defined cerebrovascular disease group [78]. Moreover, a study comparing cognitive profiles in MCI subjects with different etiologies reported no differences of memory or executive function between the vascular and non-vascular types of MCI [27]. These inconsistencies hint at the limitation of using neuropsychological patterns only as dissociable features for vascular injury/dementia [78].

With the advent of MRI, multiple modalities such as arterial spin labeling (ASL), structural and diffusion MRI have been utilized to investigate the vascular pathology on brain anatomy. It is widely accepted that vascular disease or risk factors are associated with an accelerated rate of cerebral atrophy [79–81] and decreased glucose metabolism [82], while the information as to whether these associations are independent of MCI is minimal. In the handful of studies investigating vascular pathology in the context of MCI or AD, mixed results were reported. Specifically, a volume based T1-weighted MRI study on CC in AD, VaD as well as mild ambiguous subjects reported significantly smaller anterior and posterior CC regions in the AD group, significantly smaller anterior CC regions in the VaD group and no difference in the sub-clinical dementia group as compared to controls, while no differences between VaD and AD groups were detected [83]. A region of interest-based DTI study on MCI subjects reported decreased WM integrity in selected frontal, temporal, parietal lobe regions as well as the corpus callosum in both groups of patients with and without subcortical vascular changes, while the WM alterations in the centrum semiovale and parietal lobe were believed to be more associated with the vascular pathology [84]. A T1-MRI-based study on cortical thickness and grey matter (GM) volume in MCI subjects with different levels of cardiovascular profile observed an association between elevated vascular risk factors and atrophy in the temporal and parietal lobe – the same regions affected by AD [10].

Difficulties inherent in diagnosing AD and VaD, and different inclusion criteria of vascular diseases clouded the interpretation of these studies. Moreover, the limited statistical power of volume-based methodologies and measurements focusing on single modality measurement further reduced the sensitivity of these studies to the potential neuroanatomical alterations lurking in pre-clinical stages. Therefore, in-vivo measurements with higher sensitivity are highly desired to further explore the concurrent but possibly distinct effects of CVD and MCI on the brain. In this work, we focused on neurodegenerative patterns in pre-dementia stages and excluded compounding conditions such as stroke that directly alter brain anatomy. In the present study, the joint T1 and DTI measurement in 3D CC successfully pinpointed dorsal anterior CC regions that significantly differed between the MCI-l and MCI-h groups of subjects. These findings provided new anatomical evidence for the distinctive impact of vascular pathology before clinical magnification of dementia and are thus of great importance in early preventive intervention and in guiding therapeutic design. The sensitivity of our methodology and the relevance of detected anatomic alterations and the corresponding neuroanatomic and functional implications will be described in detail in the next sections.

3.4.2 Methodological Considerations

Postmortem and probabilistic tractography studies have shown that the CC is not a homogenous structure, in terms of fiber composition [85] and topographical distribution [86]. Group differences of brain WM, including the CC, are typically analyzed based on whole structure volume or some anatomically motivated partitions, voxels, midlines or midplanes. The whole volume-based method facilitates an intuitive and coarse estimation of CC anatomy [31, 32] but has been ineffective in detecting subtle anatomical changes. Studies based on subdivisions of the CC are more tuned to the heterogeneity of CC but may easily be biased due to inconsistent classification (i.e. partitioning into three, five, or seven compartments), as well as arbitrary delineation of subdivisions [42,

43, 87]. Voxel-based methods give poor localization of differences in anatomical regions compared to surface-based ones and may be contaminated by differently oriented tracts [88], while midline- or midplane-based methods rely on assumptions that WM perpendicular to the mid-line or the mid-plane is uniformly distributed. The method introduced in this chapter uses clearly defined CC regions traced in T1 images, that are largely preserved within tract information projected onto the surface of the corpus callosa. The 3D representations may better localize injury in heterogeneous CC and may have higher statistical detection power to identify the neuro-circuit alterations underlying the observed anatomical alterations.

The vulnerability of the CC in MCI has been reported in both structural and diffusion studies [33, 34, 44, 45, 47, 48, 51]. Structural MRI is a typical choice and has been effective in deciphering brain parenchyma loss [22, 29, 30], while DTI has been promising in characterizing white matter microstructure alterations [44, 45, 47, 48, 51, 89]. These previous studies have been analyzing the brain parenchyma or its diffusion properties on their own [44, 45, 51] or by comparing them side-by-side [33, 48, 87]. To the best of our knowledge, none have tried to truly combine these two features into one statistical analysis.

As shown in the presented study, measurements in both structural and diffusion aspects have shown significant between-group differences, confirming the concomitantly occurring parenchyma and diffuse injuries in CC. Group analyses based on structural information (det J and $(s1, s2, s3)$) have successfully detected alterations in the mid-body to the posterior end, while group analyses based on diffusion information (mean FA and (λ_1, λ_2)) are more sensitive to alterations in the anterior and the posterior ends of CC. Here, for the first time, we fuse the T1-based morphometry information and DTI-based diffusion information into one, single analysis. In all three group-wise analyses, the fused method successfully outperforms analyses based on structural information or diffusion information alone. Moreover, in group comparisons between MCI-h and MCI-l group, only the fusion method reached overall significance, while no significance is detected if T1 and DTI measures are considered separately. These results show the feasibility of using the T1 and DTI fusion method to increase detection power.

3.4.3   Anatomic and Functional Implications

The corpus callosum spans the midline of the brain and possesses numerous connections to surrounding structures. At its most anterior end, the genu, WM tracts innervate the frontal lobes, and infarction of the genu has been reported to result in frontal lobe dysfunction [90–92]. The splenium, at the posterior end, lies in close proximity to the hippocampus through the amygdala [93], and alterations of the splenium are often associated with impairments in memory and visual perception [94, 95]. In our cohort of subjects, these anatomic-functional relationships have been further validated in the correlation analysis of regional CC FA values with five neuropsychological tests. As shown in Figures 3.5 and 3.6, executive functioning, verbal memory and global cognitive profile, which are higher brain functions that are extensively involved frontal networks [96, 97], showed significant associations with dorsal anterior CC, while nonverbal memory, which is highly correlated with hippocampus functioning [98], selectively correlates with the ventral posterior CC.

Thus, the anatomy of the dorsal anterior CC is more predictive of frontal lobe involved executive and verbal memory functions, while the posterior CC is more associated with temporal and parietal nonverbal memory. Taken together, the constellation of group-wise analysis and anatomical-neurophysiological correlations imply a main effect of MCI on medial to posterior cortex involved memory functions, while CVD and its risk factors add to the symptoms through frontal connections.

Comparing to controls, CCs in the MCI-h group presented similar but more extensive alterations than those from the MCI-l group, suggesting an 'interactive' impact of the vascular and neurodegenerative factors on brain morphometry. These are generally in line with anatomical findings showing associations between vascular brain injuries or risk factors and aggregated brain atrophy, especially in the parietal and temporal lobe [10, 99]. In terms of the comparison within MCI subgroups, significant differences resided in the dorsal anterior CC, implying an 'additive' effect of vascular pathology on the brain frontal network that is differentiable from the non-vascular neurodegeneration influences. The frontal lobe hypo-perfusion detected by ASL-MRI has been reported to be associated with worse executive and memory function [100]. Vascular risk factors measured by FCRP and high-density lipoprotein cholesterol are found to link with thinner frontotemporal cortex [99]. In the presented study, the distinctive impact of the cardiovascular factor in the genu of CC is consistent with the anatomic relationship between cognitive profile and the frontal lobe [99] and provides further neuroanatomic evidence supporting the neuropsychological findings of the selective executive dysfunction of vascular pathology [25, 26, 75].

However, the implications for vascular associated frontal lobe dominated cognitive functions need to be interpreted with caution. As detected by the T1 and DTI joint analysis, broader areas including anterior, mid-body and posterior CC showed different levels of alterations in the MCI-h group, while only the anterior regions reached group-wise significance. Hence, the significant alterations detected in genu between MCI with high and low vascular types does not mean that the anterior CC is the only region involved in vascular pathology. For instance, the conclusion in [99] is that vascular risk factors 'interact' with neurodegeneration factors in the temporal and posterior lobes and cause an additional adverse effect on the frontal lobe. Further, reduced executive or global cognitive functioning implied by more severely altered genu does not necessarily lead to the assertion that cardiovascular factors impair executive functioning more severely than nonverbal memory. Previously, to validate the executive predominance model of vascular pathology, [78] hypothesized a lower executive performance than episodic memory performance in cases with autopsy-defined cerebrovascular diseases but failed to detect statistically significant differences between the two tests. The presented study provides a potential interpretation of previous results that vascular pathology may accelerate the deterioration of multiple cognitive domain, with its influence on the frontal involved network especially dissociable from non-vascular neurodegeneration factors.

Our current study extends the database of vascular pathology on the brain in pre-dementia stages and suggests an 'additive', albeit not 'dominant', effect of vascular associated impact on the frontal lobe, which may eventually lead to the refinement of the widely accepted frontal predominancy theory. Our findings provide a new neuroanatomical substrate of vascular contributions to cognitive impairment before the magnification of dementia, which may serve as a new biomarker that helps clinical diagnostic and therapeutic design.

3.4.4 Limitations

There are also several limitations of the presented study. First, due to the limited size of the cohort, subjects with high FCRP and subjects with histories of myocardial infarction were merged into one single group – the MCI-h group. This vascular model shall be refined when more subjects are enrolled in the future. Second, due to the large age range within the cohort, linear regression was used to factor out the effect of age. The effect of age on brain anatomy in elderly subjects has been widely accepted. As shown in Figure 3.5, a significant linear relationship between age and the surface diffusion indices can be seen. However, the use of linear regression does not rule out the possible existence of a nonlinear relationship between age and brain anatomy. Third, here only the most representative univariate or multivariate measurements in our statistical analyses were included. Nonetheless, the T1 and DTI fusion method can also be applied to other shape or diffusion measurements like thickness, axial diffusivity and so on, as well as combinations among these. Fourth, it would be desirable to include other factors, like gender, gene, education, and ethnicity to derive a more comprehensive model.

3.5 Future Directions

In the future, this method could be extended to functionally or anatomically relevant CC subdivisions, to further strengthen the interpretation of our results. For example, a probabilistic tractography from CC to the cortex or functional-based partition could be used on the CC subregions [86], especially where showed significant group differences, to investigate the association between regional CC alterations with disturbances in specific cortical domains. In addition, it would be important to track the mental status of patients, to see whether any of them transform into clinically diagnosed dementia. This may provide additional insight into the contribution of the vascular component in the conversion to AD or other types of dementia.

3.6 Acknowledgement

We thank the families in this study for their participation. This work was supported by the National Institutes of Health through NIH grants 5P01AG012435-18, P50-AG05142-30 and NIH P01 AG06572.

References

1. D. A. Evans, H. H. Funkenstein, M. S. Albert, P. A. Scherr, N. R. Cook, M. J. Chown, L. E. Hebert, C. H. Hennekens, and J. O. Taylor, "Prevalence of Alzheimer's disease in a community population of older persons: Higher than previously reported," *JAMA*, 262(18), pp. 2551–2556, 1989.

2. A. S. Kelley, K. McGarry, R. Gorges, and J. S. Skinner, "The burden of health care costs for patients with dementia in the last 5 years of life," *Annals of Internal Medicine*, 163(10), pp. 729–736, 2015.

3. A. Wimo, L. Jonsson, J. Bond, M. Prince, B. Winblad, and Alzheimer Disease International, "The worldwide economic impact of dementia 2010," *Alzheimer's and Dementia: The Journal of the Alzheimer's Association*, 9(1), pp. 1–11, 2013.

4. F. Mangialasche, A. Solomon, B. Winblad, P. Mecocci, and M. Kivipelto, "Alzheimer's disease: Clinical trials and drug development," *The Lancet Neurology*, 9(7), pp. 702–716, 2010.

5. H. Hampel, D. Prvulovic, S. Teipel, F. Jessen, C. Luckhaus, L. Frölich, M. W. Riepe, R. Dodel, T. Leyhe, L. Bertram, et al., "The future of Alzheimer's disease: The next 10 years," *Progress in Neurobiology*, 95(4), pp. 718–728, 2011.

6. S. E. Vermeer, N. D. Prins, T. den Heijer, A. Hofman, P. J. Koudstaal, and M. M. Breteler, "Silent brain infarcts and the risk of dementia and cognitive decline," *The New England Journal of Medicine*, 348(13), pp. 1215–1222, 2003.

7. A. B. Newman, A. L. Fitzpatrick, O. Lopez, S. Jackson, C. Lyketsos, W. Jagust, D. Ives, S. T. DeKosky, and L. H. Kuller, "Dementia and Alzheimer's disease incidence in relationship to cardiovascular disease in the cardiovascular health study cohort," *Journal of the American Geriatrics Society*, 53(7), pp. 1101–1107, 2005.

8. P. B. Gorelick, A. Scuteri, S. E. Black, C. DeCarli, S. M. Greenberg, C. Iadecola, L. J. Launer, S. Laurent, O. L. Lopez, D. Nyenhuis, et al., "Vascular contributions to cognitive impairment and dementia a statement for healthcare professionals from the American Heart Association/American Stroke Association," *Stroke*, 42(9), pp. 2672–2713, 2011.

9. R. N. Kalaria, R. Akinyemi, and M. Ihara, "Does vascular pathology contribute to Alzheimer changes?." *Journal of the Neurological Sciences*, 322(1), pp. 141–147, 2012.

10. V. A. Cardenas, B. Reed, L. L. Chao, H. Chui, N. Sanossian, C. C. DeCarli, W. Mack, J. Kramer, H. N. Hodis, M. Yan, et al., "Associations among vascular risk factors, carotid atherosclerosis, and cortical volume and thickness in older adults," *Stroke*, 43(11), pp. 2865–2870, 2012.

11. C. Iadecola, "The pathobiology of vascular dementia," *Neuron*, 80(4), pp. 844–866, 2013.

12. J. A. Luchsinger, M.-X. Tang, Y. Stern, S. Shea, and R. Mayeux, "Diabetes mellitus and risk of Alzheimer's disease and dementia with stroke in a multiethnic cohort," *American Journal of Epidemiology*, 154(7), pp. 635–641, 2001.

13. H. B. Posner, M.-X. Tang, J. Luchsinger, R. Lantigua, Y. Stern, and R. Mayeux, "The relationship of hypertension in the elderly to ad, vascular dementia, and cognitive function," *Neurology*, 58(8), pp. 1175–1181, 2000.

14. M. Garcia-Alloza, J. Gregory, K. V. Kuchibhotla, S. Fine, Y. Wei, C. Ayata, M. P. Frosch, S. M. Greenberg, and B. J. Bacskai, "Cerebrovascular lesions induce transient β-amyloid deposition," *Brain*, 134(12), pp. 3694–3704, 2011.

15. M. Mather, and C. W. Harley, "The locus coeruleus: Essential for maintaining cognitive function and the aging brain," Trends in Cognitive Sciences, 20(3), pp. 214–226, 2016.

16. H. C. Chui, "Vascular cognitive impairment: Today and tomorrow," *Alzheimer's and Dementia*, 2(3), pp. 185–194, 2006.

17. J. B. Buse, H. N. Ginsberg, G. L. Bakris, N. G. Clark, F. Costa, R. Eckel, V. Fonseca, H. C. Gerstein, S. Grundy, R. W. Nesto, et al., "Primary prevention of cardiovascular diseases in people with diabetes mellitus: A scientific statement from the American Heart Association and the American Diabetes Association," *Circulation*, 115(1), pp. 114–126, 2007.

18. J. Stewart, G. Manmathan, and P. Wilkinson, "Primary prevention of cardiovascular disease: A review of contemporary guidance and literature," *JRSM Cardiovascular Disease*, 6, p. 2048004016687211, 2017.

19. N. C. Fox, W. R. Crum, R. I. Scahill, J. M. Stevens, J. C. Janssen, and M. N. Rossor, "Imaging of onset and progression of Alzheimer's disease with voxel-compression mapping of serial magnetic resonance images," *The Lancet*, 358(9277), pp. 201–205, 2001.

20. S. Oddo, A. Caccamo, J. D. Shepherd, M. P. Murphy, T. E. Golde, R. Kayed, R. Metherate, M. P. Mattson, Y. Akbari, and F. M. LaFerla, "Triple-transgenic model of Alzheimer's disease with plaques and tangles: Intracellular Aβ and synaptic dysfunction," *Neuron*, 39(3), pp. 409–421, 2003.

21. S. Kovacevic, M. S. Rafii, and J. B. Brewer, "High-throughput, fully-automated volumetry for prediction of MMSE and CDR decline in mild cognitive impairment," *Alzheimer Disease and Associated Disorders*, 23(2), p. 139, 2009.

22. L. Serra, M. Cercignani, D. Lenzi, R. Perri, L. Fadda, C. Caltagirone, E. Macaluso, and M. Bozzali, "Grey and white matter changes at different stages of Alzheimer's disease," *Journal of Alzheimer's Disease*, 19(1), pp. 147–159, 2010.

23. U. Ekman, "Functional brain imaging of cognitive status in Parkinson's disease," PhD thesis, Umeå University, 2014.

24. X. Tang, D. Holland, A. M. Dale, L. Younes, M. I. Miller, and Alzheimer's Disease Neuroimaging Initiative, "Shape abnormalities of subcortical and ventricular structures in mild cognitive impairment and Alzheimer's disease: Detecting, quantifying, and predicting," *Human Brain Mapping*, 35(8), pp. 3701–3725, 2014.

25. K. Hayden, L. Warren, C. Pieper, T. Østbye, J. Tschanz, M. Norton, J. Breitner, and K. Welsh-Bohmer, "Identification of vad and ad prodromes: The cache county study," *Alzheimer's and Dementia*, 1(1), pp. 19–29, 2005.

26. A. Nordlund, S. Rolstad, O. Klang, K. Lind, S. Hansen, and A. Wallin, "Cognitive profiles of mild cognitive impairment with and without vascular disease," *Neuropsychology*, 21(6), p. 706, 2007.

27. D. A. Loewenstein, A. Acevedo, J. Agron, R. Issacson, S. Strauman, E. Crocco, W. W. Barker, and R. Duara, "Cognitive profiles in Alzheimers disease and in mild cognitive impairment of different etiologies," *Dementia and Geriatric Cognitive Disorders*, 21(5–6), pp. 309–315, 2006.

28. J. E. Bogen, E. Fisher, and P. Vogel, "Cerebral commissurotomy: A second case report," *JAMA*, 194(12), pp. 1328–1329, 1965.

29. M. Zhu, W. Gao, X. Wang, C. Shi, and Z. Lin, "Progression of corpus callosum atrophy in early stage of Alzheimers disease: MRI based study," *Academic Radiology*, 19(5), pp. 512–517, 2012.

30. S. J. Teipel, W. Bayer, G. E. Alexander, D. Teichberg, L. Kulic, M. B. Schapiro, H.-J. Moller, S. I. Rapoport, and H. Hampel, "Progression of corpus callosum atrophy in Alzheimer disease," *Archives of Neurology*, 59(2), pp. 243–248, 2002.

31. M. Zhu, X. Wang, W. Gao, C. Shi, H. Ge, H. Shen, and Z. Lin, "Corpus callosum atrophy and cognitive decline in early Alzheimer's disease: Longitudinal MRI study," *Dementia and Geriatric Cognitive Disorders*, 37(3–4), pp. 214–222, 2013.

32. B. A.Ardekani, A. H. Bachman, K. Figarsky, and J. J. Sidtis, "Corpus callosum shape changes in early Alzheimers disease: An MRI study using the oasis brain database," *Brain Structure and Function*, 219(1), pp. 343–352, 2014.

33. M. Di Paola, F. Di Iulio, A. Cherubini, C. Blundo, A. Casini, G. Sancesario, D. Passafiume, C. Caltagirone, and G. Spalletta, "When, where, and how the corpus callosum changes in MCI and AD: A multimodal MRI study," *Neurology*, 74(14), pp. 1136–1142, 2010.

34. Z. Hu, L. Wu, J. Jia, and Y. Han, "Advances in longitudinal studies of amnestic mild cognitive impairment and Alzheimers disease based on multi-modal MRI techniques," *Neuroscience Bulletin*, 30(2), pp. 198–206, 2014.

35. J. L. Yuan, S. K. Wang, X. J. Guo, and W. L. Hu, "Acute infarct of the corpus callosum presenting as alien hand syndrome: Evidence of diffusion weighted imaging and magnetic resonance angiography," *BMC Neurology*, 11(1), p. 142, 2011.

36. L. Delano-Wood, M. W. Bondi, A. J. Jak, N. R. Horne, B. C. Schweinsburg, L. R. Frank, C. E. Wierenga, D. C. Delis, R. J. Theilmann, and D. P. Salmon, "Stroke risk modifies regional white matter differences in mild cognitive impairment," *Neurobiology of Aging*, 31(10), pp. 1721–1731, 2010.

37. R. A. Gons, L. J. van Oudheusden, K. F. de Laat, A. G. van Norden, I. W. van Uden, D. G. Norris, M. P. Zwiers, E. van Dijk, and F.-E. de Leeuw, "Hypertension is related to the microstructure of the corpus callosum: The run dmc study," *Journal of Alzheimer's Disease*, 32(3), pp. 623–631, 2012.

38. J. I. Friedman, C. Y. Tang, H. J. de Haas, L. Changchien, G. Goliasch, P. Dabas, V. Wang, Z. A. Fayad, V. Fuster, and J. Narula, "Brain imaging changes associated with risk factors for cardiovascular and cerebrovascular disease in asymptomatic patients," *JACC: Cardiovascular Imaging*, 7(10), pp. 1039–1053, 2014.

39. E. Farkas, G. Donka, R. A. de Vos, A. Mihaly, F. Bari, and P. G. Luiten, "Experimental cerebral hypoperfusion induces white matter injury and microglial activation in the rat brain," *Acta Neuropathologica*, 108(1), pp. 57–64, 2004.

40. Q. Liu, S. He, L. Groysman, D. Shaked, J. Russin, S. Cen, and W. J. Mack, "White matter injury due to experimental chronic cerebral hypoperfusion is associated with C5 deposition," *PLoS One*, 8(12), p. e84802, 2013.

41. Y. Wang, Y. Song, P. Rajagopalan, T. An, K. Liu, Y.-Y. Chou, B. Gutman, A. W. Toga, P. M. Thompson, and Alzheimer's Disease Neuroimaging Initiative, "Surface-based TBM boosts power to detect disease effects on the brain: An N= 804 ADNI study," *Neuroimage*, 56(4), pp. 1993–2010, 2011.

42. K. S. Frederiksen, E. Garde, A. Skimminge, C. Ryberg, E. Rostrup, W. F. Baaré, H. R. Siebner, A.-M. Hejl, A.-M. Leffers, and G. Waldemar, "Corpus callosum atrophy in patients with mild Alzheimers disease," *Neurodegenerative Diseases*, 8(6), pp. 476–482, 2011.

43. A. H.Bachman, S. H. Lee, J. J. Sidtis, and B. A. Ardekani, "Corpus callosum shape and size changes in early Alzheimer's disease: A longitudinal MRI study using the OASIS brain database," *Journal of Alzheimer's Disease*, 39(1), pp. 71–78, 2014.

44. Y. Zhang, N. Schuff, G.-H. Jahng, W. Bayne, S. Mori, L. Schad, S. Mueller, A.-T. Du, J. Kramer, K. Yaffe, et al., "Diffusion tensor imaging of cingulum fibers in mild cognitive impairment and Alzheimer disease," *Neurology*, 68(1), pp. 13–19, 2007.

45. M. Ukmar, E. Makuc, M. Onor, G. Garbin, M. Trevisiol, and M. Cova, "Evaluation of white matter damage in patients with Alzheimers disease and in patients with mild cognitive impairment by using diffusion tensor imaging," *La radiologia medica*, 113(6), pp. 915–922, 2008.

46. T.-F. Chen, C.-C. Lin, Y.-F. Chen, H.-M. Liu, M.-S. Hua, Y.-C. Huang, and M.-J. Chiu, "Diffusion tensor changes in patients with amnesic mild cognitive impairment and various dementias," *Psychiatry Research: Neuroimaging*, 173(1), pp. 15–21, 2009.

47. L. Zhuang, W. Wen, W. Zhu, J. Trollor, N. Kochan, J. Crawford, S. Reppermund, H. Brodaty, and P. Sachdev, "White matter integrity in mild cognitive impairment: A tract-based spatial statistics study," *Neuroimage*, 53(1), pp. 16–25, 2010.

48. S. J. Teipel, T. Meindl, L. Grinberg, M. Grothe, J. L. Cantero, M. F. Reiser, H.-J. Möller, H. Heinsen, and H. Hampel, "The cholinergic system in mild cognitive impairment and Alzheimer's disease: An in vivo MRI and DTI study," *Human Brain Mapping*, 32(9), pp. 1349–1362, 2011.

49. S. Dimitra, D. Verganelakis, E. Gotsis, P. Toulas, J. Papatriantafillou, C. Karageorgiou, T. Thomaides, E. Kapsalaki, G. Hadjigeorgiou, and A. Papadimitriou, "Diffusion tensor imaging (DTI) in the detection of white matter lesions in patients with mild cognitive impairment (MCI)," *Acta neurologica Belgica*, 113(4), pp. 441–451, 2013.

50. Y. Liu, G. Spulber, K. K. Lehtimäki, M. Könönen, I. Hallikainen, H. Gröhn, M. Kivipelto, M. Hallikainen, R. Vanninen, and H. Soininen, "Diffusion tensor imaging and tract-based spatial statistics in Alzheimer's disease and mild cognitive impairment," *Neurobiology of Aging*, 32(9), pp. 1558–1571, 2011.

51. Y. Zhang, N. Schuff, M. Camacho, L. L. Chao, T. P. Fletcher, K. Yaffe, S. C. Woolley, C. Madison, H. J. Rosen, B. L. Miller, et al., "MRI markers for mild cognitive impairment: Comparisons between white matter integrity and gray matter volume measurements," *PLoS One*, 8(6), p. e66367, 2013.

52. P. W. Wilson, R. B. D'Agostino, D. Levy, A. M. Belanger, H. Silbershatz, and W. B. Kannel, "Prediction of coronary heart disease using risk factor categories," *Circulation*, 97(18), pp. 1837–1847, 1998.

53. D. Mungas, B. R. Reed, and J. H. Kramer, "Psychometrically matched measures of global cognition, memory, and executive function for assesment of cognitive decline in older persons." *Neuropsychology*, 17(3), p. 380, 2003.

54. M. Jenkinson, P. Bannister, M. Brady, and S. Smith, "Improved optimization for the robust and accurate linear registration and motion correction of brain images," *Neuroimage*, 17(2), pp. 825–841, 2002.

55. Y. Wang, Y. Song, P. Rajagopalan, T. An, K. Liu, Y.-Y. Chou, B. Gutman, A. W. Toga, P. M. Thompson, and Alzheimer's Disease Neuroimaging Initiative, "Surface-based TBM boosts power to detect disease effects on the brain: An N= 804 ADNI study," *Neuroimage*, 56(4), pp. 1993–2010, 2011.

56. N. Toussaint, J. Souplet, and P. Fillard, "MedINRIA: Medical image navigation and research tool by INRIA," In: *Proceedings of the of MICCAI'07 Workshop on Interaction in Medical Image Analysis and Visualization*, 2007.

57. V. Arsigny, O. Commowick, X. Pennec, and N. Ayache, "A log-euclidean framework for statistics on diffeomorphisms." In: *Medical Image Computing and Computer-Assisted Intervention-MICCAI 2006*. Springer, pp. 924–931, 2006.

58. F. Lepore, C. Brun, Y.-Y. Chou, M.-C. Chiang, R. A. Dutton, K. M. Hayashi, E. Luders, O. L. Lopez, H. J. Aizenstein, A. W. Toga, et al., "Generalized tensor-based morphometry of HIV/AIDS using multivariate statistics on deformation tensors," *IEEE Transactions on Medical Imaging*, 27(1), pp. 129–141, 2008.

59. J. Shi, P. M. Thompson, B. Gutman, Y. Wang, and Alzheimer's Disease Neuroimaging Initiative, "Surface fluid registration of conformal representation: Application to detect disease burden and genetic influence on hippocampus," *NeuroImage*, 78, pp. 111–134, 2013.

60. Y. Lao, L.-A. Dion, G. Gilbert, M. F. Bouchard, G. Rocha, Y. Wang, N. Leporé, and D. Saint-Amour, "Mapping the basal ganglia alterations in children chronically exposed to manganese," *Scientific Reports*, 7, p. 41804, 2017.

61. Y. Lao, M. Law, J. Shi, N. Gajawelli, L. Haas, Y. Wang, and N. Leporé, "A T1 and DTI fused 3D corpus callosum analysis in pre-vs. post-season contact sports players." In: *Tenth International Symposium on Medical Information Processing and Analysis*. International Society for Optics and Photonics, p. 928700, 2015.

62. T. E. Nichols, and A. P. Holmes, "Nonparametric permutation tests for functional neuroimaging: A primer with examples," *Human Brain Mapping*, 15(1), pp. 1–25, 2001.

63. Y. Benjamini, and Y. Hochberg, "Controlling the false discovery rate: A practical and powerful approach to multiple testing," *Journal of the Royal Statistical Society: Series B* (Methodological), 57(1), pp. 289–300, 1995.

64. Y. Wang, T. F. Chan, A. W. Toga, and P. M. Thompson, "Multivariate tensor-based brain anatomical surface morphometry via holomorphic one-forms." In: *International Conference on Medical Image Computing and Computer-Assisted Intervention*. Springer, pp. 337–344, 2009.

65. V. Rajagopalan, J. Scott, P. A. Habas, K. Kim, F. Rousseau, O. A. Glenn, A. J. Barkovich, and C. Studholme, "Mapping directionality specific volume changes using tensor based morphometry: An application to the study of gyrogenesis and lateralization of the human fetal brain," *Neuroimage*, 63(2), pp. 947–958, 2012.

66. J. Shi, Y. Wang, R. Ceschin, X. An, Y. Lao, D. Vanderbilt, M. D. Nelson, P. M. Thompson, A. Panigrahy, and N. Lepore, "A multivariate surface-based analysis of the putamen in premature newborns: Regional differences within the ventral striatum," *PLoS One*, 8(7), p. e66736, 2013.

67. G. J. Szekely, and M. L. Rizzo, "Brownian distance covariance," *The Annals of Applied Statistics*, 3(4), pp. 1236–1265, 2009.

68. Y. Lao, Y. Wang, J. Shi, R. Ceschin, M. D. Nelson, A. Panigrahy, and N. Lepore, "Thalamic alterations in preterm neonates and their relation to ventral striatum disturbances revealed by a combined shape and pose analysis," *Brain Structure and Function*, 221(1), pp. 487–506, 2016.

69. M. K. Chung, S. M. Robbins, K. M. Dalton, R. J. Davidson, A. L. Alexander, and A. C. Evans, "Cortical thickness analysis in autism with heat kernel smoothing," *NeuroImage*, 25(4), pp. 1256–1265, 2005.

70. A. Ott, M. Breteler, F. Van Harskamp, J. J. Claus, T. J. Van Der Cammen, D. E. Grobbee, and A. Hofman, "Prevalence of Alzheimer's disease and vascular dementia: Association with education. The Rotterdam study," *BMJ*, 310(6985), pp. 970–973, 1995.

71. J. S. Meyer, G. Xu, J. Thornby, M. H. Chowdhury, and M. Quach, "Is mild cognitive impairment prodromal for vascular dementia like Alzheimers disease?" *Stroke*, 33(8), pp. 1981–1985, 2002.

72. M. Stampfer, "Cardiovascular disease and Alzheimer's disease: Common links," *Journal of Internal Medicine*, 260(3), pp. 211–223, 2006.

73. R. F. de Bruijn, and M. A. Ikram, "Cardiovascular risk factors and future risk of Alzheimers disease," *BMC Medicine*, 12(1), p. 130, 2014.

74. J. L. Ingles, D. C. Boulton, J. D. Fisk, and K. Rockwood, "Preclinical vascular cognitive impairment and Alzheimer disease neuropsychological test performance 5 years before diagnosis," *Stroke*, 38(4), pp. 1148–1153, 2007.

75. C. Marra, M. Ferraccioli, M. Gabriella Vita, D. Quaranta, and G. Gainotti, "Patterns of cognitive decline and rates of conversion to dementia in patients with degenerative and vascular forms of MCI," *Current Alzheimer Research*, 8(1), pp. 24–31, 2011.

76. N. Graham, T. Emery, and J. Hodges, "Distinctive cognitive profiles in Alzheimers disease and subcortical vascular dementia," *Journal of Neurology, Neurosurgery and Psychiatry*, 75(1), pp. 61–71, 2004.

77. O. Nystrom, A. Wallin, and A. Nordlund, "MCI of different etiologies differ on the cognitive assessment battery," *Acta Neurologica Scandinavica*, 132(1), pp. 31–36, 2015.

78. B. R. Reed, D. M. Mungas, J. H. Kramer, W. Ellis, H. V. Vinters, C. Zarow, W. J. Jagust, and H. C. Chui, "Profiles of neuropsychological impairment in autopsy-defined Alzheimer's disease and cerebrovascular disease," *Brain*, 130(3), pp. 731–739, 2007.

79. R. Kloppenborg, P. Nederkoorn, A. Grool, K. Vincken, W. Mali, M. Vermeulen, Y. van der Graaf, M. Geerlings, and SMART Study Group, "Cerebral small-vessel disease and progression of brain atrophy the SMART-MR study," *Neurology*, 79(20), pp. 2029–2036, 2012.

80. H. M. Jochemsen, M. Muller, F. L. Visseren, P. Scheltens, K. L. Vincken, W. P. Mali, Y. van der Graaf, and M. I. Geerlings, "Blood pressure and progression of brain atrophy: The SMART-MR study," *JAMA Neurology*, 70(8), pp. 1046–1053, 2013.

81. J. Barnes, O. T. Carmichael, K. K. Leung, C. Schwarz, G. R. Ridgway, J. W. Bartlett, I. B. Malone, J. M. Schott, M. N. Rossor, G. J. Biessels, et al., "Vascular and Alzheimer's disease markers independently predict brain atrophy rate in Alzheimer's Disease Neuroimaging Initiative controls," *Neurobiology of Aging*, 34(8), pp. 1996–2002, 2013.

82. G. Chetelat, B. Landeau, E. Salmon, I. Yakushev, M. A. Bahri, F. Mezenge, A. Perrotin, C. Bastin, A. Manrique, A. Scheurich, et al., "Relationships between brain metabolism decrease in normal aging and changes in structural and functional connectivity," *Neuroimage*, 76, pp. 167–177, 2013.

83. B. J. Hallam, W. S. Brown, C. Ross, J. G. Buckwalter, E. D. Bigler, J. T. Tschanz, M. C. Norton, K. A. Welsh-Bohmer, and J. C. Breitner, "Regional atrophy of the corpus callosum in dementia," *Journal of the International Neuropsychological Society*, 14(3), pp. 414–423, 2008.

84. Y. S. Shim, B. Yoon, Y.-M. Shon, K.-J. Ahn, and D.-W. Yang, "Difference of the hippocampal and white matter microalterations in MCI patients according to the severity of subcortical vascular changes: Neuropsychological correlates of diffusion tensor imaging," *Clinical Neurology and Neurosurgery*, 110(6), pp. 552–561, 2008.

85. F. Aboitiz, A. B. Scheibel, R. S. Fisher, and E. Zaidel, "Fiber composition of the human corpus callosum," *Brain Research*, 598(1), pp. 143–153, 1992.

86. H.-J. Park, J. J. Kim, S.-K. Lee, J. H. Seok, J. Chun, D. I. Kim, and J. D. Lee, "Corpus callosal connection mapping using cortical gray matter parcellation and DT-MRI," *Human Brain Mapping*, 29(5), pp. 503–516, 2008.

87. H. D. Rosas, S. Y. Lee, A. C. Bender, A. K. Zaleta, M. Vangel, P. Yu, B. Fischl, V. Pappu, C. Onorato, J.-H. Cha, et al., "Altered white matter microstructure in the corpus callosum in Huntington's disease: Implications for cortical disconnection," *Neuroimage*, 49(4), pp. 2995–3004, 2010.

88. L. J. O'Donnell, C.-F. Westin, and A. J. Golby, "Tract-based morphometry for white matter group analysis," *Neuroimage*, 45(3), pp. 832–844, 2009.

89. G. Karas, J. Sluimer, R. Goekoop, W. Van Der Flier, S. Rombouts, H. Vrenken, P. Scheltens, N. Fox, and F. Barkhof, "Amnestic mild cognitive impairment: Structural MR imaging findings predictive of conversion to Alzheimer disease," *American Journal of Neuroradiology*, 29(5), pp. 944–949, 2008.

90. S. Buklina, "The corpus callosum, interhemisphere interactions, and the function of the right hemisphere of the brain," *Neuroscience and Behavioral Physiology*, 35(5), pp. 473–480, 2005.

91. B. L. Miller, and J. L. Cummings *The Human Frontal Lobes: Functions and Disorders*. Guilford Press, 2007.

92. K. Krupa, and M. Bekiesinska-Figatowska, "Congenital and acquired abnormalities of the corpus callosum: A pictorial essay," *BioMed Research International*, 2013, 265619, 2013.

93. H.-J. Kretschmann, W. Weinrich, and W. Fiekert, *Neurofunctional Systems: 3D Reconstructions with Correlated Neuroimaging*. University of California Press, 1998.

94. P. Rudge, and E. K. Warrington, "Selective impairment of memory and visual perception in splenial tumours," *Brain* , 114(1), pp. 349–360, 1991.

95. M. G. Knyazeva, "Splenium of corpus callosum: Patterns of interhemispheric interaction in children and adults," *Neural Plasticity*, 2013, 639430, 2013.

96. J. L. Cummings, "Frontal-subcortical circuits and human behavior," *Archives of Neurology*, 50(8), p. 873, 1993.

97. M. Hoffmann, "The human frontal lobes and frontal network systems: An evolutionary, clinical, and treatment perspective," *ISRN Neurology*, 2013, 892459, 2013.

98. A. Bonner-Jackson, S. Mahmoud, J. Miller, and S. J. Banks, "Verbal and non-verbal memory and hippocampal volumes in a memory clinic population," *Alzheimer's Research and Therapy*, 7(1), pp. 1–10, 2015.

99. S. Villeneuve, B. R. Reed, C. M. Madison, M. Wirth, N. L. Marchant, S. Kriger, W. J. Mack, N. Sanossian, C. DeCarli, H. C. Chui, et al., "Vascular risk and Aβ interact to reduce cortical thickness in AD vulnerable brain regions," *Neurology*, 83(1), pp. 40–47, 2014.

100. M. L. Alosco, J. Gunstad, B. A. Jerskey, X. Xu, U. S. Clark, J. Hassenstab, D. M. Cote, E. G. Walsh, D. R. Labbe, R. Hoge, et al., "The adverse effects of reduced cerebral perfusion on cognition and brain structure in older adults with cardiovascular disease," *Brain and Behavior*, 3(6), pp. 626–636, 2013.

Chapter 4

Automated Diagnosis and Prediction in Cardiovascular Diseases Using Tomographic Imaging

Lisa Duff and Charalampos Tsoumpas

4.1   Introduction

Medical imaging is well established in clinical routine to help diagnose and monitor a range of conditions. Medical images are often evaluated qualitatively so the amount of data extracted is limited and is at risk of inter-observer variability. Medical images contain vast amounts of unused data which, if extracted, may help better understand and stratify the patient's disease. The process that extracts the entire measurement as captured by the medical imaging modality and makes use of higher-order correlations existing within the medical imaging dataset is known as 'Radiomics'.

Traditionally, humans determine relevant features (e.g. shape, size, texture) in the images, using practices such as radiomics. They then encode them into a machine learning algorithm. Machine learning is a type of artificial intelligence where systems learn from examples rather than being explicitly programmed to conduct a task [1]. In the case of medical image analysis the examples are the images, and an example of the task is classification by diagnosis, outcome or response to treatment. Deep learning is a subset of machine learning that determines the relevant features without the programmer needing to include them in the algorithm. Deep learning is a powerful tool in the field of radiomics that can be used to analyse intelligently patient datasets and offer diagnostic and prognostic information to the clinicians. This recently established methodology can be combined with additional information (e.g. genomics and/or phenomics) empowering patient stratification. As a result clinicians can plan treatment with higher efficacy and prevent unnecessarily adverse experiences for the patient.

The fields of deep learning and radiomics have recently gained astonishing popularity (Figure 4.1), and they have been extensively utilised in oncology where they have proven to be successful. The field of radiomics is hard to trace backwards as there were several terms for quantifying imaging metrics for describing diseases before this generic term was commonly used. Many reviews cite El Naqa et al. as the first radiomics paper which, in 2009, looked at predicting cancer outcomes with image features [2] but also credit Gillies et al. for first proposing the term radiomics the following year [3]. Likewise the popularity of deep learning is difficult to track as it can go by different

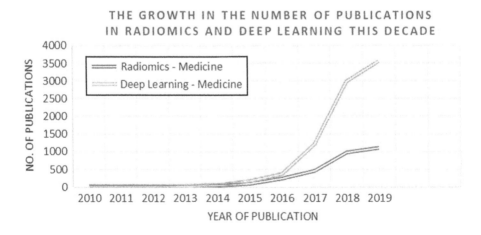

FIGURE 4.1:   The growth in the number of medical publications per year returned on Scopus when searching the term radiomics and deep learning. (as of 14th August 2019.)

names, and the technique has evolved to resemble what it is today. The number of publications per year using the term deep learning, and included on the Scopus database, only exceeded 100 in 2010 with the majority of medical image studies utilising deep learning being published post-2016 [4]. This surge has partly been driven by the ability to gather large sets of data and partly by the access to HPC (high performance computing) which has helped provide the computing power to construct and train complex algorithms [5]. The field of oncology has contributed most of the published studies, and the use of deep learning in cardiovascular diseases (Figure 4.2) is still in its infancy in comparison [6].

The establishment of these two research fields can help with laborious tasks such as organ segmentation, and they have been used to determine the causes of illness along with several other applications. In cardiac imaging, the left ventricle and the blood vessels are the most commonly analysed (Figure 4.3). This book chapter will focus on radiomics and deep learning and will attempt to describe how they are used in imaging of the cardiovascular system. The advantages and disadvantages of each technique will be discussed alongside the challenges they face on their route to clinical adoption.

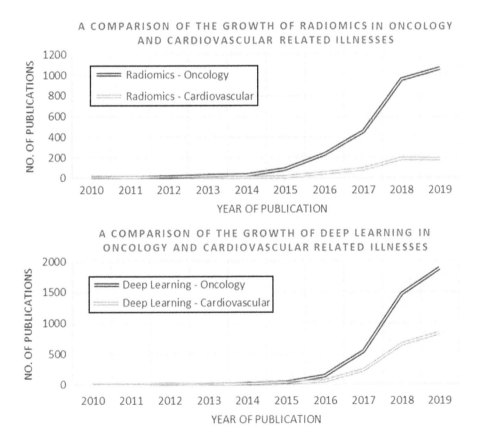

FIGURE 4.2: A comparison between the number of oncology- and cardiovascular-based radiomics (top) and deep learning (bottom) studies per year. (According to Scopus as of 14th August 2019.)

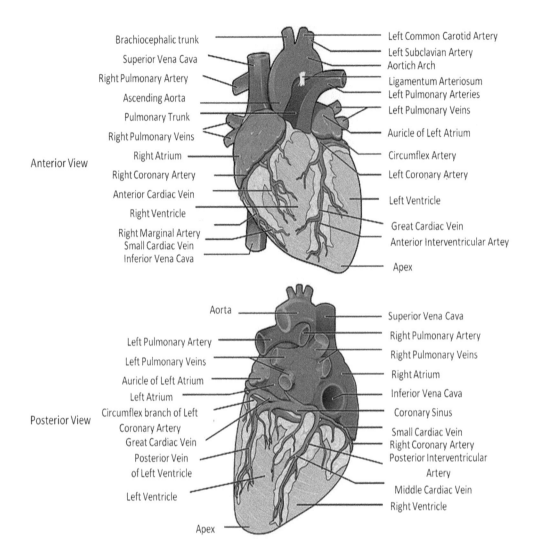

FIGURE 4.3: 'Cardiac Anatomy' by OpenStax College. Available at: https://radiopaedia.org/cases/cardiac-anatomy-creative-commons-illustration. Licensed under Creative Commons Attribution 3.0 Generic (CC BY 3.0) Full Terms at https://creativecommons.org/licenses/by/3.0/Modification: Labels replaced to make more legible.

4.2 Cardiovascular Imaging

Cardiovascular imaging is a vital part of the diagnosis and monitoring of cardiovascular diseases, e.g. coronary artery disease, cardiomyopathy and congenital heart disease. The modality selected depends on the suitability for a given or suspected condition, the accessibility and the attributes of the patient, e.g. pre-existing conditions [7]. This section covers four important imaging modalities and summarises their main characteristics, applications, advantages and disadvantages.

4.2.1 Computed Tomography Angiography

Computed tomography angiography (CTA), works similarly to conventional CT but utilises an injected contrast agent to enhance the visualisation of the blood vessels as shown in Figure 4.4 and illustrated by several investigations [8]. By doing so the physical changes of the blood vessels can be detected such as narrowing and blockages, aneurysms (bulges) or plaque build-up [9, 10]. One of the primary uses of CTA is to detect coronary artery disease [7, 9], but it is also helpful in detecting and monitoring other conditions such as vasculitis [10].

CTA is less invasive than conventional angiography as there is no need for a small catheter to be guided through an artery up to the heart. It also involves less radiation exposure, and it is often considered safer [7, 9]. However, it does not indicate hemodynamic significance, which requires a technique that measures the heart function [7].

4.2.2 Magnetic Resonance Imaging

Magnetic resonance angiography (MRA) and cardiac magnetic resonance imaging (CMR) are often utilised in the diagnosis and treatment of patients with cardiovascular diseases due to its high soft tissue contrast (Figure 4.5). MRA is used to image blood vessels while CMR looks at the heart and larger vessels.

There are a range of different pulse sequences that can be used to acquire MRA and CMR images. The pulse sequence dictates the magnitude and timing of the pulses produced by the corresponding MRI scanner and in turn alter the image produced. The optimal pulse sequence is determined by what is required from the image. If anatomical information is sought after, sequences such as spin echo (SE), fast spin echo (FSE) and turbo spin echo (TSE) will be selected. On the other hand, if cardiac function is being evaluated sequences such as gradient echo (GE) or steady-state free precession (SSFP) would be a better choice [11, 12].

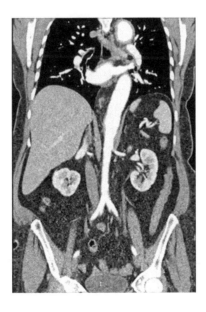

FIGURE 4.4: CTA scan of the aorta.

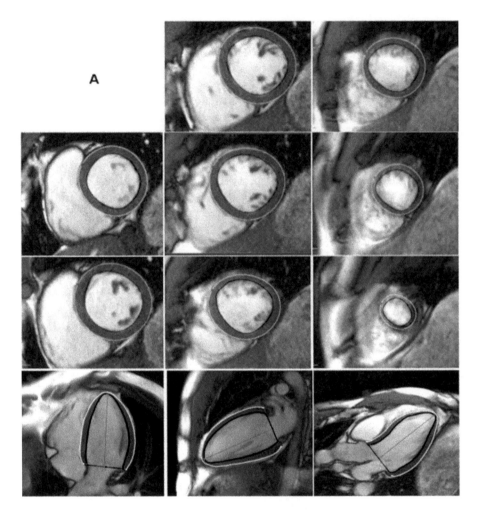

FIGURE 4.5: Epicardial borders defined on CMR 'Steps taken for 3D FTCMR' by Boyang Liu, Ahmed M. Dardeer, William E. Moody, Manvir K. Hayer, Shanat Baig, Anna M. Price, Francisco Leyva, Nicola C. Edwards, Richard P. Steeds. Available at: https://research.birmingham.ac.uk/portal/files/49118040/Liu_et_al_Reference_ranges_ IJCI.pdf. Licensed under Creative Commons Attribution License (CC BY) Full Terms at http://creativecommons.org/licenses/by Modification: Only first step of a process shown.

MRA can be performed without the injection of a contrast agent, but the occurrence of artefacts makes acquisition and analysis more difficult. Several techniques are available to take MRA scans without contrast agent and each are better suited to different applications and vascular territories. For example, SSFP can be used to acquire high resolution images of blood vessels such as the aorta and renal arteries while ECG-gated FSE is used to look at peripheral arteries [12]

CMR can be used to image the structure of the heart so as to detect problems such as congenital heart disease, damage after a heart attack and heart valve degeneration. It can also evaluate how well it pumps blood and the blood flow to the heart, e.g. perfusion MRI [13]. The adaptability and versatility of CMR mean it is used for a range of conditions such as cardiomyopathy, myocarditis, myocardial oedema and coronary heart disease [7, 14]. Besides the high tissue contrast and versatility of magnetic resonance imaging (MRI), one of its benefits is the lack of ionising radiation exposure especially in cases that require repeated or regular imaging [7, 11]. It also offers better spatial and temporal resolution than radionuclide imaging. However, several problems are faced when patients have metal implants or implanted medical devices, e.g. pacemakers, because MRI relies on a strong magnet to acquire images. In medical devices such as pacemakers the magnet can exert mechanical forces on the components, influence the operation of the device, e.g. pacing, or induce current [15]. Recently, implants and devices are designed to be MRI compatible so this is becoming less of an issue [7, 13, 14]. Overall, MRI is safe but if a contrast agent is used there are some associated risks. Firstly, some patients may experience an allergic reaction to the agent, and secondly, patients with kidney damage may suffer further damage by the contrast agent [13, 14]. Finally, MRI faces technical challenges, one of which is motion from both the heart and respiratory action. This can be tackled with techniques such as cardiac gating, breath holding and optimised pulse sequences [11].

4.2.3   Radionuclide Imaging

Positron emission tomography (PET) and single photon emission computed tomography (SPECT) are commonly used in cardiology. Radioactive tracers are injected into the body, or in some cases swallowed or inhaled, and the emitted gamma radiation is detected by the scanner. Every tracer serves a different purpose, and its associated process in the body will be highlighted by the emitted radiation. For example, FluoroDeoxyGlucose ([^{18}F]FDG) gathers in areas of high glucose uptake and so highlights area of inflammation or cancer (Figure 4.6). In cardiology, radionuclide

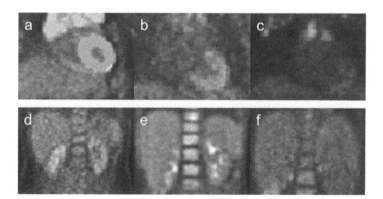

FIGURE 4.6: FDG uptake in (a) left ventricular and left atrium, (b) left ventricle uptake, (c) no uptake, (d) uptake in the cortex and collecting system (kidneys), (e) collecting system (kidneys) and (f) no uptake (kidneys). 'Various tissue uptake of heart (a–c) and kidney (d–f) in FDG-PET' by Chih-Yang Hsu, Mike Doubrovin, Chia-Ho Hua, Omar Mohammed, Barry L. Shulkin, Sue Kaste, Sara Federico, Monica Metzger, Matthew Krasin, Christopher Tinkle, Thomas E. Merchant and John T. Lucas Jr. Available at www.nature.com/articles/s41598-018-22319-4 Licensed under Creative Commons Attribution 4.0 International License (CC BY 4.0) Full Terms at http://creativecommons.org/licenses/by/4.0/.

imaging can visualise blood flow, damaged muscle tissue and inflammation of the blood vessels [16–18].

PET and SPECT have lower spatial resolution than CT or MRI, but they are often combined with the latter two modalities to improve anatomical localisation of disease. The combination of both helps physicians diagnose and treat patients more effectively. Acquiring both components at the same time means patients will be in the same position and the same protocol will have been used. In addition, the CT component of PET-CT or SPECT-CT provides attenuation correction [19]. PET and SPECT can be combined with MR imaging, but such scanners are more expensive to purchase and maintain, so as a result are limited in availability and dedicated protocols are still under development [20]. SPECT and PET each have their own advantages and disadvantages in cardiac imaging. SPECT is more widely available and uses cheaper equipment and tracers but takes longer to acquire. PET is higher resolution giving it better diagnostic capabilities, but tracers are much harder to produce or source and usually have shorter half-lives [21].

One common application of radionuclide imaging in cardiology is nuclear stress perfusion, which looks at heart pumping and blood flow. It is used to assess patients with previous myocardial infarction or artery bypass surgery, chest pain due to blockages and when echocardiographic images are unsuitable [7, 22]. The technique is limited however by the drawbacks of perfusion tracers. PET tracers, such as 15O water, 13N ammonia and 82Rb, require onsite or close-by cyclotrons or generators. Recently, interest has been building around 18F labelled perfusion tracers, 18F flurpiridaz in particular, due to its longer half-life (108 minutes) meaning regional cyclotrons could be used and there is time for use in exercise-based tests [23, 24].

Overall, radionuclide imaging is safe, but it involves some exposure to radiation [16, 17]. SPECT and PET have low image resolution so they are not often the first techniques used when diagnosing or treating a patient. In some cases, though, they are the most appropriate [7].

4.2.4 Echocardiography

Echocardiography or echocardiograms are ultrasounds of the heart and can either be taken from outside the body or by passing the probe through the oesophagus. The ultrasound image is produced in real time and can give information about the structure and performance of the heart, e.g. how well it pumps blood [7, 25]. The transthoracic echocardiogram (TTEs) is the most commonly used version of this imaging technique and is taken from outside the body by placing the probe against the patient's chest. Other types of echocardiogram include taking 3D imaging of the heart, Doppler ultrasound (used to visualise blood flow) and stress tests (for example during exercise) [25]. Sometimes contrast agents are used which enhance image quality and help assess perfusion [26]. Echocardiography is used in the diagnosis of cardiac failure, identification of congenital heart disease and detecting pulmonary arterial hypertension among other applications (Figure 4.7) [7].

Echocardiography can often provide all required information and does not involve exposing the patient to radiation. Images can be obtained in real time; they are widely accessible and relatively inexpensive. In some patients however it is not as effective, for example if they have lung disease or obesity the transmission of sound waves is compromised [7, 25].

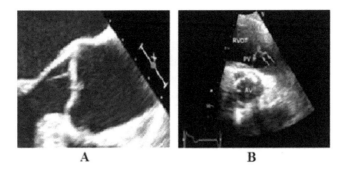

FIGURE 4.7: 'Shows fibrous tags on the (A) aortic valve and (B) pulmonary valve' by Shaimaa A. Mostafa. Available at www.omicsonline.org/open-access/ech ocardiographic-criteria-of-the-elderly-2329-9517-1000192.php?aid=40993#tables_fi gures. Licensed under Creative Commons Attribution License Full Terms at http://creativecommons.org/licenses/by/4.0/.

4.2.5 Conclusion

In summary, there are multiple approaches to cardiac imaging, and the more suitable depends on the patient's condition and situation and the availability of the technique. CTA provides anatomical information such as detecting narrowing of the blood vessels. It requires a contrast agent to make the vessels easier to observe and does expose the patient to some radiation. CMR or MRA also record anatomical information but are better suited to soft tissues. They can be optimised by altering the pulse sequence but sometimes still require a contrast agent. Radionuclide imaging is more suitable to functional assessments of body processes such as perfusion or inflammation. Echocardiography uses ultrasound to image the heart, and it is cheap and accessible. The choice of modality depends on the clinical situation and protocols.

4.3 Radiomics

Radiomics is the practice of mining quantitative parameters (often referred to as features) from medical images and using those features to aid clinical decision making. With the recent growing interest in personalised and precision medicine, where small differences in disease characteristics can result in drastically different clinical decisions, the field of radiomics has gained a lot of interest [27–29].

Some radiomic features are currently well known, e.g. blood vessel diameter, but some are less so, e.g. entropy. Several features involve a lot of detail and require complex calculations so they are not extractable purely from observation. In cardiovascular medicine various imaging biomarkers have already been discovered that correlate with specific disease outcomes, but as they are usually qualitative they are less robust due to inter- and intra-observer variations. Radiomic workflows can be established to gather values for a large range of features, regardless of a radiologist's experience with them or the amount of detail required, and measure the values in a systematic way to reduce variations [28].

The aim of radiomics is to use images that would be acquired as part of normal clinical practice. Once these images are acquired and pre-processed the region of interest (ROI) is defined and relevant tissue is segmented. The features are then extracted from the ROI and analysed alongside the clinical data to determine any important correlations (Figure 4.8). In this section we will look at the most commonly used radiomic features in cardiac imaging as well as discussing the difficulties radiomics faces in translating into routine clinical practice.

4.3.1 Radiomic Features

The radiomic features discussed below have shown some success in cardiac imaging. Some others have been developed for different fields of medicine, in most cases oncology, but have not yet proven useful in cardiology. It is worth noting that there are thousands of candidate radiomic features, but only the most representative will be mentioned in this chapter. The variations are usually small changes to the algorithm used to calculate them, a combination of a few different features or a change to the way the feature is applied.

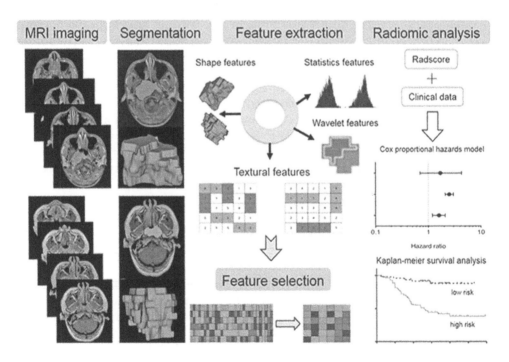

FIGURE 4.8: 'The image post-processing workflow' by Fu-Sheng Ouyang, Bao-Liang Guo, Bin Zhang, Yu-Hao Dong, Lu Zhang, Xiao-Kai Mo, Wen-Hui Huang, Shui-Xing Zhang and Qiu-Gen Hu. Available at: www.oncotarget.com/index.php?journal=oncotarget &page=article&op=view&path[]=20423&path[]=65107 Licensed under the Creative Commons Attribution Licence CC BY 3.0 Full Terms at https://creativecommons.org/licenses/by/3.0/.

4.3.1.1 Conventional Features

The term conventional feature refers to the standardised parameters that are currently measured in various medical images. Some of these features may be used in radiomics as well, either on their own or in conjunction with other features. While the established conventional parameters may have been successful in diagnosing and monitoring several conditions, other radiomic parameters may be more effective and reliable [27].

Geometric parameters, for example the vessel diameter, are conventionally used to identify stenosis, dilation and aneurysm [30–32]. This is a good diagnostic feature in anatomical imaging techniques that is easily readable from most images provided they have sufficient resolution. However, it still gives a limited amount of information.

An example of a feature used commonly in PET is the standardised uptake value (SUV) which is defined as the ratio of the radioactivity concentration in an ROI to the radioactivity concentration in the body (Equation 4.1). It is used to identify areas of abnormal uptake such as malignant or inflamed tissue (Figure 4.9) [33].

$$\text{SUV} = \frac{\text{concentration of activity in tissue}}{\text{injected activity/body weight}} \tag{4.1}$$

Over an ROI the SUV is normally quoted as either SUV_{mean} or SUV_{max}. The mean value is vulnerable to the segmentation of the ROI but avoids issues with noise while the issues with the maximum and minimum values are vice versa. SUV is a useful parameter as it is straightforward to calculate and it accounts for the patient's body weight thus removing some variability. It is also proven to be reproducible in quantifying glucose uptake in the ascending aorta and vena cava [34, 35]. However it is affected by respiratory motion, the blood glucose level of patients and the body fat percentage.

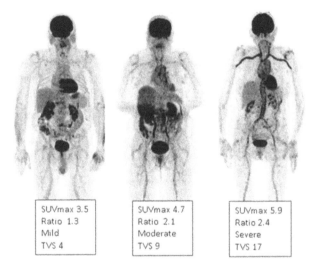

FIGURE 4.9: From left to right mild, moderate and severe cases of Large Vessel Vasculitis indicated by increasing SUV in FDG-PET images. 'FDG-PET' by Riemer H. J. A. Slart. Available at www.ncbi.nlm.nih.gov/pmc/articles/PMC5954002/. Licensed under the Creative Commons Attribution Licence CC BY 4.0 Full Terms at https://creativecommons.org/licenses/by/4.0/.

Technical factors such as image processing and scanner variability can also influence results [36]. In cardiac imaging SUV is used to evaluate the response to steroid therapy in some inflammatory diseases [37] but could also be used as a radiomic feature to predict cardiovascular disease [34].

4.3.1.2 First Order Statistics

First order statistics are derived from voxel intensity values and not the spatial relationships between voxels. Voxel intensity values are extracted from an ROI and are usually represented as a histogram binned according to intensity value. Several first order statistics are derived from the intensity distribution on the histogram, so are referred to as textural features. This term also encompasses higher order statistics that evaluate intensity variation as well. Although first order statistics have proven to be successful in providing diagnostic and prognostic information, they discard information concerning the spatial relationship between voxels, so are inherently limited [28]. Due to their reliance on intensity alone they are also extremely vulnerable to image processing techniques, inter-scanner variability and the number and size of bins in histogram analysis [38].

4.3.1.2.1 Averages, Variance and Standard Deviation

Both the mean and median intensities are often recorded, but average intensity is rarely used as a parameter on its own. It is often quoted alongside variance, standard deviation or another first order parameter for example in coronary plaque characterisation [39].

4.3.1.2.2 Skewness and Kurtosis

Skewness refers to the asymmetry of the histogram and its divergence from a normal distribution. A skewness of zero is symmetrical, less than zero is left skewed and more than zero is right skewed (Equation 4.2). Kurtosis describes the sharpness of the histogram peak. If the kurtosis is equal to three it is peaked like a normal distribution, if it is higher than three it is sharper, and if it is less than three the peak is flatter (Equation 4.3) [40]. Skewness and kurtosis have proven to be useful in evaluating patients with left ventricular hypertrophy and cardiomyopathies and stratifying the group by the cause of the condition [41, 42].

$$\text{Skewness} = \frac{\frac{1}{E}\sum_i \left(\text{HISTO}(i) - \overline{\text{HISTO}}\right)^3}{\left(\sqrt{\frac{1}{E}\sum_i \left(\text{HISTO}(i) - \overline{\text{HISTO}}\right)^2}\right)^3} \tag{4.2}$$

$$\text{Kurtosis} = \frac{\frac{1}{E}\sum_i \left(\text{HISTO}(i) - \overline{\text{HISTO}}\right)^4}{\left(\frac{1}{E}\sum_i \left(\text{HISTO}(i) - \overline{\text{HISTO}}\right)^2\right)^2} \tag{4.3}$$

$\text{HISTO}(i)$ = number of voxels with intensity i
$\overline{\text{HISTO}}$ = average of grey levels in histogram
E = total number of voxels

4.3.1.2.3 Energy and Entropy

Energy represents the uniformity of the histogram distribution (Equation 4.4) while entropy represents its randomness (Equation 4.5) [43, 44]. Energy has been shown to correlate with extracellular volume in a study investigating left ventricle remodelling [45] while entropy could be used to stratify patients by the cause of left ventricle hypertrophy [41].

$$\text{Energy} = \sum_i p(i)^2 \tag{4.4}$$

$$\text{Entropy}_{\log 10} = \sum_i p(i) \cdot \log_{10}\left(p(i) + \varepsilon\right) \tag{4.5}$$

$p(i)$ = probability of a voxel having intensity i
ε $\quad = 2 \times 10^{-16}$

4.3.1.3 Second and Higher Order Statistics

Second and higher order statistics look at the patterns in an image and the relationship between voxels rather than the voxel values themselves. Second order statistics describe the relationship between two voxels while higher order statistics describe the relationship between three or more voxels. On one hand it is important to include spatial information when looking at cardiovascular-related diseases as spatial relationships have been proven to be of relevance when assessing the severity of some illnesses, e.g. coronary plaques [46]. However as coronary lesions are fairly small relative to tumours the practicality of doing so has limited the use of these parameters in cardiac radiology [28]. As these statistics utilise intensity they are also vulnerable to similar factors as first order statistics such as image processing and image acquisition protocol.

4.3.1.3.1 Grey-Level Co-occurrence Matrices

Grey-level co-occurrence matrices (GLCMs) are second order statistics meaning they express the relationship between two voxels rather than the voxel values themselves. They are matrices that express how often an intensity value i (columns) occurs in a neighbouring voxel to intensity value j (rows) (Figure 4.10). Only four

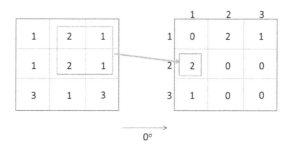

FIGURE 4.10: (Left) Example intensity values for nine neighbouring voxels. (Right) The 0 degrees raw GLCM expressing how often a given intensity value i sits at 0 degrees (indicated by the blue arrow) to intensity value j. For example, the red boxes demonstrate that as intensity = 1 sits at 0° to intensity = 2 two times in the left-hand array the 1,2 matrix component in the right hand array should equal two.

directions are required when looking at the relationship in 2D ($0°$, $45°$, $90°$ and $135°$) as the remaining directions are accounted for by the same angles in another voxel. To create symmetrical matrices the transpose is added to the original GCLM, and to create rotationally invariant matrices the four directions are accounted for and then averaged.

4.3.1.3.2 Higher Order Statistics

One type of higher order statistic is a grey-level run length matrix (GLRLM) which assesses how many voxels are next to each other with the same value (run length). The rows of a GLRLM represent the intensities while the columns express the run length. While the GLRLM is the most commonly used, several other higher order statistic matrices have been derived to express the spatial relationship between voxel intensities, e.g. the grey-level zone length matrix (provides information of the size of homogenous zones).

4.3.1.3.3 Matrix-Based Statistics

While the matrices mentioned above contain a lot of information some parameters have been derived to either summarise the matrices or pull out desired information. Some are based on first order statistics such as the energy and entropy of the matrix [43, 44] while others were created solely for this purpose, e.g. statistics that emphasise the number of short or long run lengths in a GLRLM [28].

4.3.1.3.4 Use of Second and Higher Order Statistics

As mentioned earlier the small size of coronary lesions in comparison to tumours has limited the use of these matrices in cardiology [28]. However, there are a few cases where they have proven to be successful. For example, GLCMs and GLRLMs were more effective than first order statistics at identifying coronary plaques with the Napkin-Ring Sign (a proven prognostic biomarker for severe cardiac events) [27]. Furthermore, a 3D version of the GLCM was used in one study to evaluate heart muscle structure [47].

4.3.1.4 Shape-Based Features

Shape-based metrics, also referred to as geometry-based features, are fairly well established and can be one, two or three dimensional. One common use of geometry-based features, more specifically lesion length and volume, is to evaluate the severity of coronary plaque build-up in coronary heart disease [27, 48]. The sphericity of the plaque volume is used in some studies as well [49]. The curvature of the aorta is another geometric parameter and has been shown to assess and predict the risk of aneurysm [50–52]. Some shape-based parameters are used in conjunction with other radiomic features such as total lesion glycolysis (TLG) which is the product of the volume and SUV_{mean} [43, 44]. TLG was used successfully to group large vessel vasculitis patients into favourable and unfavourable outcomes [53].

As mentioned earlier, some geometric features are already in clinical practice. The simplicity and the comprehensibility of shape-based parameters make them easier to use in clinical workflows. As they are less reliant on intensity, they are less affected by external factors such as image processing [38]. However, shape-based features can be complicated to measure as the heart and the blood vessel geometries contribute to the shape of a given lesion [28, 54].

4.3.2 Translating Radiomics into Clinical Practice

Despite showing great promise radiomics has yet to be adopted in general medical practice. This can be attributed to several factors. First, several radiomic features are reported to be vulnerable to image reconstruction, image acquisition and segmentation methods. Second, the process is complex resulting in low reproducibility and is difficult to fit into clinical practice. Some methods require human input introducing bias and measurement variability [55]. Radiomics also requires large datasets, multi-centre studies, and it has a high false positive rate [56]. Beichel et al. state that PET is especially at risk of confounding factors in radiomics studies due to low resolution, high noise and a high variety of reconstruction options [55], but the problem is found in all commonly used imaging modalities.

Most publications cite as one of their weaknesses their 'small cohort size' even if they had a larger number of participants relative to others. Orlhac et al. stated that 77% of PET radiomic studies had less than 100 participants when they conducted their study earlier that year (2018) and only three had more than 200 [57]. The most commonly given reason for small sample size was the difficulty in finding a homogenous cohort with the same or similar disease, stage, treatment and imaging protocol [41], [57]. This is closely tied to the lack of standardisation in image acquisition, reconstruction and analysis discussed earlier. As a result most studies are single-centre, and there is not a significant amount of evidence from multi-centre studies that would prove the transferability of radiomics. If this was achieved it could be shown that radiomics is a useful tool in clinical practice.

4.3.2.1 Image Processing

The use of different image processing methods, including but not limited to image reconstruction methods, is the most commonly reported source of variation among radiomic features with several studies having to consider the effects when designing their experiment [27, 28, 57].

The magnitude of the effect varied massively, with van Velden et al., Gallivanone et al. and Altazi et al. finding 97%, 26% and 1% of features were stable when the reconstruction method varied respectively [58–60]. This shows that the effect is feature dependent which is reinforced by the study from Shiri et al. who found geometry-based features were unaffected but that this was only true for 44% of intensity-based features and 41% of texture-based features [38]. Several papers suggest that the reliance on intensity for intensity- and texture-based features makes them unreliable due to the lack of standardised quantification and image reconstruction which limit their use in patient management [28, 61, 62].

Image reconstruction was not the only processing step to have an effect on radiomic features. Schofield et al. investigated the influence of different filters on image smoothing and found a 3 mm filter to be the most reproducible in a test–retest [41]. Similarly, Desseroit et al. found image quantisation methods made feature repeatability highly variable when looking at the reliability of shape and heterogeneity features in both the PET and CT components of PET–CT [63].

4.3.2.2 Image Acquisition

While not as commonly discussed it is widely accepted that the image acquisition process needs to be standardised if radiomics will ever be applied in clinical practice [57]. One study demonstrated this when using two different scanners made 4 out of 11 features significantly different [64]. Likewise, Beichel et al. found that using different scanners gave a coefficient of variation (COV) of 42.5%, varying acquisition times gave a COV of 5.3% and differences in institutional approaches gave a COV of 26.8% [55]. Several papers made a point of designing their experiments to ensure that the images were acquired the same way [27, 41] however many assumed if they were from the same centre the acquisition would have been the same. This assumption is not necessarily valid as many experiments were conducted over years, and software as well as the scanner will eventually change.

4.3.2.3 Segmentation

The extraction and analysis of radiomic features is usually a complex process offering multiple opportunities for a variety of approaches which may potentially lead to different results. As segmentation can define the region to be analysed it makes sense that if a different region is defined the radiomic analysis will give a different result. There are also several different manual and automated segmentation methods, and without standardisation of the methods, the results will remain irreproducible. Using a manual method is slow, laborious and vulnerable to inter/intra-observer variation, making it difficult to reproduce. However, it can be more accurate than automatic methods, especially if low resolution or non-contrast enhanced imaging was used, or if the shape and structure of the ROI is significantly altered due to pathological reasons. Automatic methods are much faster and more reproducible which allows much larger datasets to be analysed and more significant results to be determined. However, they often struggle with images that do not conform to normal anatomy and can be highly inaccurate if the images are not high resolution, partially corrupted or not of high contrast. A compromise can be found with semi-automatic methods in some cases.

For example, Gallivanone et al. found in their PET–CT study using a phantom that only 20% of features were stable when the segmentation method was different [58]. Similarly, Altazi et al. demonstrated that 13% of features were insensitive to whether a computed or manual segmentation method was used [59]. However, van Velden et al. found that 75% of features were insensitive to segmentation methods [60].

4.3.2.4 Reproducibility

Reproducibility of radiomic features was often studied in its own right to investigate natural variations between experiments. Figueroa et al. [34] and Rudd et al. [35] demonstrated that SUV_{max} and SUV_{mean} on the ascending aorta and vena cava were reproducible in PET imaging.

Gallivanone et al. found in their PET-CT study that only 53% of their features were reproducible when a test–retest was conducted [58]. Leijenaar et al. were more successful when testing feature stability and achieved 71% reproducibility with a test retest and 91% reproducibility with an inter-observer test [65].

One factor that must be considered in radiomics reproducibility however is that many papers are irreproducible to others [66]. Beichel et al. attributed this to complex methods that allow for error but are often not set out in full in publications [55]. This reinforces the need for standardised methods.

4.3.2.5 Outlook

Despite the obstacles radiomics faces in achieving clinical adoption there are several factors working in its favour. The importance of standardisation has been recognised by both the Radiological Society of North America (RSNA) and the European Society of Radiology (ESR) who have set up the Quantitative Imaging Biomarkers Alliance (QIBA) and the European Imaging Biomarkers Alliance (EIBALL) respectively. Part of their purpose is to work towards standardisation and optimisation of quantitative imaging which will facilitate the use of radiomics in clinical practice. However, these organisations are not new and standardisation has yet to be adopted so some progress still needs to occur.

Scientifically progress has been made in harmonising data collected from different departments or centres. Orlhac et al. have produced a method that standardises features from PET by removing the centre effect but keeping the patient effects [57]. The method expresses each feature using Equation 4.6.

$$y_{ij} = a + X_{ij}\beta + \gamma_i + \delta_i\varepsilon_{ij} \tag{4.6}$$

a = average value for feature
X_{ij} = design matrix for the covariates of interest
β = vector of regression coefficients for each covariate
γ_i = additive effect of scanner i
δ_i = multiplicative scanner effect
ε_{ij} = error term
y_{ij} = value of each feature y measured in VOI (Volume of Interest) j and scanner i

It aims to estimate γ_i and δ_i using empirical Bayes estimates. The authors used data from two departments (Department A and B) and made a third dataset by Gaussian smoothing Department A's data. Tumours were segmented to define an ROI, and similarly a spherical ROI was defined in healthy liver. Three SUV values were calculated along with six textural features using the LifeX software [43, 44], and feature distributions in each department were compared before and after harmonisation. In the healthy liver four out of nine features were significantly different between Department A and B and six out of nine features between Department A's original data and the smoothed data. After the harmonisation there was no significant difference between any set. In tumour images the trend was mostly the same. This method was previous used in genomics which faced a similar problem in its early stages called the Batch Effect. Like in radiomics this was due to technical differences in settings used to acquire the data making the solution easily transferable to radiomics.

Finally, large cohort studies in radiomics have proven to be feasible. Bohnen et al. investigated the potential of a large range of radiomic features for risk stratification for major cardiac diseases using a pilot cohort of 200 participants out of 45,000 participants from the Hamburg City Health Study (HCHS) [67]. Their investigation was successful, and they claimed this demonstrated that HCHS would facilitate radiomics. However, a standardised CMR workflow was established for the HCHS which provides a further example that standardisation across all aspects of image acquisition, reconstruction and analysis is helpful for radiomics to be used effectively in clinical practice. If this is not established then more work needs to be done in harmonising data from different centres, in a similar vein to Orlhac et al. [57] but across all modalities, with large patient cohorts and with larger numbers of radiomic features to prove robustness. Besides standardisation, software with a user-friendly interface will be very helpful, so that it does not create a burden for radiologists [28].

While standardisation is required for radiomics to reach its full potential, some features appear to be mostly unaffected by the factors discussed. In total, 47% of the features investigated by Shiri et al. had a small (<5%) COV, meaning they could be candidates for larger scale multi-centre studies [38]. Although this may be a helpful practical approach for multi-centre studies, this limits the information radiomics could provide.

4.3.3 Conclusion

In summary, radiomics can be used to extract large amounts of data from medical images, some of which correlate with clinical data and can be used as a diagnostic or predictive marker. First order statistics express the average intensity and the distribution of intensity in an ROI but do not involve the spatial relationships between voxels. Second or higher order statistics express the spatial relationships between voxels but have limited use in small coronary lesions. Shape-based parameters have proven quite successful and are simple and easily comprehensible. Radiomic features are vulnerable to several external factors such as image processing techniques, image acquisition protocols and the segmentation of the ROI. For radiomics to reach its full potential in clinical practice standardised imaging workflows need to be established.

4.4 Deep Learning

Machine learning is a type of artificial intelligence where computational systems (i.e. artificial neural networks (ANNs)) learn from examples rather than being explicitly programmed to conduct a task [1]. In the case of medical image analysis, the examples are the images and an example of a task could be segmentation or classification by diagnosis or prognosis. Traditionally, humans determine relevant features (e.g. shape, size, texture) in the images using practices such as radiomics. They then encode them into the machine learning algorithm.

Deep learning is a subset of machine learning where the ANN may be able to determine the optimal features for classification directly from the inputted data and/or images [68]. The advantage of this approach is that it can potentially eradicate the need for humans to extract features themselves, but its downside is that the ANN requires a significant amount of input data.

Deep learning makes use of ANNs. These networks consist of several interconnected layers made of processing units, called by definition artificial neurons [6, 69]. In the case of medical image analysis a specific type of ANN, the convolutional neural networks (CNNs) are the most commonly used. CNNs are designed to mimic the visual cortex and are often used to analyse images.

Several different CNN architectures have been created using different arrangements of the main types of layers – input layer, convolution layer, pooling layer, fully connected layer and output layer (Figure 4.11). The input layer in this case is the medical image and the convolution layer learns the features present in an image. The next layer is a pooling layer which downsamples the output from the convolution layer and makes the network less sensitive to a features position. Once several layers of convolution and pooling layers have been applied a fully connected layer receives every value

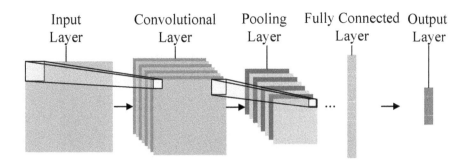

FIGURE 4.11: 'Structure of a Convolutional Artificial Neural Network (CNN)' by Min Peng, Chonyang Wang, Tong Chen and Guangyan Liu. Available at: www.mdpi. com/2078-2489/7/4/61. Licensed under the Creative Commons Attribution Licence CC BY 4.0 Full Terms at https://creativecommons.org/licenses/by/4.0/

in the final array, sets them into a 1D array and every value receives a vote of the final classification [70]. The network is trained by determining the error for images with a known classification and altering the parameters in the network to minimise the error.

The number of layers required increases depending on the complexity of the task and the level of accuracy required as shown in Table 4.1. In this section the three most common applications of deep learning in cardiology will be discussed – segmentation, diagnosis and prediction. Alongside this the strengths and weaknesses of deep learning will be explored.

4.4.1 Segmentation Using Tomographic Imaging

Anatomical segmentation is a key application of deep learning in medical imaging. There are several reasons for its emergence; in some cases, such as ventricle segmentation in MRI, the manual process is relatively simple but laborious and an automated segmentation process would improve the clinical workflow [74]; in other cases, such as cranial vasculature segmentation in CTA imaging, the process is difficult as the vessels have a similar intensity to nearby tissues [75]. Figure 4.12 demonstrates a manual segmentation of the aorta. While several automated methods have been attempted to tackle these problems, deep learning has proven to be superior in accuracy [76].

The majority of work in this field has been in left ventricle (LV) segmentation [74, 77–85] as LV structure and function are commonly used to evaluate heart health and to diagnose several cardiovascular conditions [74]. Right ventricle segmentation has only recently become an area of interest after its prognostic value was determined [76, 86, 87]. Notable methods that improve ventricle segmentation include: converting

TABLE 4.1: A Comparison of Three Well-Known CNNs, Their Error Rates and Their Number of Layers

Name of CNN	Number of Layers	Top Five Test Error Rates (%)
AlexNet [71]	8	15.3
GoogleNet [72]	22	6.67
ResNet [73]	152	3.57

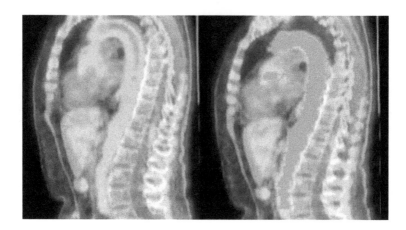

FIGURE 4.12:　Manual segmentation of the aorta.

from semi-automated to fully automated segmentation as shown by Tan et al. [74, 88] albeit only with a small improvement; combining deep learning with deformable models to help deal with edge slices [82]; and using the level set method to deal with shape variations and small data size [77]. While all of these investigations state that their method improves the accuracy of the deep learning model, it is hard to evaluate these claims due to different methods of measuring accuracy and variations in the purpose of the study. Many papers mentioned that they placed highly or would have performed favourably in the segmentation challenges run by MICCAI (Medical Image Computing & Computer Assisted Intervention) conferences. This along with the fact that the models did perform well by the metrics used by the author (with many, but not all, approximate accuracies >90%) means that the methods are still worth considering when planning to use deep learning in medical image segmentation.

Also worth noting is the left ventricle segmentation method used by Yang et al. [89] where Multi-Atlas Segmentation (MAS) and deep learning were combined to overcome challenges that MAS alone presented. MAS uses atlases – sets of images segmented by experts – to conjecture the segmentation of a new image. The authors admitted that while their method performed well, Ngo et al. [77] had published a more accurate model. They attributed this to Ngo et al. using a semi-automated rather than a fully automated segmentation method, meaning they required more manual input which is known for being more accurate but is more time consuming and at a higher risk of subjective errors [79].

The pay off between human intervention and fully automated methods was a common theme in several papers. For example, Chen et al. [75] constructed a deep learning network to segment cranial vasculature in CTA imaging, but to avoid the need for human annotation they utilised segmented MRA images of the same area to train the network. By building the network and training procedure to account for the different training data, they achieved a fairly complete segmentation of the vasculature. The segmentation may have been improved if training was conducted using fully annotated CTA images, but this would have required a significant amount of work and time from an experienced clinician. Overall the decision between using semi-automated methods for higher accuracy or fully automated methods for efficiency should come down to the purpose of the model. If a highly accurate segmentation method is required then the method should reflect it, but if only a rough segmentation is required or deep learning is employed due to the large amount of images needing annotation then a more efficient method may be a better fit. As López-Linares et al. [90] point out, in many cases a good estimate that can then be improved upon by a clinician in a much shorter time than they would have previously needed to invest would still have a positive influence on the clinical workflow.

Literature concerning blood vessel segmentation in tomographic imaging using deep learning is currently lacking in comparison to ventricle methods despite being essential for stenosis detection. However, the convolutional neural networks built by Nasr-Esfahani et al. [91, 92] to segment vessels in X-ray angiograms may form a good basis for any future work in the area. By adding an image enhancement step prior to the CNNs, they achieved an accuracy of 93.5% and 97.9%, respectively. The authors concluded that improvement between the papers was due to a second CNN being added to the method to improve vessel border region detection.

In summary, using models that account for heart movement and image enhancement helps improve the segmentation output of a deep learning model. The amount of human involvement required by a network depends on the performance criteria of the model. For example, if deep learning is being employed to ease the workload on clinicians then a high involvement does not meet the requirements. On the other hand if high accuracy is the main purpose of the model, more human involvement will help meet that aim. The trade-off between human labour and time efficiency should be considered when designing the model.

4.4.2　Diagnosis, Classification and Detection Using Tomographic Imaging

There is a large variety of deep learning applications in diagnostics. For example, Silva et al. [93] produced a network that classifies transthoracic echocardiographic exams (TTE) by their ejection fraction, a measurement that can evaluate left ventricle function and hence is used in cardiovascular diagnosis. The authors proposed using TTE exams as this bypassed the need for left ventricle segmentation which, as mentioned earlier, is time consuming if carried out manually. They achieved an accuracy of 78% which is not sufficient to base a diagnosis upon but can still provide decision support for clinicians.

Deep learning has also been applied to ultrasound images to detect congenital heart disease and other abnormalities in foetal hearts [94]. This approach was adopted in an attempt to overcome the difficulties in identifying a pathological condition early which were caused by the small size of foetal hearts and ultrasound artefacts. The authors made several variations of their network with the best performing having an error rate of 23.48%. This is not exceptionally high when compared with the performance of several other deep learning diagnostic networks, but there is space for improvement. The authors suggested incorporating the temporal information from ultrasound into the neural network to improve accuracy. Another potential improvement would be to increase the number of subjects as they only used ten.

The accuracy of a model can be improved when additional processing steps are added before or after the CNN. Identification of cerebral microbleeds (CMBs) is essential for diagnosing dementia and several other vascular diseases, but like many manual tasks concerning medical imaging it is time consuming and is not highly reproducible. Chen et al. [95] presented a method for detection using three steps: potential CMB located by statistical thresholding, CNN for feature representation and a support vector machine classifier to reduce false positives. When compared with the random forest technique (another form of deep learning) and other CNNs, the authors' method had about half as many false positives and was only 2–3% more sensitive.

One noticeable gap in the previous work in the area is a comparison between the accuracy of diagnostic deep learning methods and a clinician's diagnostic accuracy. Several studies used images labelled by an expert as their ground truth for comparison, but these tend to be specialists, and often only one or two experts are used rather than an average of several leading to a bias. A good evaluation of the methods may be to compare them to the accuracy of several different clinicians with different experience levels as Becker et al. [96] did to demonstrate the use of the model as either decision support or as a training aid. Although ANNs may not be systematically more accurate than the experts manually annotating the images, in clinical practice the ANN may still perform better once fatigue and human error are more prevalent. In comparison to similar studies in oncology, there is very little discussion concerning how the methods would fit into the clinical workflow – for example how fast the networks can produce an answer and their ease of use – but as significantly more work has been conducted in oncology; it may just take more time for cardiovascular applications to reach that stage.

4.4.3   Prediction and Prognosis Using Tomographic Imaging

Deep learning can also be used for risk-stratification and to predict therapeutic outcomes allowing for more efficient treatment management. One example application is in stroke management. Vincent et al. [97] produced tissue hyperperfusion (increased cerebral blood flow) measurements using arterial spin labelling (ASL) MRI. Hyperperfusion in turn is useful for predicting intracerebral haemorrhage (a type of stroke). When compared to manually annotated images the predictions were 97.45% accurate making it a potentially powerful tool in clinical management.

There are clearly fewer studies using deep learning for outcome prediction in cardiovascular-related diseases in comparison to segmentation or diagnostics; however, more studies are likely to be published in the near future as deep learning for prediction has proven successful in other fields.

4.4.4   Using Deep Learning with Radionuclide Imaging

Most of the examples of deep learning in cardiological and cardiovascular examinations described in the previous sections used either CTA or MRA images. Examples using radionuclide imaging are less common in literature. One possible cause of this is the difficulty in gathering a large number of images required for deep learning due to the relatively smaller number of scans with these techniques. However, a promising result from Betancur et al. demonstrates that deep learning can be used for SPECT in cardiovascular diseases. They produced a deep learning method for predicting obstructive disease using Fast Myocardial Perfusion SPECT, and it was shown to perform better than the current gold standard [98]. PET and SPECT deep learning models in oncology are becoming more common in comparison, and a lot can be learned from the progress made in that field [99–101].

4.4.5 Discussion

The use of deep learning in cardiovascular-related diseases is still a young field, and the full potential is yet to be realised. When using ANNs to segment cardiovascular anatomy and regions of interest the main barrier faced is the movement of the heart due to respiration or cardiac contraction. This can be overcome to some degree with deformable models and the level set method. The potential of deep learning for diagnosis has been studied extensively, but it is yet to bridge the translational gap to be used in clinical practice even in a trial version. Going forward more work needs to be conducted to establish how the neural networks hold up in a clinical setting. There is insufficient literature to evaluate predictive ANNs for cardiovascular diseases.

Beside segmentation, diagnosis and prediction, there are several other applications of deep learning in medical imaging that have started to emerge recently. For example, deep learning is often used in image acquisition and reconstruction to improve image quality [102], [103]. Other exciting applications are treatment response assessment [104], adaptive radiotherapy [105] and transplant acceptance assessment [106].

The deep learning studies often face a barrier which is to find sufficiently large high-quality datasets [77, 97]. Studies aim to have thousands of images and at least several hundred per category (e.g. healthy/not healthy) [107]. However, several only manage to acquire tens or hundreds, but in some instances this was proven to be sufficient [97, 108]. The sample size required grows with the complexity of the deep learning network, so while simple classifications may only require a few hundred images, more complex methods may require thousands to be adequately accurate [6]. If models are trained on insufficient data, outliers and noise in the data become problematic and overfitting becomes an issue. Overfitting is when a model used to describe a set of data is very specific to the training data and does not generalise well to other datasets (Figure 4.13). This is often a result of an unnecessarily complex or flexible model being created to account for noise or outliers that are not part of the true underlying relationship [109]. Overfitting can be hard to detect, so a validation dataset is often held back to test the model after training. Similarly, cross-validation works by holding back a portion of the data (for example 20%), training it on the remaining data, validating with the held back set and then repeating the process holding back a different set of data each time. This also allows for a final test with an unseen dataset. Using a validation dataset is a good technique for detecting overfitting, but it does not necessarily work every time. Regularisation is a technique used to force a model to become simpler. In the case of neural networks one example is the dropout method that randomly drops neurons from the ANN during training to prevent them from adapting very well to the input data [109]. As with any simplification method, there is the risk of producing an inflexible and oversimplified model.

There are several techniques to improve deep learning accuracy when a small training set is unavoidable, such as dividing images into smaller patches [110] or manipulating images to look slightly different thus allowing them to be used more than once [111]. Transfer learning is a popular option which involves transferring knowledge from a previously trained model into a new model without needing to retrain the new model from scratch [112]. The first layers of a deep neural network extract general features which are easily transferable even between models with completely different purposes. The final few layers are very task specific and are generally not transferable. The more similar the purpose of the models the easier it is to transfer knowledge between them.

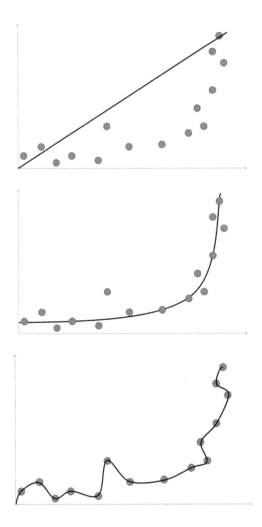

FIGURE 4.13:   Model underfits the data and is oversimplified (top). Model correctly fits the data (middle). Model overfits the data and is more complex than required (bottom).

The exact point a layer changes from general to specific is hard to determine, but some authors such as Yosinski et al. have proposed methods to achieve this [113]. In some cases though, deep learning is not feasible with a small amount of data, but techniques such as traditional machine learning may still be possible.

Another commonly discussed topic in deep learning is deciding whether to choose a semi-automated deep learning method (some human contribution) or fully automated. Semi-automated methods are more accurate than fully automated methods, but the human contribution is time consuming. Deep learning is often employed to save time in analysis, diagnosis and treatment planning in clinical workflow so a significant amount of human contribution is undesirable. Some argue that a less accurate output can easily be refined by the clinician afterwards in much less time. There is not a method that is clearly better, and the most appropriate one should be considered on a case-by-case basis.

In summary, deep learning and more specifically CNNs are an extremely useful tool for medical image analysis. In cardiac imaging specifically there have been several successful attempts to use them as segmentation and diagnostic tools, but there are not as many predictive models produced yet. The largest barrier to using deep learning is the need for extremely large amounts of data, but there are techniques to overcome this difficulty. The amount of human involvement in a deep learning model is a common point for debate, and the correct decision depends on the requirements of the model. Overall, deep learning is an exciting and promising tool that could allow clinicians to gain much more information from medical images.

4.5   Concurrent Research

The fields of both radiomics and deep learning progress quickly and throughout the process of editing this chapter, several other relevant studies were published. These articles are not be incorporated or fully discussed in the same way as previous research in the field but the most relevant are summarised here.

In radiomics, three key papers were published. Sun et al. used CMR and echocardiography to assess the severity of mitral regurgitation. Their radiomic model performed well with a sensitivity and specificity of 85.7% and 94.1%, respectively [114]. They also demonstrated that combining imaging modalities is a worthwhile approach. Gould et al. used radiomics to help with treatment selection and found that scar heterogenity measured using CMR mean entropy of the relevant tissue was a good indicator for the appropriate implantable cardioverter-defibrillator therapy [115]. Manabe et al. found that PET-CT textural analysis could be of use in cardiac sarcoidosis diagnosis [116]. Oikonomou et al. used radiomic features from tissue CT to improve cardiac risk prediction and found several significant results including indicators for inflammation, fibrosis and vascularity as well as developing a machine learning algorithm that improved upon traditional risk stratification [117].

Several deep learning papers related to cardiovascular diseases were published but the one of the most relevant was a continuation of the work by Betancur et al. discussed earlier [118]. Diller et al. also published two papers using deep learning for decision making in adult congentital heart disease, including a large single centre study with 10,019 patients [119, 120]. Lastly, there were several examples of more work being published using deep learning for image segmentation [121, 122].

4.6   Conclusion

In this book chapter, we extensively discussed the arising fields of radiomics and deep learning focusing on how they can be used to advance cardiac and cardiovascular imaging. The advantages and disadvantages of each technique were explored alongside the challenges they face on their route to clinical adoption. In summary, radiomics allows us to extract large amounts of previously unused data from medical images, and in many cases these data will correlate with other types of clinical data so to strengthen their diagnostic or predictive potential. There are several types of features each with their own strengths and weaknesses. Features based on first order statistics are simple and commonly used. However, they discard information about the spatial relationships between voxels and are vulnerable to external factors such as image processing. Second and higher order statistics express the spatial relationships between voxels but have relatively limited use in small coronary lesions and are also vulnerable to the same external factors. Shape-based parameters have proven somewhat successful and are simple and comprehensible. They overlap significantly with conventional parameters as they are easy enough to include in some clinical workflows. Most, if not all, radiomics features can be influenced by external factors so standardised imaging workflows need to be established for radiomics to work on a larger scale. Deep learning is a useful supplementary technique to radiomics, and perhaps it will eventually become a necessary add-in for analysing future medical imaging datasets. The ANN mines features from medical images to help classify them by disease or outcome without the need for a human to tell it what to look for. As a result, more relationships between image features and clinical data can be determined. In conclusion, radiomics and deep learning are promising techniques that could allow us to gain much more information from medical images and provide better treatment for patients.

Acknowledgements

This work was supported by the EPSRC (Engineering and Physical Sciences Research Council) [Grant number EP/L014823/1]. Dr Tsoumpas is supported by a Royal Society Industry Fellowship (IF170011).

References

1. G. Langs et al., "Machine learning: From radiomics to discovery and routine," *Radiologe*, vol. 58, no. 1, pp.1–6, 2018.

2. I. El Naqa et al., "Exploring feature-based approaches in PET images for predicting cancer treatment outcomes," *Pattern Recognit.*, vol. 42, no. 6, pp. 1162–1171, 2009.

3. R. J. Gillies, A. R. Anderson, R. A. Gatenby, and D. L. Morse, "The biology underlying molecular imaging in oncology: From genome to anatome and back again.," *Clin. Radiol.*, vol. 65, no. 7, pp. 517–21, 2010.

4. G. Litjens et al., "A survey on deep learning in medical image analysis," *Med. Image Anal.*, vol. 42, pp. 60–88, 2017.

5. B. Van Essen, H. Kim, R. Pearce, K. Boakye, and B. Chen, "LBANN: Livermore big artificial neural network HPC toolkit," in *Proceedings of MLHPC 2015: Machine Learning in High-Performance Computing Environments - Held in conjunction with SC 2015: The International Conference for High Performance Computing, Networking, Storage and Analysis*, Austin, TX, 2015.

6. S. Gandhi, W. Mosleh, J. Shen, and C.-M. Chow, "Automation, machine learning, and artificial intelligence in echocardiography: A brave new world," *Echocardiography*, vol. 35, no. 9, pp. 1402–1418, 2018.

7. M. B. Stokes and R. Roberts-Thomson, "The role of cardiac imaging in clinical practice," *Aust. Prescr.*, vol. 40, no. 4, pp. 151–155, 2017.

8. R. Blankstein, "Introduction to noninvasive cardiac imaging," *Circulation*, vol. 125, no. 3, pp. 267–271, 2012.

9. R. K. Pai, E. G. Thompson, M. J. Gabica, A. Husney, G. J. Philippides "Computed Tomography Angiogram (CT Angiogram) | Michigan Medicine," *Univeristy of Michigan*, 2018. [Online]. Available: https://www.uofmhealth.org/health-library/bo1097. [Accessed: 01-Dec-2018].

10. A. Dehaene, A. Jacquier, C. Falque, G. Gorincour, and J. Y. Gaubert, "Imaging of acquired coronary diseases: From children to adults," *Diagn. Interv. Imaging*, vol. 97, pp. 571–580, 2016.

11. D. T. Ginat, M. W. Fong, D. J. Tuttle, S. K. Hobbs, and R. C. Vyas, "Cardiac imaging: Part 1, MR pulse sequences, imaging planes, and basic anatomy," *Am. J. Roentgenol.*, vol. 197, no. 4, pp. 808–815, 2011.

12. J. J. Westenberg et al., "Vessel diameter measurements in gadolinium contrast-enhanced three-dimensional MRA of peripheral arteries," *Magnetic Resonance Imaging*, vol. 18, no. 1, pp. 13–22, 2000.

13. British Heart Foundation (BHF), "MRI Scans - Tests for Heart Conditions - British Heart Foundation," *BHF*, 2018. [Online]. Available: https://www.bhf.org.uk/informationsupport/tests/mri-scans. [Accessed: 02-Dec-2018].

14. National Heart, Lung, and Blood Institute (NHLBI), "Cardiac MRI ," *U.S. Department for Health and Human Services*, 2018.[Online]. Available: https://www.nhlbi.nih.gov/health-topics/cardiac-mri. [Accessed: 02-Dec-2018].

15. R. Kalin and M. S. Stanton, "Current clinical issues for MRI scanning of pacemaker and defibrillator patients," *Pacing Clin. Electrophysiol.*, vol. 28, no. 4, pp. 326–328, 2005.

16. National Heart, Lung, and Blood Institute (NHLBI), "Nuclear Heart Scan | National Heart, Lung, and Blood Institute (NHLBI)," *U.S. Department for Health and Human Services*, 2018. [Online]. Available: https://www.nhlbi.nih.gov/health-topics/nuclear-heart-scan. [Accessed: 30-Nov-2018].

17. American Heart Association (AHA), "Single Photon Emission Computed Tomography (SPECT)," *AHA*, 2015. [Online]. Available: http://www.heart.org/en/health-topics/heart-attack/diagnosing-a-heart-attack/single-photon-emission-computed-tomography-spect. [Accessed: 01-Dec-2018].

18. Z. Li, A. A. Gupte, A. Zhang, and D. J. Hamilton, "Pet imaging and its application in cardiovascular diseases," *Methodist Debakey Cardiovasc. J.*, vol. 13, no. 1, pp. 29–33, 2017.

19. R. Rocha and D. Fornell, "Advances in Cardiac Nuclear Imaging," *DAIC*, 2014. [Online]. Available: https://www.dicardiology.com/article/advances-cardiac-nuclear-imaging. [Accessed: 30-Nov-2018].

20. E. C. Ehman et al., "PET/MRI: Where might it replace PET/CT?" *J. Magn. Reson. Imaging*, vol. 46, no. 5, pp. 1247–1262, 2017.

21. G. Angelidis et al., "SPECT and PET in ischemic heart failure," *Heart Fail. Rev.*, vol. 22, no. 2, pp. 243–261, 2017.

22. British Heart Foundation (BHF), "Myocardial Perfusion Scan -British Heart Foundation," *BHF*, 2018. [Online]. Available: https://www.bhf.org.uk/informationsupport/tests/myocardial-perfusion-scan. [Accessed: 01-Dec-2018].

23. J. Maddahi and R. R. S. Packard, "Cardiac PET perfusion tracers: Current status and future directions," *Semin. Nucl. Med.*, vol. 44, no. 5, pp. 333–343, 2014.

24. S. Dorbala et al., "Prognostic value of stress myocardial perfusion positron emission tomography: Results from a multicenter observational registry," *J. Am. Coll. Cardiol.*, vol. 61, no. 2, pp. 176–84, 2013.

25. I. H. O. [Internet], "What Is an Echocardiogram?" 2016. [Online]. Available: https://www.ncbi.nlm.nih.gov/books/NBK395556/?report=reader#_NBK395556_pubdet_. [Accessed: 29-Nov-2018].

26. R. Senior et al., "Clinical practice of contrast echocardiography: Recommendation by the European Association of Cardiovascular Imaging (EACVI) 2017," *Eur. Heart J. Cardiovasc. Imaging*, vol. 18, no. 11, p. 1205, 2017.

27. M. Kolossváry et al., "Radiomic features are superior to conventional quantitative computed tomographic metrics to identify coronary plaques with napkin-ring sign," *Circ. Cardiovasc. Imaging*, vol. 10, no. 12, pp. e006843, 2017.

28. M. Kolossváry, M. Kellermayer, B. Merkely, and P. Maurovich-Horvat, "Cardiac computed tomography radiomics," *J. Thorac. Imaging*, vol. 33, no. 1, pp. 26–34, 2018.
29. Computational Imaging & Bioinformatics Lab – Harvard Medical School, "Radiomics," *Harvard*, 2017. [Online]. Available: http://www.radiomics.io/. [Accessed: 09-Nov-2018].
30. R. Lorbeer et al., "Reference values of vessel diameters, stenosis prevalence, and arterial variations of the lower limb arteries in a male population sample using contrast-enhanced MR angiography," *PLoS One*, vol. 13, no. 6, p. e0197559, 2018.
31. A. J. Boyd, D. C. S. Kuhn, R. J. Lozowy, and G. P. Kulbisky, "Low wall shear stress predominates at sites of abdominal aortic aneurysm rupture," *J. Vasc. Surg.*, vol. 63, no. 6, pp. 1613–1619, 2016.
32. A. Forteza et al., "Efficacy of losartan vs. atenolol for the prevention of aortic dilation in Marfan syndrome: A randomized clinical trial," *Eur. Heart J.*, vol. 37, no. 12, pp. 978–985, 2016.
33. K. Mah and C. B. Caldwell, "Biological target volume," in *PET-CT in Radiotherapy Treatment Planning*, Content Repository Only!, 2008, pp. 52–89.
34. A. L. Figueroa et al., "Measurement of arterial activity on routine FDG PET/CT images improves prediction of risk of future CV events," *JACC Cardiovasc. Imaging*, vol. 6, no. 12, pp. 1250–1259, 2013.
35. J. H. F. Rudd et al., "18Fluorodeoxyglucose positron emission tomography imaging of atherosclerotic plaque inflammation is highly reproducible," *J. Am. Coll. Cardiol.*, vol. 50, no. 9, pp. 892–896, 2007.
36. M. C. Adams, T. G. Turkington, J. M. Wilson, and T. Z. Wong, "A systematic review of the factors affecting accuracy of SUV measurements," *Am. J. Roentgenol.*, vol. 195, no. 2, pp. 310–320, 2010.
37. S.-A. Chang et al., "[18 F]Fluorodeoxyglucose PET/CT predicts response to steroid therapy in constrictive pericarditis," *J. Am. Coll. Cardiol.*, vol. 69, no. 6, pp. 750–752, 2017.
38. I. Shiri, A. Rahmim, P. Ghaffarian, P. Geramifar, H. Abdollahi, and A. Bitarafan-Rajabi, "The impact of image reconstruction settings on 18F-FDG PET radiomic features: Multiscanner phantom and patient studies," *Eur. Radiol.*, vol. 27, no. 11, pp. 4498–4509, 2017.
39. G. Y. Kim, J. H. Lee, Y. N. Hwang, and S. M. Kim, "A novel intensity-based multi-level classification approach for coronary plaque characterization in intravascular ultrasound images.," *Biomed. Eng. Online*, vol. 17, no. Suppl 2, p. 151, 2018.
40. G. J. R. Cook, M. Siddique, B. P. Taylor, C. Yip, S. Chicklore, and V. Goh, "Radiomics in PET: Principles and applications," *Clin. Transl. Imaging*, vol. 2, no. 3, pp. 269–276, 2014.
41. R. Schofield et al., "Texture analysis of cardiovascular magnetic resonance cine images differentiates aetiologies of left ventricular hypertrophy," *Clin. Radiol.*, vol. 74, no. 2, pp. 1–10, 2018.
42. C.-W. Wu et al., "Histogram analysis of native T1 mapping and its relationship to left ventricular late gadolinium enhancement, hypertrophy, and segmental myocardial mechanics in patients with hypertrophic cardiomyopathy," *J. Magn. Reson. Imaging*, vol. 49, no. 3, pp. 668–677, 2018.
43. LIFEx, "LIFEx," *LIFEx*, 2018. [Online]. Available: https://www.lifexsoft.org/. [Accessed: 17-Dec-2018].
44. C. Nioche et al., "LIFEx: A freeware for radiomic feature calculation in multimodality imaging to accelerate advances in the characterization of tumor heterogeneity," *Cancer Res.*, vol. 78, no. 16, pp. 4786–4789, 2018.
45. A. Esposito et al., "Assessment of remote myocardium heterogeneity in patients with ventricular tachycardia using texture analysis of Late Iodine Enhancement (LIE) Cardiac Computed Tomography (cCT) images," *Mol. Imaging Biol.*, vol. 20, no. 5, pp. 816–825, 2018.
46. R. Virmani, A. P. Burke, A. Farb, and F. D. Kolodgie, "Pathology of the vulnerable plaque," *J. Am. Coll. Cardiol.*, vol. 47, no. 8, pp. C13–C18, 2006.
47. N. M. Salih and D. E. Octorina Dewi, "Structural characterization of heart muscle in coronary artery CT image using extended 3D GLCM: Preliminary study on normal systole and diastole phases," *Int. J. Cardiol.*, vol. 249, pp. S14–S15, 2017.

48. R. Nakanishi et al., "Changes in coronary plaque volume: Comparison of serial measurements on intravascular ultrasound and coronary computed tomographic angiography," *Texas Hear. Inst. J.*, vol. 45, no. 2, pp. 84–91, 2018.

49. H. E. Barrett, J. J. Mulvihill, E. M. Cunnane, and M. T. Walsh, "Characterising human atherosclerotic carotid plaque tissue composition and morphology using combined spectroscopic and imaging modalities," *Biomed. Eng. Online*, vol. 14, no. Suppl 1, p. S5, 2015.

50. S. Ruiz De Galarreta, R. Anton, A. Cazon, and A. Pradera-Mallabiabarrena, "Influence of the local mean curvature on the abdominal aortic aneurysm stress distribution," *J. Mech. Med. Biol.*, vol. 17, no. 08, p. 1750106, 2017.

51. R. C. L. Schuurmann et al., "Aortic curvature as a predictor of intraoperative type Ia endoleak," *J. Vasc. Surg.*, vol. 63, no. 3, pp. 596–602, 2016.

52. H. B. Alberta, J. L. Secor, T. C. Smits, M. A. Farber, W. D. Jordan, and J. S. Matsumura, "Differences in aortic arch radius of curvature, neck size, and taper in patients with traumatic and aortic disease," *J. Surg. Res.*, vol. 184, no. 1, pp. 613–618, 2013.

53. L. Dellavedova et al., "The prognostic value of baseline18F-FDG PET/CT in steroid-naïve large-vessel vasculitis: Introduction of volume-based parameters," *Eur. J. Nucl. Med. Mol. Imaging*, vol. 43, no. 2, pp. 340–348, 2016.

54. C. J. Slager et al., "The role of shear stress in the generation of rupture-prone vulnerable plaques," *Nat. Clin. Pract. Cardiovasc. Med.*, vol. 2, no. 8, pp. 401–407, 2005.

55. R. R. Beichel et al., "Multi-site quality and variability analysis of 3D FDG PET segmentations based on phantom and clinical image data," *Med. Phys.*, vol. 44, no. 2, pp. 479–496, 2017.

56. M. Vallieres, A. Zwanenburg, B. Badic, C. Cheze-Le Rest, D. Visvikis, and M. Hatt, "Responsible radiomics research for faster clinical translation," *J. Nucl. Med.*, vol. 59, no. 2, pp. 189–193, 2018.

57. F. Orlhac et al., "A post-reconstruction harmonization method for multicenter radiomic studies in PET," *J. Nucl. Med.*, vol. 59, no. 8, pp. 1321–1328, 2018.

58. F. Gallivanone, M. Interlenghi, D. D'Ambrosio, G. Trifirò, and I. Castiglioni, "Parameters influencing PET imaging features: A phantom study with irregular and heterogeneous synthetic lesions," *Contrast Media Mol. Imaging*, vol. 2018, Article ID 5324517, 12 pages, 2018.

59. B. A. Altazi et al., "Reproducibility of F18-FDG PET radiomic features for different cervical tumor segmentation methods, gray-level discretization, and reconstruction algorithms," *J. Appl. Clin. Med. Phys.*, vol. 18, no. 6, pp. 32–48, 2017.

60. F. H. P. van Velden et al., "Repeatability of radiomic features in non-small-cell lung cancer [18F]FDG-PET/CT studies: Impact of reconstruction and delineation," *Mol. Imaging Biol.*, vol. 18, no. 5, pp. 788–795, 2016.

61. H. Precht et al., "Influence of adaptive statistical iterative reconstruction on coronary plaque analysis in coronary computed tomography angiography," *J. Cardiovasc. Comput. Tomogr.*, vol. 10, no. 6, pp. 507–516, 2016.

62. T. A. Fuchs et al., "CT coronary angiography: Impact of adapted statistical iterative reconstruction (ASIR) on coronary stenosis and plaque composition analysis," *Int. J. Cardiovasc. Imaging*, vol. 29, no. 3, pp. 719–724, 2013.

63. M.-C. Desseroit et al., "Reliability of PET/CT shape and heterogeneity features in functional and morphologic components of non–small cell lung cancer tumors: A Repeatability analysis in a prospective multicenter cohort," *J. Nucl. Med.*, vol. 58, no. 3, pp. 406–411, 2017.

64. S. Reuzé et al., "Prediction of cervical cancer recurrence using textural features extracted from18F-FDG PET images acquired with different scanners," *Oncotarget*, vol. 8, no. 26, pp. 43169–43179, 2017.

65. R. T. H. Leijenaar et al., "Stability of FDG-PET Radiomics features: An integrated analysis of test-retest and inter-observer variability," *Acta Oncol. (Madr).*, vol. 52, no. 7, pp. 1391–1397, 2013.

66. J. P. A. Ioannidis, "Why most published research findings are false," *PLoS Med.*, vol. 2, no. 8, p. e124, 2005.

67. S. Bohnen et al., "Cardiovascular magnetic resonance imaging in the prospective, population-based, Hamburg City Health cohort study: Objectives and design," *J. Cardiovasc. Magn. Reson.*, vol. 20, no. 1, p. 68, 2018.

68. G. Chartrand et al., "Deep learning: A primer for radiologists," *RadioGraphics*, vol. 37, no. 7, pp. 2113–2131, 2017.

69. L. D. Jones, D. Golan, S. A. Hanna, and M. Ramachandran, "Artificial intelligence, machine learning and the evolution of healthcare: A bright future or cause for concern?" *Bone Jt. Res.*, vol. 7, no. 3, pp. 223–225, 2018.

70. Y. LeCun, Y. Bengio, and G. Hinton, "Deep learning," *Nature*, vol. 521, no. 7553, pp. 436–444, 2015.

71. A. Krizhevsky, I. Sutskever, and G. E. Hinton, "ImageNet classification with deep convolutional neural networks," *Advances in Neural Information Processing Systems*, pp. 1097–1105, 2012.

72. C. Szegedy et al., "Going deeper with convolutions," *Proceedings of the IEEE Conference on Computer Vision and Pattern Recognition*, pp. 1–9, 2015.

73. K. He, X. Zhang, S. Ren, and J. Sun, "Deep residual learning for image recognition," *Proceedings of the IEEE Conference on Computer Vision and Pattern Recognition*, pp. 770–778, 2016.

74. L. K. Tan, R. A. McLaughlin, E. Lim, Y. F. Abdul Aziz, and Y. M. Liew, "Fully automated segmentation of the left ventricle in cine cardiac MRI using neural network regression," *J. Magn. Reson. Imaging*, vol. 48, no. 1, pp. 140–152, 2018.

75. X. Chen et al., "Train a 3D U-Net to segment cranial vasculature in CTA volume without manual annotation," in *Proceedings - International Symposium on Biomedical Imaging*, 2018, vol. 2018, April, pp. 559–563.

76. M. R. Avendi, A. Kheradvar, and H. Jafarkhani, "Automatic segmentation of the right ventricle from cardiac MRI using a learning-based approach," *Magn. Reson. Med.*, vol. 78, no. 6, pp. 2439–2448, 2017.

77. T. A. Ngo, Z. Lu, and G. Carneiro, "Combining deep learning and level set for the automated segmentation of the left ventricle of the heart from cardiac cine magnetic resonance," *Med. Image Anal.*, vol. 35, pp. 159–171, 2017.

78. O. Emad, I. A. Yassine, and A. S. Fahmy, "Automatic localization of the left ventricle in cardiac MRI images using deep learning," in *Proceedings of the Annual International Conference of the IEEE Engineering in Medicine and Biology Society, EMBS*, 2015, vol. 2015, November, pp. 683–686.

79. G. Luo, S. Dong, K. Wang, and H. Zhang, "Cardiac left ventricular volumes prediction method based on atlas location and deep learning," in *Proceedings - 2016 IEEE International Conference on Bioinformatics and Biomedicine, BIBM 2016*, 2017, pp. 1604–1610.

80. G. Luo, G. Sun, K. Wang, S. Dong, and H. Zhang, "A novel left ventricular volumes prediction method based on deep learning network in cardiac MRI," *Comput. Cardiol.*, vol. 43, pp. 89–92, 2016.

81. S. Dong, G. Luo, G. Sun, K. Wang, and H. Zhang, "A combined multi-scale deep learning and random forests approach for direct left ventricular volumes estimation in 3D echocardiography," *Comput. Cardiol.*, vol. 43, pp. 889–892, 2016.

82. M. R. Avendi, A. Kheradvar, and H. Jafarkhani, "A combined deep-learning and deformable-model approach to fully automatic segmentation of the left ventricle in cardiac MRI," *Med. Image Anal.*, vol. 30, pp. 108–119, 2016.

83. C. Xu et al., "Direct delineation of myocardial infarction without contrast agents using a joint motion feature learning architecture," *Med. Image Anal.*, vol. 50, pp. 82–94, 2018.

84. G. Yang et al., "A fully automatic deep learning method for atrial scarring segmentation from late gadolinium-enhanced MRI images," in *Proceedings - International Symposium on Biomedical Imaging*, 2017, pp. 844–848.

85. L. V. Romaguera, F. P. Romero, C. F. Fernandes, and M. G. Fernandes Costa, "Myocardial segmentation in cardiac magnetic resonance images using fully convolutional neural networks," *Biomed. Signal Process. Control*, vol. 44, pp. 48–57, 2018.

86. F. Haddad, R. Doyle, D. J. Murphy, and S. A. Hunt, "Right ventricular function in cardiovascular disease, part II: Pathophysiology, clinical importance, and management of right ventricular failure," *Circulation*, vol. 117, no. 13, pp. 1717–1731, 2008.

87. G. Luo, R. An, K. Wang, S. Dong, and H. Zhang, "A deep learning network for right ventricle segmentation in short-axis MRI," *Comput. Cardiol.*, vol. 43, pp. 485–488, 2016.

88. L. K. Tan, Y. M. Liew, E. Lim, and R. A. McLaughlin, "Convolutional neural network regression for short-axis left ventricle segmentation in cardiac cine MR sequences," *Med. Image Anal.*, vol. 39, pp. 78–86, 2017.

89. H. Yang, J. Sun, H. Li, L. Wang, and Z. Xu, "Neural multi-atlas label fusion: Application to cardiac MR images," *Med. Image Anal.*, vol. 49, pp. 60–75, 2018.

90. K. López-Linares et al., "Fully automatic detection and segmentation of abdominal aortic thrombus in post-operative CTA images using Deep Convolutional Neural Networks," *Med. Image Anal.*, vol. 46, pp. 202–214, 2018.

91. E. Nasr-Esfahani et al., "Vessel extraction in X-ray angiograms using deep learning," in *Proceedings of the Annual International Conference of the IEEE Engineering in Medicine and Biology Society, EMBS*, 2016, vol. 2016, October, pp. 643–646.

92. E. Nasr-Esfahani et al., "Segmentation of vessels in angiograms using convolutional neural networks," *Biomed. Signal Process. Control*, vol. 40, pp. 240–251, 2018.

93. J. F. Silva, A. Guerra, S. Matos, and C. Costa, "Ejection fraction classification in transthoracic echocardiography using a deep learning approach," in *Proceedings - IEEE Symposium on Computer-Based Medical Systems*, 2018, vol. 2018, June, pp. 123–128.

94. V. Sundaresan, C. P. Bridge, C. Ioannou, and J. A. A. Noble, "Automated characterization of the fetal heart in ultrasound images using fully convolutional neural networks," in *Proceedings - International Symposium on Biomedical Imaging*, 2017, pp. 671–674.

95. H. Chen, L. Yu, Q. Dou, L. Shi, V. C. T. Mok, and P. A. A. Heng, "Automatic detection of cerebral microbleeds via deep learning based 3D feature representation," in *Proceedings - International Symposium on Biomedical Imaging*, 2015, vol. 2015, July, pp. 764–767.

96. A. S. Becker, M. Mueller, E. Stoffel, M. Marcon, S. Ghafoor, and A. Boss, "Classification of breast cancer from ultrasound imaging using a generic deep learning analysis software: A pilot study," *Br. J. Radiol.*, vol. 91, p. 20170576, 2018.

97. N. Vincent, N. Stier, S. Yu, D. S. S. Liebeskind, D. J. J. Wang, and F. Scalzo, "Detection of hyperperfusion on arterial spin labeling using deep learning," in *Proceedings - 2015 IEEE International Conference on Bioinformatics and Biomedicine, BIBM 2015*, 2015, pp. 1322–1327.

98. J. Betancur et al., "Deep learning for prediction of obstructive disease from fast myocardial perfusion SPECT. A multicenter study," *JACC Cardiovasc. Imaging*, vol. 11, no. 11, pp. 1654–1663, 2018.

99. Z. Zhong et al., "3D fully convolutional networks for co-segmentation of tumors on PET-CT images," in *2018 IEEE 15th International Symposium on Biomedical Imaging (ISBI 2018)*, 2018, pp. 228–231.

100. H.-H. Tseng, Y. Luo, S. Cui, J.-T. Chien, R. K. Ten Haken, and I. El Naqa, "Deep reinforcement learning for automated radiation adaptation in lung cancer," *Med. Phys.*, vol. 44, no. 12, pp. 6690–6705, 2017.

101. H. Wang et al., "Comparison of machine learning methods for classifying mediastinal lymph node metastasis of non-small cell lung cancer from 18F-FDG PET/CT images," *EJNMMI Res.*, vol. 7, no. 1, p. 11, 2017.

102. Y. Zhang and M. An, "Deep learning- and transfer learning-based super resolution reconstruction from single medical image," *J. Healthc. Eng.*, vol. 2017, no. 2, pp. 1–20, 2017.

103. X. Yang, R. Kwitt, M. Styner, and M. Niethammer, "Fast predictive multimodal image registration," in *Proceedings - International Symposium on Biomedical Imaging*, 2017, pp. 858–862.

104. K. H. Cha et al., "Bladder cancer treatment response assessment in CT using radiomics with deep-learning," *Sci. Rep.*, vol. 7, no. 1, p. 8738, 2017.

105. H.-H. Tseng, Y. Luo, S. Cui, J.-T. Chien, R. K. Ten Haken, and I. El Naqa, "Deep reinforcement learning for automated radiation adaptation in lung cancer," *Med. Phys.*, vol. 44, no. 12, pp. 6690–6705, 2017.

106. F. Khalifa et al., "A generalized MRI-based CAD system for functional assessment of renal transplant," in *Proceedings - International Symposium on Biomedical Imaging*, 2017, pp. 758–761.

107. M. Treder, J. L. Lauermann, and N. Eter, "Automated detection of exudative age-related macular degeneration in spectral domain optical coherence tomography using deep learning," *Graefe's Arch. Clin. Exp. Ophthalmol.*, vol. 256, no. 2, pp. 259–265, 2018.

108. C. S. Lee, D. M. Baughman, and A. Y. Lee, "Deep learning is effective for classifying normal versus age-related macular degeneration OCT images," *Kidney Int. Reports*, vol. 2, no. 4, pp. 322–327, 2017.

109. N. Srivastava, G. Hinton, A. Krizhevsky, and R. Salakhutdinov, "Dropout: A simple way to prevent neural networks from overfitting," *J. Mach. Learn. Res.*, vol. 15, no. 1, pp. 1929–1958, 2014.

110. C. Lam, C. Yu, L. Huang, and D. Rubin, "Retinal lesion detection with deep learning using image patches," *Investig. Ophthalmol. Vis. Sci.*, vol. 59, no. 1, pp. 590–596, 2018.

111. C.-L. Chin et al., "An automated early ischemic stroke detection system using CNN deep learning algorithm," in *Proceedings - 2017 IEEE 8th International Conference on Awareness Science and Technology, iCAST 2017*, 2018, vol. 2018, January, pp. 368–372.

112. H. Yu, Z. Luo, and Y. Tang, "Transfer learning for face identification with deep face model," in *2016 7th International Conference on Cloud Computing and Big Data (CCBD)*, 2016, pp. 13–18.

113. J. Yosinski, J. Clune, Y. Bengio, and H. Lipson, "How transferable are features in deep neural networks?" *Advances in Neural Information Processing Systems*, pp. 3320–3328, 2014.

114. X. Sun, Z. Feng, X. Yuan, W. Zhang, P. Rong, "Radiomics strategy based on cardiac magnetic resonance imaging cine sequence for assessing the severity of mitral value regurgitation," *Zhong nan da xue xue bao. Yi xue ban= Journal of Central South University. Medical Sciences*, vol. 44, no. 3, pp. 290–296, 2019.

115. J. Gould, B. Porter, S. Claridge, Z. Chen, B. J. Sieniewicz, B. S. Sidhu, S. Niederer et al., "Mean entropy predicts implantable cardioverter-defibrillator therapy using cardiac magnetic resonance texture analysis of scar heterogeneity," *Heart Rhythm*, vol. 16, no. 8, pp. 1242–1250, 2019.

116. O. Manabe, O. Hiroshi, H. Kenji, H. Souichiro, N. Masanao, T. Ichizo, A. Tadao et al., "Use of 18 F-FDG PET/CT texture analysis to diagnose cardiac sarcoidosis," *Eur. J. Nucl. Med. Mol. Imaging*, vol. 46, no. 6, pp. 1240–1247, 2019.

117. E. K. Oikonomou et al., "A novel machine learning-derived radiotranscriptomic signature of perivascular fat improves cardiac risk prediction using coronary CT angiography," *Eur. Heart J.*, vol. 0, pp. 1–15, 2019.

118. J. Betancur et al., "Deep learning analysis of upright-supine high-efficiency spect myocardial perfusion imaging for prediction of obstructive coronary artery disease: A multicenter study," *J. Nucl. Med.*, vol. 60, no. 5, pp. 664–670, 2019.

119. G. P. Diller et al., "Machine learning algorithms estimating prognosis and guiding therapy in adult congenital heart disease: Data from a single tertiary centre including 10 019 patients," *Eur. Heart J.*, vol. 40, no. 13, pp. 1069–1077, 2019.

120. G. P. Diller et al., "Utility of machine learning algorithms in assessing patients with a systemic right ventricle," *Eur. Heart J. Cardiovasc. Imaging*, vol. 20, no. 8, pp. 925–931, 2019.

121. T. Wang et al., "A learning-based automatic segmentation and quantification method on left ventricle in gated myocardial perfusion SPECT imaging: A feasibility study," *J. Nucl. Cardiol.*, pp. 1–12, 2019.

122. M. Khened et al., "Fully convolutional multi-scale residual DenseNets for cardiac segmentation and automated cardiac diagnosis using ensemble of classifiers," *Med. Image Anal.*, vol. 51, pp. 21–45, 2019.

Chapter 5

Big Data in Computational Health Informatics

Ruogu Fang, Yao Xiao, Jianqiao Tian, Samira Pouyanfar,
Yimin Yang, Shu-Ching Chen, and S. S. Iyengar

5.1 Introduction

Computational health informatics is an emerging multidisciplinary research topic closely related to the medical industry. It involves various methodical science studies such as biomedical, radiology, nursing, medicine, computer science, information technology, and statistics. By collecting and analyzing information from different health-related domains, health informatics enables the application of information and communication technologies (ICTs) for predicting patients' health situations. The research principle and scientific merits in health informatics are to promote health care output (HCO) and patients' safety and quality of care [1]. The emergence of novel technologies in recent years contributes to healthcare and the growth of digital healthcare data significantly. Modern applications such as capturing devices, sensors, wearable technology, mobile apps, and the Internet of Things (IoT) increase the amount of data sources and therefore the volume of data. Based on all possible sources of healthcare data, the healthcare industry will then need to process and produce a large amount of digital health data such as electronic health records (EHR)* and personal health records (PHR).†

The healthcare industry has undergone rapid growth of data volume, as shown in Figure 5.1, for both healthcare data and digital health data. In this figure, the exponential growth of the healthcare data shows that the data size level since 2011 from 150 exabytes (10^8) will surpass the zettabyte (10^{21}) level and the yottabyte (10^{24}) level in the near future [2]. By 2020, the data size level will become around six times more than the current year and hundreds of times more than 2011. The estimated digital healthcare data in 2012 was almost 500 petabytes (10^{15}), and it will increase to around 25 exabytes in 2020 [3].

Besides the large volume of digital health data, the complexity of its structure is also a problematic circumstance for traditional pathways. We summarize some of the characteristic determinants of traditional systems for failing to handle these datasets as:

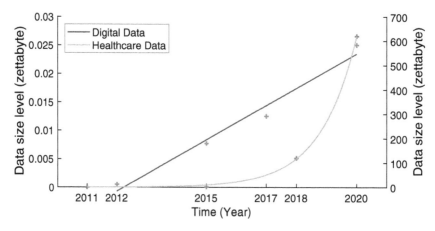

FIGURE 5.1: U.S. healthcare data growth.

---

* EHR, including electronic medical records.
† PHR, one subset of EHR including medical history, laboratory results, and medications.

- The enormous variety of structured and unstructured healthcare data, such as handwritten doctor notes, medical records, diagnostic images (MRI, CT), and radiographic films [4]
- The existence of noisy, heterogeneous, complex, longitudinal, diverse, and large datasets in healthcare informatics [3]
- The difficulties in capturing, storing, analyzing, and visualizing large and complex datasets
- The necessity of increasing the storage capacity, computation power, and the processing capability [5]
- The ability of handling medical issues, which is not adequately addressed in traditional systems. such as improving the quality of healthcare, improving sharing policy and security of patients' data, reducing the healthcare cost [6]

Therefore, seeking solutions for managing and analyzing such complicated, diversified, and gigantic datasets efficiently and with reasonable storage capacity is of vital importance. A popular term, big data analytics, is given to such large and complex datasets. Big data analytics thus performs an essential task of managing the huge amount of healthcare data and enhancing patients' quality of healthcare. It is also expected to reduce nursing costs, improve treatment methods, achieve more personalized medication, and assist doctors to make individualized decisions. According to the definition by Gartner [7], big data is "high-volume, high-velocity and high-variety information assets that demand cost-effective, innovative forms of information processing for enhanced insight and decision-making". Of course, big healthcare data are within the big data domain. A survey from McKinsey Global Institute estimates that more than $300 billion in benefits every year will be created by healthcare analytics [8].

In health informatics, big data combine the usage of technologies and procedures including the management, manipulation, and organization of vast, various, and complicated datasets that can be used to ameliorate the patient experience. Consequently, the major advantages of big data in health informatics are three-fold. First, it applies the massive volume of data to assist efficient and effective treatments. Second, based on different patients' conditions, it can assist personalized healthcare. Third, it can coordinate the medical system components such as patients, providers, payers, and administrators appropriately [3].

5.1.1 Three Scenarios of Big Data in Health Informatics

Nowadays, the healthcare industry is converting to using big data technologies to manage and enhance the medical systems. Many health-related companies and organizations are utilizing big data in health informatics for such purposes. The following shows several descriptions of health informatics scenarios to demonstrate the importance of big data in solving current medical issues and assisting treatments.

5.1.1.1 High-Risk and High-Cost Patient Management

According to the statistic report in National Health Expenditure Projections 2013, the anticipated growth of health expenditure rate will be 5.7%, faster than the annual growth rate. The expense of healthcare and related activities in the United States is around $2.5 trillion (in 2009), much higher than other developed countries [9]. Of this, approximately half of the total cost is relevant to the health expense of only 5% of the patients. Therefore, an efficient and patient-oriented healthcare management system is needed, especially for those high-risk and high-cost patients. D. Bates et al. [9] recently proposed a solution for efficiently managing those issues with big data techniques. They have considered several factors for reducing the medical cost and improving the accuracy of predicting high-risk patients. Moreover, they suggest that applying a large amount of data from high-risk patients is necessary for developing the predictive models in the analytic systems.

5.1.1.2 Risk-of-Readmission Prediction

In healthcare informatics, seeking solutions for risk prediction is considered as one of the immensely challenging tasks. Accurately predicting the patients with a high possibility of readmission is considered a practical solution for lowering healthcare costs and improving the healthcare quality. However, such predictions are complicated. It is a fact that giving predictions is associated with many factors, for instance, patients' health conditions, disease parameters, diagnosis analysis, and the quality parameters of clinic care [10]. The critical point is to extract key elements from those unstructured data like doctors' notes and various reports, where natural language processing (NLP) can be used to handle the problem. Recently, the University of North Carolina (UNC) healthcare system has started to apply IBM big data analytics to assist high-cost readmission predictions and preventable cost reduction. Besides this use-case, many readmission predicting models are being studied by using big data solutions, like K. Zolfaghar et al. [10]'s work for predicting the 30-day risk of readmission for congestive heart failure incidents.

5.1.1.3 Mobile Health for Lifestyle Recommendation

Mobile technology has the characteristics of universal and personalized design and is an ideal platform for offering fast, cheap, and easily accessible healthcare [11]. For example, it can provide patient's information and health status such as heart rate, blood pressure level, and sleep patterns [12] to the doctors at real-time without location and time limitations. However, the huge amount of generated mobile data needs effective management. Thus big data solutions are being implemented for managing all types of health information, improving healthcare, and increasing access to healthcare.

5.1.2 Related Work

To date, various research based on big data tools and approaches has been intensely studied with the focus on biomedical [2, 4, 13, 14] and computational aspects [1, 15, 16], [17].

An overview of big data in health informatics is presented by [2]; this paper emphasizes the advantages and the characteristics of big data in the area of clinical operations, public health, genomic analytics, remote monitoring, etc. What is more, it also provides several commonly used big data platforms and analysis tools. The conclusion this paper has drawn is that health informatics can lower the cost as well as improving healthcare outcomes. Costa [13] discusses the significant improvements in combining clinical health data and omics, that is, different fields of study in biology such as genomics, proteomics, and metabolomics. In this article, authors review the challenges associated with using big data in biomedicine and translational sciences. By interviewing some companies in the emerging healthcare big data ecosystem, B. Feldman et al. [4] introduce big data and the necessity of big data in health informatics. A general overview is outlined for big data analytics in health informatics and the ways that big data can help with healthcare by grouping different companies and organizations.

Most of the big data in health informatics surveys favor biomedical aspects instead of the computational viewpoint. Herland et al. [1] discuss the work in health informatics branches, such as using big data based on bioinformatics, public health informatics, and neuroinformatics. Their central focus is to gather health informatics data at the levels such as microscopic, tissue, and population levels and the related data mining methods. Nevertheless, this work only talks about a few healthcare examples but not categorized based on computational and big data approaches. Merelli et al. [15] discuss several technologies of big data in health informatics, including the solutions for big data architecture, parallel platforms, and security issues for biomedical data. Still, many of those technical issues require further study, such as data capturing, feature analysis, machine learning algorithms. In the work of [16], a brief overview of different machine learning and data mining approaches is presented with the focus on the association rules, classification, and clustering methods.

5.1.3 Contributions and Organization of This Chapter

In this chapter, we provide a structured and comprehensive overview of computational methods of big data in health informatics with the goal of connecting big data and health informatics communities. In contrast with other articles mainly focused on the high-level biomedical polities without a detailed explanation of processing steps, this chapter offers an extensive study of the computational approaches and the processing steps. It sketches out the computational methodologies by expanding the study in several directions of health informatics processing and presents a structured pipeline which details each step from the raw data to clinical diagnosis. The primary emphasis also includes the associated challenges, current big data mining techniques, strengths and limitations of current works, and future directions.

We separate out this chapter into three parts: Challenges and opportunities, healthcare informatics processing pipeline, and future directions. Section 5.2 explains the four Vs (Volume, Velocity, Variety, and Veracity) of big data and provides several examples of challenges and opportunities related to big data in health informatics. Section 5.3 discusses the pipeline of the computational process in health informatics, where Section 5.3.1 describes several data-capturing methods for effectively gathering useful healthcare data.

Section 5.3.2 and Section 5.3.3 present the ways of storing data and data sharing with other systems worldwide respectively. Section 5.3.4 covers data analyzing methods including preprocessing, feature extraction and selection, and machine learning approaches. Section 5.3.5 and Section 5.3.6 present the data searching methods and decision supports. Conclusively, Section 5.4 explores the potential directions based on big data's challenges for future work.

5.2 Challenges and Opportunities

The challenges of big data in health informatics can be described as four Vs: Volume, Variety, Velocity, and Veracity. The characteristics of big data properties in health informatics are described in Figure 5.2. Dealing with a big volume of health data is challenging and overwhelming not only because of the diversity of data sources but also because of the limited speed for generating and processing those data, not to mention to examine and verify the quality and legitimacy of those data [4].

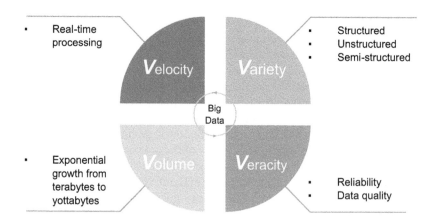

FIGURE 5.2: Four Vs of big data.

5.2.1 Volume

Global health data has been growing exponentially in recent years with the development of medical devices and information technology. The quantity unit terabyte for these data is no longer sufficient. Instead, zettabyte and yottabyte are more suitable. Each healthcare company with more than a thousand employees collects over 400 terabytes of data as reported in 2009 according to Health Catalyst [18]. This indicates that the healthcare industry is a high-data volume and big-data-driven industry, notwithstanding the high real-time streams from the web and social media data.

There are various origins of healthcare data which contribute to the tremendous volume of data. Sources include the traditional medical records from handwritten notes, personal reports, and clinical trial data to the new types of data, for example, 3D imaging and sensors [4]. We summarize different data sources in Section 5.3.1. Recently, the emerging of wearable medical devices which can continuously monitor the patient's physiological information has significantly contributed to the amount of healthcare data. The information includes biopotential, heart rate, blood pressure, and so on [19].

The high volume of healthcare data creates a big challenge, which requires scalable storage and support for distributed queries across multiple data sources. Specifically, the challenge is to be able to locate and mine specific pieces of data in an enormous, partially structured dataset. Many advanced data management techniques such as virtualization and cloud computing have been widely studied and experimented with in industrial companies. Those proposed platforms are capable of manipulating large volumes of data virtually or geographically distributed on multiple physical machines, enabling the universal sharing of information.

5.2.2 Variety

Healthcare data could be characterized by the variety of sources and the complexity of different forms of data. Generally, healthcare data could be classified as unstructured, structured, and semi-structured.

Historically, most unstructured data usually come from office medical records, handwritten notes, paper prescriptions, magnetic resonance imaging (MRI), computed tomography (CT), and so on. The structured and semi-structured data refer to electronic accounting and billings, actuarial data, laboratory instrument readings, and EMR data converted from paper records [4].

Nowadays, more and more data streams add variety to healthcare information, both structured and unstructured, including intelligent wearable, fitness devices, social media, and so on. The challenge lies in the seamless combination of old-fashioned and new forms of data, as well as the automatic transformation between the structured and unstructured data, which relies on effective distributed processing platforms, advanced data mining, and machine learning techniques. Leveraging heterogeneous datasets and securely linking them has the potential to improve healthcare by identifying the right treatment for the right individual.

5.2.3 Velocity

The increase in volume and variety of healthcare data is highly related to the velocity at which it is produced and the speed needed to retrieve and analyze the data for timely decision-making.

Compared with relatively static data such as paper files, X-ray films, and scripts, it is gradually more important and challenging to timely and accurately process real-time stream, such as various monitoring data, in order to provide the right treatment to the right patient at the right time [4]. A concrete example can be found in the prevalence of wearable monitoring devices, which provide continuous and ever-accumulating physiological data. Being able to perform real-time analytics on continuous monitoring data could help predict life-threatening pathological changes and offer appropriate treatment as early as possible. The high velocity of healthcare data poses another big challenge for big data analytics. Although the traditional database management systems (DBMS) are reported to perform well on large-scale data analysis for specific tasks [20], they simply cannot catch up with the pace of high-velocity data, not to mention the limitation of flexibility when facing multi-level nesting and hierarchies data structure with high volatility, which is a common property for healthcare data. This situation creates an opportunity for introducing high-velocity data processing tools. An initial attempt includes the utilization of the Hadoop platform for running analytics across massive volumes of data using a batch process. More recently, industrial practices have approved the convergence of traditional relational database and key-value store.

Spanner [21], Google's globally distributed database, is an example, which provides high consistency and availability.

5.2.4 Veracity

Coming from a variety of sources, the large volume of healthcare data varies in its quality and complexity. It is not uncommon that the healthcare data contain biases, noise, and abnormality, which pose a potential threat to proper decision-making processes and treatments of patients.

High-quality data can not only ensure the correctness of information but also reduce the cost of data processing. It is highly desirable to clean data in advance of analyzing it and using it to make life or death decisions. However, the variety and velocity of healthcare data raise difficulties in generating trusted information. There are mainly two types of data quality problems depending on the causes. The first type is primarily due to IT issues, e.g., data management, audit, error reporting, and compliance. The second type reveals the underlying veracity of the data, i.e., the truthfulness, relevance, and predictive value [4]. It is the second type that is of greater importance and is a bigger challenge, which potentially could be handled by using big data analytic tools.

The biggest challenge is determining the proper balance between protecting the patient's information and maintaining the integrity and usability of the data. Robust information and data governance programs will address a number of these challenges. As Techcrunch [22] points out, "while today we rely on the well-trained eye of the general practitioner and the steady hand of the surgeon, tomorrow's lifesavers will be the number-crunching data scientists, individuals with only a passing understanding of first aid".

5.2.5 Opportunities

Except for the "4Vs" that are most familiar to the readers, there are other emerging issues to be considered, such as the validity and the volatility of big data in health informatics. While validity is concerned with the correctness and accuracy of data, volatility refers to how long the data would be valid and should be kept for. All of the above big data characteristics (including all "Vs") offer great challenges to big data analytics in healthcare. At the same time, it is the very same challenges that bring unprecedented opportunities for introducing cutting-edge technologies to make sense out of large volumes of data and provide new insights for improving decision-making processes in near-real-time. The enormous scale and diversity of temporal-spatial healthcare data have created unprecedented opportunities for data assimilation, correlation, and statistical analysis [23].

By utilizing parallel computing platforms, various models and visualization techniques could be applied to accommodate the characteristics of big data analytics and take the most advantage of it. The following are some concrete examples of opportunities that are to be explored [8, 24, 2, 25, 26, 14, 23]:

- Personalized care: Create predictive models to leverage personalized care (e.g., genomic DNA sequence for cancer care) in real-time to highlight best practice treatments for patients [26]. These solutions may offer early detection and diagnosis before a patient develops disease symptoms.
- Clinical operations: Conduct comparative study to develop better ways for diagnosing and treating patients, such as mining large amounts of historical and unstructured data, looking for patterns, and modeling various scenarios to predict events before they actually happen [26, 14].
- Public health: Turn big healthcare data from a nationwide patient and treatment database into actionable information for timely detection and prevention of infectious diseases and outbreaks, thus benefiting the whole population [2].
- Genomic analytics: Add genomic analysis to the traditional healthcare decision-making process by developing efficient and effective gene sequencing technologies. Utilize high throughput genetic sequencers to capture organism DNA sequences and perform genome-wide association studies (GWAS) for human disease and human microbiome investigations [23].
- Fraud detection: Analyze large amount of claim requests rapidly by using a distributed processing platform (e.g, MapReduce for Hadoop) to reduce fraud, waste, and abuse, such as a hospital's overutilization of services or identical prescriptions for the same patient filled in multiple locations, etc. [2, 14]

• Device/remote monitoring: Capture and analyze continuous healthcare data in huge amounts from wearable medical devices both in hospital and at home, for monitoring of safety and prediction of adverse events [2].

5.3 Health Informatics Processing Pipeline

A mature computing framework of big data in health informatics involves a sequence of steps that constitute a comprehensive health informatics processing pipeline. Each step in the pipeline plays a critical role in rendering qualified and valuable outcomes of big data analytics. Specifically, the capturing, storing, and sharing of big data prepare appropriate input for the subsequent analyzing procedure, where various analytical approaches are applied to explore meaningful patterns from healthcare big data for making timely and effective decisions. This section will discuss the pipeline (as shown in Figure 5.3) in detail.

Since this is a survey paper aimed at introducing the state-of-art of health informatics in big data to a broad audience, the details of each part can be found in the references.

5.3.1 Capturing

The discovery and utilization of meaningful patterns in data have helped to drive the success of many industries. However, the usage of analytics to improve outcomes in the healthcare industry has not gained as much traction due to the difficulty of collecting large and complex datasets, which is the very first step of big data analytics. Without the ability to capture big data, healthcare providers are unable to utilize analytics to improve outcomes.

Big data in health informatics are characterized by a variety of sources (from both internal and external points of view), diverse formats (such as flat files and database records), and different locations (either physically distributed machines or multiple sites) [4]. The various sources and data types are listed in Table 5.1 [6, 25].

Recently, the introduction of EHR to U.S. hospitals led the healthcare industry into a new, high-tech age, with a high potential for the use of analytics to improve outcomes. The EHR is well known for its benefits of delivering standardized medical records and improving patient care with reduced errors and costs. With the EHR, healthcare providers are able to access information about individual patients or populations, including demographics, medical history, laboratory test results, personal statistics, etc. [27].

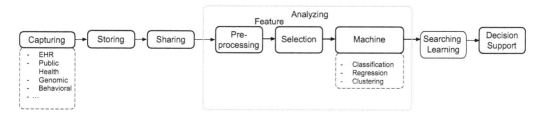

FIGURE 5.3: Health informatics processing pipeline.

TABLE 5.1: Healthcare Data Types and Sources

Data Type	Examples	Structured/ Un-structured	Data Format
Human-generated data	Physicians notes, email, and paper documents	Structured and un-structured	ASCII/text
Machine-generated data	Readings from various monitoring devices	Structured and un-structured	Relational tables
Transaction data	Billing records and healthcare claims	Semi-structured and structured	Relational tables
Biometric data	Genomics, genetics, heart rate, blood pressure, x-ray, finger prints, etc.	Structured and un-structured	ASCII/text, images
Social media data	Interaction data from social web-sites	Unstructured	Text, images, videos
Publications	Clinical research and medical reference material	Unstructured	Text

Taking advantage of big data means quickly capturing high volumes of data generated in many different formats with dynamic structures. However, there are issues like latency and scalability. Low latency is a highly desired property for stream processing as a big data technology, while scaling data integration is critical for adapting to the high-volume and high-velocity nature of big data. Apache Hadoop coupled with existing integration software and Hadoop Map/Reduce framework could provide the computing environment for parallel processing. More recently, Hadoop Spark [28], a successor system that is more powerful and flexible than Hadoop MapReduce, is getting more and more attention due to its lower latency queries, iterative computation, and real-time processing. Storm* is another scalable and fast distributed framework with a special focus on stream processing.

5.3.2 Storing

New technologies and the move to EHR are creating massive amounts of healthcare data with increased complexity, from various types of diagnostic images to physicians' notes, laboratory results, and much more. It is essential to provide efficient solutions for cost-effective storage and management. More specifically, there are several critical properties for a desired healthcare data storage and management system [29]:

- High availability: It is desirable for healthcare staff to access records quickly, securely, and reliably anywhere anytime. Reliable and rapid retrieval of patients' information saves valuable time for physicians and enables them to make responses and deliver immediate care, which can mean the difference between life and death.
- High scalability: Healthcare data storage requirements can easily reach tens or hundreds of terabytes or even petabytes, considering annual enterprise storage

* https://github.com/nathanmarz/storm

needs. To accommodate this explosive growth of data, the storage platforms should be incrementally scalable.

- Cost-effective: Storing and managing high volumes of healthcare data could be redundant and complex. An efficient storage infrastructure should reduce the cost and complexity and provide protection to the data without compromising performance.

Databases are the basic components of a storage infrastructure. In general, there are a number of storage options for analyzing big data. Compared with the traditional relational database management systems (RDBMS), the analytical RDBMSs can scale to handle big data applications. However, they are more appropriate for structured data. Hadoop technology expands the opportunity to work with a broader range of content with Hadoop Distributed File System (HDFS) and the Map/Reduce programming framework [30]. Depending on what you are analyzing, the options vary. For example, Hadoop storage options are often associated with poly-structured data such as text, while NoSQL databases are more appropriate for data with a multitude of representations [31].

As mentioned earlier, electronic images and reports are among the major sources of healthcare big data. To accommodate the special needs of storing and managing these types of data, the picture archiving and communication system (PACS) technology has been developed, which has the advantages of economical storage and convenient access [32]. With the emergence and advancement of big data analytics in healthcare, many industrial companies have provided enterprise-level solutions for storing and management. For example, NetApp [33] provides EHR and PACS solutions for reducing storage management and complexity, transforming clinical workflows, and lowering the total cost of ownership (TCO) of the data storage environment. Moreover, Intel Distribution for Apache Hadoop software [34] provides distributed computation frameworks to store, pre-process, format, and normalize patients' free-text clinical notes, which are generally difficult and time-consuming to process and analyze. It achieves scalability by processing each patient's information separately and in parallel.

5.3.3 Sharing

After capturing patient and clinical data, the problem becomes how to securely exchange healthcare information with scientists and clinicians across a given healthcare organization over institutional, provincial, or even national jurisdictional boundaries [35]. There are a number of challenges associated with the sharing of big data in health informatics, for example,

- The ad hoc use of a variety of data formats and technologies in different systems and platforms
- The assurance of the controlled sharing of data by using secure systems

Cloud computing is one of the main reasons why big data has been so ubiquitous in recent years. By ensuring storage capacity, server management, network power, and bandwidth utilities, cloud computing will help synchronize data storage with devices so that all of the information being generated automatically streams into internal systems, enabling the sharing of information. It means that security and management will be centralized and made more effective. To summarize, cloud deployments share common characteristics as follows [36]:

- They involve an optimized or virtualized infrastructure, leverage the Internet for shared access, and charge for use based on actual consumption.
- The hardware is distributed and fault tolerant, satisfying privacy and data security requirements.
- They can enhance collaboration to mine patient and claims data.
- They allow a shared pool of computing and storage resources on a pay-as-you-go basis.

A successful use of a cloud platform in healthcare is the "PACS-on-demand" systems. By storing and sharing medical images with cloud infrastructure, it reduces the need to invest in IT capacity as well as allowing efficient and secure collaboration with radiology specialists and affiliated practices [36]. By using a cloud platform, it enables data sharing between healthcare entities (such as providers and payers) which often have disparate data systems that are unable to bring together different types of data to make healthcare decisions. The research society also discusses secure sensitive data sharing on big data platforms, which will help enterprises reduce the cost of providing users with personalized services and provide value-added data services [37].

5.3.4 Analyzing

Data mining approaches have been widely used for analyzing healthcare data. The major steps of the analyzing procedure, including preprocessing, feature extraction/ selection, and machine learning, are addressed in the following sections.

5.3.4.1 Preprocessing

Real world healthcare data are noisy, skewed, and heterogeneous in nature. It is impractical to directly apply analytical algorithms to the raw healthcare data due to their variety and complexity. Therefore, in order to improve data quality and prepare them for further analysis, dealing with noise and missing values in large-scale healthcare datasets through pre-processing becomes a necessity.

A typical healthcare data pre-processing procedure usually includes the following steps depending on the source and format of the data [38, 39]:

- Data cleaning: This step involves the removal of noise in healthcare data, which includes artifacts [40, 41] and frequency noise in clinical data [39]. For example, thresholding methods are used to remove incompliant measurements [42, 43] and low-pass/high-pass filtering tools are usually applied to remove frequency noise in sensor signals [44, 45, 46]. In [47], an improved version of the SURF technique is proposed to cleanse healthcare data generated by radio frequency identification (RFID) readers.
- Missing value interpolation: Missing values may be caused by unintentional reasons (e.g., sensor failures) or intentional reasons (e.g., transportation of patients) [48]. Usually, single missing values caused by sensor failures are interpolated by their previous and following measurements [43]. Data missing for an intentional

reason or because of irrelevancy to the current clinical problem are considered non-recoverable and thus deleted [48].

- Data synchronization: Sensor data are reported at different rates with time-stamps based on their internal clocks [39]. However, the clocks across sensors are often not synchronized. It is necessary to make reasonable assumptions and derive alignment strategies to synchronize sensor data [49, 50].
- Data normalization: This step is often required to cope with differences in the data recording process [39]. For example, a daily heart rate may represent a daily average heart rate or a measurement during a specific time range. Furthermore, a normalization step is usually performed to transform the original feature set into a comparable format by adopting and mapping standardized terminologies and code sets.
- Data formatting: Each analytical approach (such as data mining algorithms) requires data to be submitted in a specified format [51]. Therefore, there is a necessity to transform the original healthcare data into machine understandable format. For example, the data should be stored in the Attribute-Relation File Format (.ARFF), and the data type of the attributes must be declared in a recognizable manner in the WEKA tool [52]. Sometimes, a discretization process should be carried out to convert the original numerical values to nominal ones for a specific algorithm. It is worth mentioning that the discretization process may cause information loss and thus impact data quality. This is another challenge for big healthcare data analysis.

Performing preprocessing for healthcare data is a challenging task, which requires an efficient and effective big data framework. One of the solutions, Hadoop platform, is designed to store and process extremely large datasets [10]. To simulate a scalable data warehouse and make the data extraction process more effective, big data tools such as Hive [53] and Cassandra are coupled with the MapReduce framework, built on top of Hadoop, to provide a powerful distributed data management system with flexibility. Specifically, first raw healthcare data will be stored as flat files on various nodes in Hadoop, which will later be loaded into HDFS. Later, Hive commands will be invoked to create appropriate tables and develop schema to structurize the data to be queried.

5.3.4.2 Feature Extraction/Selection

The process of extracting and selecting a subset of important and relevant features to form a large set of measured data is called feature selection, also known as attribute selection, or variable selection. In general, input data may include redundant features, which can be transformed into a small set of relevant features using dimensional reduction algorithms. By applying feature selection algorithms, complex and large datasets, which are computationally expensive and need large amounts of memory, are transformed to a small, relevant feature set with sufficient accuracy. Medical big and high dimensional data may cause inefficiency and low accuracy. To overcome this issue, many researchers utilize feature extraction algorithms in healthcare informatics [54].

Principal component analysis (PCA), linear discriminant analysis (LDA), and independent component analysis (ICA) are the most commonly used algorithms for dimension reduction and feature selection in the healthcare domain [38]. In PCA, d-dimensional data is transformed into a lower-dimensional space to reduce complexities. Pechenizkiy et al. [55] evaluate several feature transformation strategies based on PCA in healthcare diagnostics. Five medical datasets are used for evaluating PCA feature transformation: diabetes, liver disorders, heart disease, thyroid gland, and breast cancer. Experimental results show that PCA is appropriate for certain problems related to medical diagnostics. In [56], the effects of various statistical feature selection approaches such as PCA, LDA, and ICA on the medical domain are discussed.

Fialho et al. [48] apply a tree-based feature selection algorithm combined with a fuzzy modeling approach to a large ICU database to predict ICU readmissions. Tree search feature selection algorithm builds a tree to order all the possible feature combinations. This approach appeals to clinicians due to its graphical representation and simplicity. In this paper, sequential forward selection (SFS) and sequential backward elimination (SBE) [57] are applied to the ICU datasets as a tree search technique. Sun et al. [58] also apply a greedy forward selection approach to the EHR dataset. The proposed method includes a two-level feature reduction which selects predictive features based on Information Gain (IG). In every step, best performing features from one concept combine with the features selected from the next proper concept. The combination process will be continued until prediction performance fails to improve. Although the greedy tree search approach is simple and easy to interpret, it may get trapped at a local optimum [48]. Correlation feature selection (CFS) is a measure which evaluates the merit of feature subsets based on the following hypothesis: "Good feature subsets contain features highly correlated with the classification, yet uncorrelated to each other" [59]. To predict five-year life expectancy in older adults in [60], CFS combined with greedy stepwise search is used as a feature selection strategy to find the feature subset with best average competency. Table 5.2 summarizes some feature extraction/selection algorithms in healthcare informatics, as well as their pros and cons.

Several feature selection algorithms have been recently proposed, dealing with big data. Unlike conventional feature selection algorithms, online learning, specifically online feature selection, is suitable for large-scale real-world applications in such a way that each feature is processed upon its arrival and each time the best feature set is maintained from all seen features [61]. Yu et al. [62] proposes a scalable and accurate online approach (SAOLA) to select important features from large datasets with high dimensionality. Using a sequential scan, the SAOLA overcomes critical challenges in online feature selection, including the computational cost of online processing specifically for large datasets, which keeps growing on and on. Tan et al. [63] also presents a new feature selection method for extremely high dimensional datasets on big data. In this paper, the problem is transformed into a convex semi-infinite programming (SIP) issue, which is solved by a new feature generating machine (FGM). FGM repeatedly extracts the most relevant features using a reduced and primal form of the multiple kernel learning (MKL) subproblems. FGM is appropriate for feature selection on big data due to its subproblem optimization which involves a small subset of features with reduced memory overhead.

5.3.4.3 Machine Learning

Machine learning is an area of computer science which explores the creation of algorithms that can automatically learn from data and improve through experience. Nowadays, machine learning techniques have been applied in a variety of applications including audio processing, autonomous vehicles, detection of fraudulent credit card activity, to name a few [64]. Computerization in healthcare is growing day in, day out, which leads to complex and large medical databases. Machine learning algorithms are

TABLE 5.2: Summary of Feature Extraction/Selection Algorithms in the Healthcare Informatics

Feature Extraction/ Selection Algorithm	Healthcare Examples	Pros	Cons
PCA [55], [56]	Diabetes, heart disease, breast cancer	Simple, non-parametric, spread out data in the new basis, useful in unsupervised learning, most used algorithm in healthcare informatics	Non-statistical method, difficult to interpret, linear combinations of all input variables
LDA [56]	Hepatitis diagnosis, coronary artery disease	Multiple dependent variables, reduced error rates, easier interpretation	Parametric method (assumes unimodal Gaussian likelihood), mean is the discriminating factor not variance, extremely sensitive to outliers, produces limited feature projections
ICA [56]	Heart disease, genetic disease	Finds uncorrelated and independent components, generalization	Inappropriate for small training data, permutation and stability ambiguity
Tree based [48], [58]	ICU readmission, EHR and MRI datasets	Graphical representation, simple	Traps to local optimum, redundant features might be selected
CFS [59]	Breast, colon, and prostate cancer	Fast, reduced error rate, better results on small datasets	Ignores interaction with classifier, less scalable

able to automatically manage such large databases. A general overview of machine learning techniques in healthcare informatics is presented in the following.

Generally, machine learning algorithms are subdivided into two categories: Supervised learning (or predictive learning) and unsupervised (or descriptive learning) [16]. In supervised learning, both input and their desired outputs are presented to the machine for it to learn a general prediction rule that maps inputs to outputs. In other words, prediction rules are acquired from training data to predict unseen data labels. Classification and regression are the two major categories of supervised learning studied in this survey. The classification algorithms surveyed in this paper are as follows: first, the decision tree is introduced. It is a simple and easy-to-implement classifier, which is useful for physicians who want to easily understand and interpret the classification results. However, it may not be applicable for very large datasets with high dimensional features due to its space limitation and overfitting problem. Another classifier presented is support vector machine (SVM), which is widely applied on image-based classification and large medical datasets, in spite of its slow training and expensive computational time. Neural network (NN) classifier is also presented, which has the same weaknesses as SVM as well as its black-box nature and difficulty in interpretation. Although NN is broadly used for medical applications, its modified version, called deep learning, has better capabilities in dealing with big data issues such as volume and velocity. Deep learning algorithms are usually used as classifiers in image-based medical research such as neuroimaging applications. In addition, different types of sparse classifiers and ensemble algorithms are introduced in the following sections, which are widely used in the big data medical datasets to overcome imbalanced data and the overfitting problem, respectively. However, most of these algorithms are computationally expensive and are not easy to interpret. The regression algorithms, another type of supervised learning, are also widely used in healthcare applications such as brain image analysis, CT scans of different organs, and battery health monitoring.

In unsupervised learning, there are no labels, and the goal is to find the structure of unknown input data by searching the similarity between records. Clustering is one type of unsupervised learning approach. Three different types of clustering algorithms are covered in this survey as follows: Partitioning algorithm is presented as it is simple, fast, and capable of handling large datasets; however it is not recommended for noisy datasets including lots of outliers. Hierarchical algorithm is another clustering method discussed in this survey due to its visualization capability, which is requested by many physicians. Though it is sometimes not appropriate for big data due to the space and time limitations. The last algorithm presented in this section is density-based clustering, which handles non-static and complex datasets and detects outliers and arbitrary shapes specifically in biomedical images. However, it is slow for large datasets, like hierarchical clustering.

Until now, various research studies have been conducted on machine learning and data mining that outline advantages and disadvantages of multiple machine learning algorithms in healthcare informatics [65, 66, 38]. It is worth mentioning that there is a significant overlap among machine learning and data mining as both are used in data analysis research and include supervised and unsupervised algorithms. Machine learning is used to extract models and patterns in data. However, data mining, a combination of machine learning and statistics, mainly deals with existing large datasets and analyzes massive, complicated, and structured/unstructured data. Data mining is a more general concept, which has gradually merged with and been used in database management systems. Therefore, data mining algorithms should be more scalable to discover rich knowledge from large datasets.

5.3.4.3.1 Classification

Classification is the problem of identifying the category of new observation records based on the training data whose categories are known. Automatic classification can be used in diagnosis systems to help clinicians with disease detection. In the following, several widely used classification algorithms and their applications in healthcare and medical informatics are discussed.

5.3.4.3.1.1 Decision Tree and Rule-Based Algorithms Decision tree is a simple and widely used classifier. It classifies instances by sorting them in a tree which can be re-represented as sets of if-then rules [64].

This learning method has been successfully used in a wide variety of medical diagnostic systems [66, 38].

Zhang et al. [67] introduce a real-time prediction and diagnosis algorithm. This prediction algorithm is mainly based on very fast decision tree (VFDT) [68]. VFDT is a decision tree algorithm form on Hoeffding tree which can control large and continuous data streams by remembering the mapping connection among leaf nodes and the history entries. Zhang et al. discuss that VFDT outperforms traditional decision tree algorithms, although it is not able to predict the patient illness using only the current situation on its own. Therefore, they modified VFDT as follows: several pointers of leaf node are added in the training phase of learning. Then, a mapping table is designed to store medical data, address, and its pointer. Therefore, whenever VFDT sends a stream to a leaf node, the corresponding pointer is used to search the table and return similar history medical records. These similarities can be used for medical prediction and help physicians to better treat their patients.

Fuzzy decision tree (FDT) is compared with three other classifiers in [69], which demonstrates the performance superiority of FDT in the brain MRI classification problem. Unlike traditional decision trees, FDT has the ability to handle fuzzy data. In this study, an automatic classification algorithm is proposed, which includes the preprocessing step (wavelet decomposition of the image) and feature selection algorithm in order to extract a few morphological features, as well as FDT classifier. The authors conclude that their method can efficiently classify patients into the three levels of Alzheimer (close to 90% efficiency, which is better than human experts) using a few morphological features.

Other utilizations of decision trees in medical and healthcare informatics are presented in [38]. Although decision tree techniques are simple and easy to implement, they have space limitations. Furthermore, if the dataset contains many features, it may be inefficient to create a tree. To overcome this issue, and also dataset overfitting [70], pruning algorithms are used in decision trees.

5.3.4.3.1.2 Support Vector Machine SVM is another supervised learning method used for both classification and regression [71]. As a classifier, SVM constructs hyperplanes in a multi-dimensional space to classify samples with different labels. In the SVM models, several kernel functions can be used including polynomial, sigmoid, Gaussian, and radial basis function (RBF).

In [72], children's health is studied, and specifically, the effect of socio-economic status on educational attainment is addressed. In order to reduce dimensionality and identify the important features, SVM-based classifier is optimized by the particular swarms optimization (PSO) to update the hyper-parameters in an automatic manner and simultaneously identify the important features via entropy regularization. To evaluate the proposed method, 21 features are extracted from 3792 data samples which are divided into training data (for SVM model construction), validation data (for model selection by PSO), and testing data. The method is compared with LDA and multilayer perceptron (MLP) [73], which are linear and nonlinear classification algorithms, respectively.

5.3.4.3.1.3 Artificial Neural Networks "Neural network" is a family of statistical learning and artificial intelligence approaches widely used for classification. This algorithm is inspired by biological neural networks of animals and humans, particularly the brain and central nervous system. NN includes neurons which can be trained to compute values from inputs and predict a corresponding class of test data. Until now, there has been a wide variety of decision-making processes and predictions using NN in the healthcare domain.

Vu et al. [74] presents an online three-layer NN to detect heart rate variability (HRV) patterns. Specifically, HRV related to Coronary Heart Disease (CHD) risk is recognized using ECG sensors. When a sample enters, the hidden layer nodes find the excessive similarity between nodes and the input in a competitive manner. The proposed method outperforms other NN algorithms such as MLP, growing neural gas (GNG) [75], and self-organizing map (SOM) [76].

Yoo et al. [16] discuss several disadvantages of NN in healthcare domain as follows: First, NN requires numerous parameters which are very critical for the classification result. Second, the training phase of NN is computationally expensive and time consuming. Furthermore, it lacks model transparency due to the black-box nature of the NN system, and thus it is difficult for medical experts to understand its structure in order to gain knowledge from it. Finally, the accuracy is usually lower than other algorithms such as random forest and SVM, to name a few.

5.3.4.3.1.4 Deep Learning With the tremendous growth of data, deep learning is playing an important role in big data analysis. One notable success of deep learning for big data is the use of a large number of hidden neurons and parameters, involving both large models and large-scale data [77], to model high-level abstraction in data [78]. To date, various deep learning architectures including deep belief networks [79], deep neural networks [80], deep Boltzmann machine [81], and deep convolution neural network [82] have been applied in areas such as speech recognition, audio processing, computer vision, and healthcare informatics.

Liang et al. [83] present a healthcare decision-making system using multiple layer neural network deep learning to overcome the weaknesses of conventional rule-based models. The deep model combines features and learning in a unified model to simulate the complex procedures of the human brain and thought. This system is achieved by applying a modified version of the deep belief networks to two large-scale healthcare datasets.

Deep learning models have shown success in neuroimaging research. Plis et al. [84] use deep belief networks and restricted Boltzmann machine to the functional and structural MRI data. The results are validated by examining whether deep learning models are effective compared with representative models of the class label of the dataset used in training the models, examining the depth parameter in the deep learning analysis for this specific medical data, and determining if the proposed methods can discover the unclear structure of large datasets.

Li et al. [85] also leverage a deep learning-based framework to estimate the incomplete imaging data from the multi-modality database. They take the form of convolution neural networks, where the input and output are two volumetric modalities. They evaluate this deep learning-based method on the Alzheimer's Disease Neuroimaging Initiative (ADNI) database, where the input and output are MRI and PET images respectively.

5.3.4.3.1.5 Sparse Representation Sparse representation classification (SRC) of signals has been attracting a great deal of attention in recent years. SRC is the problem of finding the most compact signal representation using atoms linear combination in a given over-complete dictionary [86].

In the healthcare informatics area, several research studies have been done using SRC to improve the classification results. Marble [87] is a sparse non-negative tensor factorization method for count data. It is used to fit EHR count data which are not always correctly mapped to phenotypes. In this model, the sparsity constraints are imposed by decreasing the unlikely mode elements. To solve the optimization problem, a periodic minimization approach (cycling through each mode as fixing others) is used.

Sparse representation has also been an effective approach for machines to learn characteristic patterns from the medical data for image restoration, denoising, and super-resolution. Sparse representation has been effective in medical image denoising and fusion using group-wise sparsity [88] and image reconstruction [89, 90]. Recently, coupled with dictionary learning, Fang et al. [91] restore the hemodynamic maps in the low-dose computed tomography perfusion by learning a compact dictionary from the high-dose data, with improved accuracy and clinical value using tissue-specific dictionaries [92] and applying to various types of medical images [93]. The sparsity property in the transformed domain has also been important in restoring the medical information by combining with the physiological models [94].

5.3.4.3.1.6 Ensemble Ensemble is a supervised learning algorithm which combines different classification algorithms to increase the performance of single classifiers. In other words, instead of using an individual classifier, ensemble learning can be used to aggregate the prediction results of several classifiers. Ensemble learning improves the generalization and predictive performance.

Random forest [95] is one type of ensemble learning algorithm which constructs multiple trees at training time. This algorithm overlaps the overfitting problem of decision trees by averaging multiple deep decision trees. Recently, various researches have been applying the random forest algorithm to the bioinformatics domain. Díaz-Uriarte et al. [96] apply random forest for classification of microarray data. They also apply random forest for a gene selection task. In this research, ten DNA microarray datasets focusing on several parts of the body are used.

Rotation forest ensemble (RFE) [97] with alternating decision tree (ADT) [98] is a modified version of the decision tree technique that is used as a classifier in the paper by Mathias et al [60]. For this purpose, 980 features are extracted from EHR data. Afterwards, the greedy stepwise algorithm is used for feature selection. Finally, RFE with ADT is applied to predict the five-year mortality rate. Liu et al. [99] propose ensemble learning using a sparse representation algorithm to classify Alzheimer's disease from medical images such as MRI. For this purpose, a random patch-based subspace ensemble classification is proposed. This technique builds several individual classifiers using various subsets of local patches and finally combines weak classifiers to improve performance results. Experimental results show the high performance of the proposed method on Alzheimer's MRI data.

5.3.4.3.1.7 Other Classification Algorithms In this section, other classification algorithms used in healthcare informatics are introduced.

Hidden Markov model (HMM) is a statistical model representing probability distribution over the sequences of observations [100]. This model uses the Markov chain to model signals in order to calculate the occurrence probability of states. Cooper [101] applies structured and unstructured HMM to analyze hospital infection data. Structured HMM is more parsimonious and can estimate important epidemiological parameters from time series data. Another work on HMM addresses the detection of anomalies in measured blood glucose levels [102]. Based on the experimental results, the HMM technique is accurate and robust in the presence of moderate changes.

Gaussian mixture model (GMM) is one more statistical model widely used as a classifier in pattern recognition tasks. It consists of a number of Gaussian distributions in the linear way [103]. Giri et al. [104] present a method which uses GMM for the automatic detection of normal and coronary artery disease conditions with electrocardiogram (ECG) signals. In this research, various feature selection and classification algorithms are applied to heart rate signals. The GMM classifier combined with the ICA reduction algorithm results in the highest accuracy compared to other techniques such as LDA and SVM.

Bayesian networks, also known as belief networks, are another probabilistic model corresponding to a graphical model called directed acyclic graph (DAG) [105]. Each node of the graph represents a random variable, and the edge between the nodes indicates probabilistic dependencies among the related random variables. A Bayesian network, for instance, could be applied to measure the probabilities of the existence of various diseases and find the relationship between diseases and symptoms.

Furthermore, some researchers have developed more complex classification algorithms in the healthcare domain, including collateral representative subspace projection modeling (C-RSPM) [106] and multi-layer classification framework, for discovering the temporal information of biological images [107].

5.3.4.3.2 Regression

Regression is a supervised learning algorithm used to model relationships between objects and targets. The difference between regression and classification is that in regression, the target is continuous, while in the latter, it is discrete. In other words, regression is the problem of approximating a real-valued target function [64].

Yoshida et al. [108] propose a radial basis function-sparse partial least squares (RBF-sPLS) regression and apply it to high-dimensional data including MRI brain images. The sPLS regression [109] reduces dimension and selects features simultaneously using sparse and linear combination of the explanatory variables. The proposed method, a combination of sPLS with basis expansion, is applicable to real data including MRI brain images with large-scale characteristics from chronic kidney disease patients. The authors evaluate the performance of RBF-sPLS by comparing it with the method without basis expansion. Saha et al. [110] introduce a Bayesian regression estimation algorithm implemented as a relevance vector machine (RVM) via particle filters (PF). This algorithm is used to integrate monitoring, diagnosis, and prediction of battery health. RVM is a Bayesian form obtaining solutions for regression and probabilistic classification [111], which represents a generalized linear form of the SVM. To estimate state dynamically, a general framework is provided using PF. The results show the advantage of proposed methods over conventional methods of battery health monitoring.

Regression forests is another supervised algorithm, which is used for anatomy localization and detection in [112]. The main goal of this algorithm is to train a non-linear mapping from a complicated input to continuous parameters (from voxels to the location and size of organ), and its difference with other forests classifiers is that it is utilized to predict multivariate and continuous outputs using a tree-based regression method. To evaluate the proposed regression algorithm, it is applied on a database including 400 three-dimensional CT scans with high variety images, which shows the robustness and the accuracy of the trained model compared to the conventional methods.

5.3.4.3.3 Clustering

Clustering [113] is the task of categorizing a group of objects into sub-groups (called clusters) in such a way that objects in the same category are more similar to each other compared with those in other categories. Clustering is a technique of unsupervised learning and statistical analysis which is applicable in many fields. The main difference between classification and clustering is that clustering does not use labels and finds natural grouping based on the structure of objects. Several clustering algorithms applied in healthcare data are discussed in the following. Further information regarding clustering algorithms for healthcare is discussed in [114].

5.3.4.3.3.1 Partitioning Algorithms A partitioning clustering algorithm divides a set of data objects into various partitions (clusters) in such a way that each object is placed in exactly one partition [115]. K-means is a known partitioning clustering algorithm used in various areas such as computer vision, data mining, and market segmentation. It partitions objects into k clusters, computes centroids (mean points) of the clusters, and assigns every object to the cluster that has the nearest mean in an Expectation-Maximization fashion.

Zolfaghar et al. [10] present a big data-driven solution for predicting Risk of Readmission (RoR) of congestive heart failure patients. To achieve this, they apply data mining models to predict risk of readmission. K-means is used to segment a national inpatient sample (NIS) dataset. Using K-means, the average income of each patient (used as a predictor variable for ROR) is calculated to map each record of a dataset to the closest cluster based on Euclidean distance.

Graph-partitioning based algorithm is another popular clustering method which is applicable for partitioning a graph G into sub-components with specific properties. Yuan et al. [116] apply a new skin lesion segmentation algorithm using narrow-band graph-partitioning (NBGP) and region fusion to improve efficiency and performance of segmentation algorithm. The proposed method can produce skin lesion segmentation well even for weak edges and blurred images.

A modified version of K-means recently utilized is called subspace clustering. It partitions a dataset along with the various subsets of dimension instead of whole space, to overcome the challenge of dimensionality in big data analysis [117]. Hund et al. [118] apply a subspace clustering approach to real-world medical data and analyze the patient data relationship and immunization treatment. Experimental results show how subspace clustering can effectively identify grouping of patients compared with a full space analysis such as hierarchical clustering.

5.3.4.3.3.2 Hierarchical Algorithms Hierarchical algorithms build a hierarchy known as a dendrogram [119]. There are two general strategies for hierarchical clustering: The agglomerative (bottom-up) approach and the divisive (top-down) approach. Both combination (for agglomerative) and splitting (for divisive) of clusters are determined in a greedy manner in which a distance or dissimilarity measure is needed by use of an appropriate metric such as Euclidean distance or Manhattan distance.

A hybrid hierarchical clustering algorithm is proposed by Chipman et al. [120] to analyze microarray data. The proposed method utilizes mutual clusters and takes advantage of both bottom-up and top-down clustering. Belciug [121] also applies a hierarchical agglomerative clustering for categorizing and clustering the patients based on length of their stay in terms of days at the hospital.

5.3.4.3.3.3 Density-Based Algorithms Density-based clustering algorithms are extensively used to search for clusters of non-linear and arbitrary shapes based on the density of connected points. This algorithm defines clusters by a radius which contains maximum objects based on a defined threshold. One of the most popular density-based clustering approaches is called density-based clustering of applications with noise (DBSCAN) which was first introduced by Ester et al. in 1996 [122]. DBSCAN not only groups points with many close nearest neighbors and high-density areas but also detects outliers and noise points within low-density areas.

Density-based algorithm is also widely applied on healthcare and medical datasets such as biomedical images. In [123], an unsupervised region-based segmentation approach based on DBSCAN is used to detect homogeneous color regions in skin lesions images. The segmentation approach separates lesion from healthy skin and also detects color regions inside the lesion. Experimental results show that lesion borders are successfully identified in 80% of the tested biomedical images. As density-based algorithms cannot efficiently cluster high-dimensional datasets, a hierarchical density-based clustering of categorical data (HIERDENC) is proposed by Andreopoulos et al. [124] to overcome some challenges of conventional clustering approaches. HIERDENC detects clusters in a better runtime scalability which is more efficient for large datasets and big data applications. When new data are introduced, only HIERDENC index is updated, so that there is no need to repeat clustering on all data. HIERDENC is applied on several large and quickly growing biomedical datasets such as finding bicliques of biological networks (protein interaction or human gene networks), clustering of biomedical images, and retrieving clusters of force–distance curves.

A summary of the aforementioned machine learning algorithms with some healthcare examples, as well as their pros and cons, are shown in Table 5.3.

5.3.5 Searching

Specialists and physicians use analyzed data to search for systematic patterns in patients' information, which helps them in having a more precise diagnosis and treatment. Data mining is an analytic process which is designed to search and explore large-scale data (big data) to discover consistent and systematic patterns. One of the main challenges in big data mining in the medical domain is searching through unstructured and structured medical data to find a useful pattern from a patient's information.

Information retrieval (IR) is the process of extracting, searching, and the analysis of data objects based on metadata or other content-based indexing [125]. Data objects may be text documents, images, and audio. In this section, text and image retrieval in the medical domain are discussed.

Text mining refers to the process of extracting information and analyzing unstructured textual information [126]. Extracting information from textual documents is a useful information retrieval technique widely used in healthcare informatics [127, 128, 129]. Using text mining, it is possible to extract information from patient records, reports, lab results, and generally, clinical notes. The major problem is that the clinical notes are unstructured. To tackle this challenge, a wide variety of methods have been applied to analyzing these unstructured texts in the field of NLP [130].

Content-based image retrieval (CBIR) reveals its crucial role in medical image analysis by providing physicians and doctors with diagnostic aid including visualizing existing and relevant cases, together with diagnosis information. Therefore, retrieving images that can be valuable for diagnosis is a strong necessity for clinical decision-support methods including evidence-based medicine or case-based reasoning. In addition, their use will allow for the exploration of structured image databases in medical education and training. Therefore, information retrieval in medical images has been widely investigated in this community.

TABLE 5.3: Summary of Machine Learning Algorithms in the Healthcare Informatics

ML Category	Algorithm	Healthcare examples	Dataset examples	Pros	Cons
Classification	Decision tree	Brain MRI classification, medical prediction	ADNI [131], hemodialysis [132]	Simple, easy to implement	Space limitation, overfitting
	SVM	Image-based MR classification, children's health	NCCHD [133]	High accuracy	Slow training, computationally expensive
	Neural Network	Cancer, blood glucose level prediction, Heart Rate Variability recognition	Cleveland [134], Acute Nephritis Diagnosis [135]	Handles noisy data, detects nonlinear relationship	Slow, computationally expensive, black-box models, low accuracy
	Sparse	EHR count data, heartbeats classification, tumor classification, gene expression	Colon cancer, MIT-BIH ECG [136], DE-SynPUF [87]	Efficiency, handles imbalanced data, fast, compression	Computationally expensive
	Deep learning	Registration of MR brain images, healthcare decision-making, Alzheimer's diagnosis	ADNI, Huntington disease [137]	Handles large dataset, deals with deep architecture, generalization, unsupervised feature learning, supports multi-task learning and semi-supervised learning	Difficult to interpret, computationally expensive
	Ensemble	Microarray data classification, drug treatment response prediction, mortality rate prediction, Alzheimer's classification	ADNI	Overcomes overfitting, generalization, predictive, high performance	Hard to analyze, computationally expensive

(Continued)

TABLE 5.3: (*Continued*) Summary of Machine Learning Algorithms in the Healthcare Informatics

ML Category	Algorithm	Healthcare examples	Dataset examples	Pros	Cons
	Other classifiers	Hospital infection analysis, anomalies detection, health monitoring, drug reaction signal generation, health risk assessment	ECG data, ADRs [138]	Depends on method	Depends on method
Regression		Brain imaging analysis, battery health diagnosis	CKD [139], Li-ion batteries [110]	Depends on method	Depends on method
Clustering	Partitioning	Risk of readmission prediction, depression clustering	NIS and MHS [10], ADNI	Handles large datasets, fast, simple	High sensitivity to initialization, noise, and outliers
	Hierarchical	Microarray data clustering, patients grouping based on length of stay in hospital	Microarray datasets, HES [140]	Visualization capability	Poor visualization for large data, slow, uses huge amount of memory, low accuracy
	Density-based	Biomedical images clustering, finding bicliques in a network	Skin lesions images, BMC biomedical images	Detects outliers and arbitrary shapes, handles non-static and complex data	not suitable for large datasets, slow, tricky parameter selection

For example, Comaniciu et al. [141] propose a CBIR system supporting decision-making in the domain of clinical pathology, in which a central module and fast color segmenter are used to extract features such as nucleus appearance (e.g., texture, shape, and area). Performance of the system is evaluated using a classification with ten-fold cross-validation and compared with that of an individual expert on a database containing 261 digitized specimens.

CBIR has been employed for histopathological image analysis. For example, Schnorrenberg et al. [142] extend the biopsy analysis support system to consist of indexing and CBIR for retrieving biopsy slide images. A database containing 57 breast cancer cases is used for evaluation. Akakin et al. [143] propose a CBIR system using the multi-tiered approach to classify and retrieve microscopic images. To maintain the semantic consistency between images retrieved from the CBIR system, both "multi-image" query and "slide-level" image retrieval are enabled.

As emphasized in [144], scalability is the key factor in CBIR for medical image analysis. In fact, with the ever-increasing amount of annotated medical data, large-scale, data-driven methods provide the promise of bridging the semantic gap between images and diagnoses. However, the development of large- scale medical image analysis algorithms has lagged greatly behind the increasing quality and complexity of medical images.

Specifically, owing to the difficulties in developing scalable CBIR systems for large-scale datasets, most previous systems have been tested on a relatively small number of cases. With the goal of comparing CBIR methods on a larger scale, ImageCLEF and VISCERAL provide benchmarks for medical image retrieval tasks [145, 146, 147].

Recently, hashing methods have been intensively investigated in the machine learning and computer vision community for large-scale image retrieval. They enable fast approximated nearest neighbors (ANN) search to deal with the scalability issue. For examples, the locality sensitive hashing (LSH) [148] uses random projections to map data to binary codes, resulting in highly compact binary codes and enabling efficient comparison within a large database using the Hamming distance. Anchor graph hashing (AGH) [149] has been proposed to use neighborhood graphs which reveal the underlying manifold of features, leading to a high search accuracy. Recent research has focused on data-dependent hash functions, such as the spectral graph partitioning and hashing [150] and supervised hashing with kernels [151] incorporating the pairwise semantic similarity and dissimilarity constraints from labeled data. These hashing methods have also been employed to solve the dimensionality problem in medical image analysis. Particularly, Zhang et al. [152, 153] build a scalable image-retrieval framework based on the supervised hashing technique and validate its performance on several thousand histopathological images acquired from breast microscopic tissues. It leverages a small amount of supervised information in learning to compress high dimensional image feature vector into only tens of binary bits with the informative signatures preserved. The supervised information is employed to bridge the semantic gap between low-level image features and high-level diagnostic information, which is critical to medical image analysis.

In addition to hashing and searching the whole image, another approach is to segment all cells from histopathological images and conduct large-scale retrieval among cell images [154]. This enables cell-level and fine-grained analysis, achieving high accuracy. It is also possible to fuse multiple types of features in a hashing framework to improve the accuracy of medical image retrieval. Specifically, the composite anchor graph hashing algorithm [149] has been developed for retrieving medical images [155], e.g., retrieving lung microscopic tissue images for the differentiation of adenocarcinoma and squamous carcinoma. Besides hashing-based methods, vocabulary tree methods have also been intensively investigated [156] and employed for medical image analysis [157].

5.3.6 Decision Support

Clinical decision support (CDS) is a process for improving quality of healthcare. It helps physicians, doctors, and patients to make better decisions [158]. Although CDS systems have made great contributions to improve medical care and reduce healthcare errors, they do not always improve clinical decision support systems due to some technical and non-technical factors. Kawamoto et al. [159] studied the literature to recognize the specific factors and features of such systems for enhancing clinical practice. Based on the results, 68% of decision support systems enhance clinical practice remarkably, which utilize four features, including automatic provision of decision support, provision of recommendations, provision of decision support during decision-making and at its location, and computerized decision support, that are significantly correlated with system success. In addition, some direct experimental evidence proves the significance of three extra features including sharing decision support with patients, providing performance feedback periodically, and requesting documentation of reasons if system recommendations are not followed. Healthcare systems can leverage new technologies in big data to provide better clinical decision support.

Today, CDS is a hot topic and an essential system in hospitals since it improves clinical output and efficiency of healthcare. Using big data technologies, many of CDS's limitations have been broken and dynamic clinical knowledge base systems have been created to deploy more complicated models for the CDS systems. Therefore, big data make the CDS systems more credible and effective [160].

5.3.6.1 Patient Similarity

Patient similarity computation is a significant process in healthcare informatics and decision support systems, which finds patients with similar clinical characteristics. It is very helpful for decision support applications and predicting patients' future conditions. The main goal is to find the similarity between patients by extracting distance metric. Based on the IBM Patient Care and Insights solution [161], the patient similarity algorithm includes the following steps:

- Both structured (e.g., clinical factors) and unstructured data (e.g., physicians' notes) are integrated for analysis.
- Personalized healthcare delivery plans are generated based on health history of each individual patient.
- Personalized treatment plans are created using thousands of patient characteristics examined by professionals.
- Patients with similar clinical characteristics are identified which helps doctors to see what treatments were most effective.
- Based on patient–physician matching, each patient is paired with a doctor who is suitable for a specific condition.

Due to the fact that each expert has different knowledge on patient similarities, Wang et al. [162] propose an approach called Composite distance integration (Comdi) to unify the single metrics achieved for each physician into a single unified distance metric. In addition to learning a globally optimized metric, Comdi presents a technique to share knowledge of expertise without sharing private data. To achieve this, it provides the neighborhood information for each party to integrate them into a global consistent metric. Breast cancer and Pima Indian diabetes [163] datasets are used to evaluate Comdi. The results show the Comdi's leverage in comparison with other individual and shared methods such as PCA, LDA, and locally supervised metric learning (LSML).

Wang et al. [164] present a Similarity Network Fusion (SNF) to integrate data samples (e.g., patients) and construct patient networks. A patient similarity network is demonstrated as a graph where nodes correspond to the patients and edges correspond to the similarity weight between patients. To determine the weight of each edge, a scaled exponential similarity kernel using Euclidean distance is applied. In this paper, three data types including DNA methylation, mRNA expression, and microRNA for five cancer datasets are combined by SNF to compute and fuse patient similarity.

Information visualization of big data is a novel and important process in data mining, known as visual data mining or visual data exploration [165]. Tsymbal et al. [166] propose and compare three techniques for visualizing patient similarity including treemaps [167], relative neighborhood graphs [168], and combined distance-heat-maps [169], which they believe is the most promising approach in clinical workflow.

Several studies have been done recently to bring big data to personalized healthcare, specifically patient similarity. The CARE system [170], for instance, is developed to predict and manage patient-centered disease. CARE is a computational assistant for doctors and physician to assess the potential disease risks of patients. This work utilizes shared experiences and similarity among a large number of patients resulting in a personalized healthcare plan. This big data includes patient history, prognosis, treatment strategies, disease timing, disease progression, and so on.

5.3.6.2 Computer-Assisted Interventions

More and more computer-based tools and methodologies are developed to support medical interventions especially in the big data era. This particular field of research and practice is called Computer Assisted Interventions (CAI) [171]. Examples include image processing methods, surgical process modeling and analysis, intraoperative decision supports, etc.

With the development of EHR and big data analytics, algorithms and approaches for healthcare big data are proposed to mine multi-modal medical data consisting of imaging and textual information. To be specific, modern, scalable, and efficient algorithms are generalized to harvest, organize, and learn from large-scale healthcare datasets for automatic understanding of medical images and help in the decision-making process [172]. Schlegl et al. [173] are able to learn from data collected across multiple hospitals with heterogeneous medical imaging equipment and propose a semi-supervised approach based on convolutional neural networks to classify lung tissues with promising results. In other works [174], a hierarchic multiatlas-based segmentation approach is proposed and evaluated on a large-scale medical dataset, for the segmentation of multiple anatomical structures in computed tomography scans.

Recently, the field of analyses and modeling of Surgical Process (SP) has gained its popularity for obtaining an explicit and formal understanding of surgery [175]. SP models are usually described from observer-based acquisition [176] or sensor-based acquisition [177, 178, 179]. By introducing related surgical models into a new generation of CAI systems, it improves the management of complex multi-modal information, increasing the quality and efficiency of medical care.

The developments of CAI and big data technologies interact with each other and continue to assist humans in processing and acting on complex information and providing better services to patients.

5.4   Future Directions and Outlook

Despite many opportunities and approaches for big data analytics in healthcare presented in this work, there are many other directions to be explored, concerning various aspects of healthcare data, such as the quality, privacy, and timeliness, etc. This section provides an outlook on big data analytics in healthcare informatics from a broader view, which covers the topics of healthcare data characteristics (e.g., high complexity, large-scale, etc.), data analytics task (e.g., longitudinal analysis, visualization, etc.), and objectives (e.g., real-time, privacy protection, collaboration with experts, etc.).

5.4.1   Complexity and Noise

The multi-source and multi-modal nature of healthcare data results in high complexity and noise issues. In addition, there are also problems of impurity and missing values in the high-volume data. It is difficult to handle all these problems both in terms of scale and accuracy, although a number of methods have been developed to improve the accuracy and usability of data [180]. Since the quality of data determines the quality of information, which will eventually affect the decision-making process, it is critical to develop efficient big data cleansing approaches to improve data quality for making effective and accurate decisions [181].

5.4.2   Heterogeneity

Traditional healthcare data usually lack in standardization; they are often fragmented with multiple formats [2]. Therefore it is reasonable and critical to study and develop common data standards. However, it is a challenging task due to the complexity of generating common data standards. Not only are healthcare data diverse, but there are various technical issues for integrating those data for special usage [182]. Even with standardized data formats, the multi-modal nature of data creates a challenge for effective fusion [183], which requires the development of advanced analytics that deal with large amounts of multi-modal data. The integration and fusion of the multi-source and multi-modal healthcare data with increasing scale would be a great challenge [184].

5.4.3   Longitudinal Analysis

Longitudinal data refers to the collection of repeated measurements of participant outcomes and possibly treatments or exposures [185], i.e., "the outcome variable is repeatedly measured on the same individual on multiple occasions" [186]. In recent decades, longitudinal data analysis, especially the statistical analysis of longitudinal data, has attracted more and more attention. Longitudinal studies involve the characterization of normal growth and aging, the effectiveness assessment of risk factors and treatments. It plays a key role in epidemiology, clinical research, and therapeutic evaluation. With big data analytic tools, it becomes promising to apply the longitudinal analysis of care across patients and diagnoses for identifying best care approaches.

5.4.4 Scale

Healthcare data is rapidly growing by size and coverage [183]. The fact that the data volume is scaling faster than computing resources poses a major challenge for managing large amounts of data. Several fundamental shifts (from a hardware point of view) are taking place to accommodate this dramatic change [187]. First, over the past few years, the processor technology has gradually shifted the focus to parallel data processing within nodes and the packing of multiple sockets, etc. Second, the move towards cloud computing enables information sharing and the aggregation of multiple workloads into large-scale clusters. Third, the transformative change of the traditional I/O subsystem from hard disk drives (HDDs) to solid state drives (SSDs), as well as other storage technologies, is reforming the design and operation of data processing systems.

5.4.5 Real-time

The velocity characteristic of big data in health informatics not only indicates the data acquisition and processing rate but also the timeliness of responses. There are many scenarios that call for a quick decision. For example, it would be extremely desirable to monitor and analyze a person's health condition to predict potential illness in a real-time or near real-time. It would also be of great significance to raise alarm for potential outbreak of influenza through analyzing public health data. Although real-time analytic applications are still in their infancy in the big data era, it is the strongest trend and most promising direction in the future of health informatics [188]. A good example is the development of complex event processing (CEP) for handling streaming big data and fulfilling the real-time requirement.

5.4.6 Privacy

The privacy of data is another big concern of future big data analytics in healthcare informatics [189]. Although there are strict laws governing the more formalized EHRs data, special attention should be paid and rules should be enforced to regularize the usage and distribution of personal and sensitive information acquired from multiple sources. "Managing privacy is effectively both a technical and a sociological problem, which must be addressed jointly from both perspectives to realize the promise of big data" [187]. In addition to data privacy, there are a range of other issues such as data protection, data security, data safety, and protection of doctors against responsibility derived from manipulated data, which require special big data analytics to handle these complex restrictions [189, 184, 190].

5.4.7 Visualization

Visualization of healthcare data is critical for exploratory or discovery analytics, whose purpose is to explore and discover things that are undermined and encrypted in the data [191, 192]. Effective visualization tools will help clinicians and physicians to explore the data without assistance from IT [188]. Although visualization has been studied for several decades with relative maturity, there are still challenges and open issues to be addressed, especially for big data analytics in healthcare data [184].

5.4.8   Multi-Discipline and Human Interaction

Big data in health informatics is predicted to be a multi-disciplinary task that involves continuous efforts from multiple domain experts [193]. They include, but are not limited to, engineering scientists who provide basic big data infrastructure to collect, store, share, and manage big data; computer science data scientists who provide solutions for processing and analyzing high-volume high-velocity healthcare data via numerous data mining and machine learning techniques; clinicians and physicians from the medical domain who provide professional healthcare data analysis, offer personalized care, and make final decisions. Sometimes it is difficult for computer algorithms to identify patterns and analyze results; therefore it is a desirable feature for an advanced big data analysis system to be able to support input from multiple human experts, exchange of opinions, and shared exploration of results. Furthermore, in the health domain, sometimes we do not have big data – we are confronted with a small number of datasets or rare events, where for example machine learning approaches suffer from insufficient training samples. In such cases we need more than just automatic machine learning; we still need a human in the loop. In other words, interactive machine learning (iML) or "human-in-the-loop" techniques can be utilized in health informatics where automatic machine learning approaches are not able to handle rare events alone and a human expert is needed to interact in the learning process [194]. This interaction between computer algorithms and human experts can improve the learning procedure.

5.5   Summary

This article presents a comprehensive overview of the challenges, pipeline, techniques, and future directions for computational health informatics in the big data age, by providing a structured analysis of the historical and state-of-the-art methods in over 170 papers and web articles. We have summarized the challenges of big data health informatics into four Vs: Volume, Variety, Velocity, and Veracity, and emerging challenges such as Validity and Volatility. A systematic data processing pipeline is provided for generic big health informatics, covering data capturing, storing, sharing, analyzing, searching, and decision support. Computational health informatics in the big data age is an emerging and highly important research field with a potentially significant impact on the conventional healthcare industry. The future of health informatics will benefit from the exponentially increasing digital health data.

References

1. M. Herland, T. M. Khoshgoftaar, and R. Wald, "A review of data mining using big data in health informatics," *Journal of Big Data*, Springer, vol. 1, no. 1, p. 2, 2014.
2. W. Raghupathi and V. Raghupathi, "Big data analytics in healthcare: promise and potential," *Health Information Science and Systems*, vol. 2, no. 1, p. 3, 2014.
3. J. Sun and C. K. Reddy, "Big data analytics for healthcare," in *Proceedings of the 19th ACM SIGKDD International Conference on Knowledge Discovery and Data Mining, ser. KDD '13.* New York, NY, USA, ACM, 2013, pp. 1525–1525. [Online]. http://doi.acm.org/10.1145/248 7575.2506178

4. B. Feldman, E. M. Martin, and T. Skotnes, "Big data in healthcare hype and hope," Technical Report, *Dr. Bonnie* 360°, 2012.

5. Roy, "Roy tutorials," http://www.roytuts.com/, 2015, retrieved on: 2019-09-28.

6. M. Cottle, S. Kanwal, M. Kohn, T. Strome, and N. Treister, "Transforming health care through big data: strategies for leveraging big data in the health care industry," http://c4fd63cb482ce6861463-bc6183f1c18e748a49b87a25911a0555.r93.cf2.rackcdn.com/iHT2_BigData_2013.pdf, 2013, New York: Institute for Health Technology Transformation.

7. Gartner, "IT glossary: big data," http://www.gartner.com/it-glossary/big-data/, 2014, retrieved on: 2015-03-06.

8. J. Manyika, M. Chui, B. Brown, J. Bughin, R. Dobbs, C. Roxburgh, and A. H. Byers, "Big data: the next frontier for innovation, competition, and productivity," Technical Report, McKinsey Global Institute, 2011.

9. D. W. Bates, S. Saria, L. Ohno-Machado, A. Shah, and G. Escobar, "Big data in health care: using analytics to identify and manage high-risk and high-cost patients," *Health Affairs*, vol. 33, no. 7, pp. 1123–1131, 2014.

10. K. Zolfaghar, N. Meadem, A. Teredesai, S. B. Roy, S.-C. Chin, and B. Muckian, "Big data solutions for predicting risk-of-readmission for congestive heart failure patients," in *2013 IEEE International Conference on Big Data*, 2013, pp. 64–71.

11. C. W. Cotman, N. C. Berchtold, and L.-A. Christie, "Exercise builds brain health: key roles of growth factor cascades and inflammation," *Trends in Neurosciences*, vol. 30, no. 9, pp. 464–472, 2007.

12. N. Byrnes, "MIT technology review," http://www.technologyreview.com/news/529011/can-technology-fix-medicine/, 2014, retrieved on: 2015-01-20.

13. F. F. Costa, "Big data in biomedicine," *Drug Discovery Today*, vol. 19, no. 4, pp. 433–440, 2014.

14. S. E. White, "A review of big data in health care: challenges and opportunities," *Clinical, Cosmetic and Investigational Dentistry*, vol. 6, pp. 45–56, 2014.

15. I. Merelli, H. Pérez-Sánchez, S. Gesing, and D. D'Agostino, "Managing, analysing, and integrating big data in medical bioinformatics: open problems and future perspectives," *BioMed Research International*, vol. 2014, 134023, 2014.

16. I. Yoo, P. Alafaireet, M. Marinov, K. Pena-Hernandez, R. Gopidi, J.-F. Chang, and L. Hua, "Data mining in healthcare and biomedicine: a survey of the literature," *Journal of Medical Systems*, vol. 36, no. 4, pp. 2431–2448, 2012.

17. M. Herland, T. M. Khoshgoftaar, and R. Wald, "Survey of clinical data mining applications on big data in health informatics," in *12th International Conference on Machine Learning and Applications (ICMLA)*, vol. 2, IEEE, 2013, pp. 465–472.

18. J. Crapo, "Big data in healthcare: separating the hype from the reality," https://www.healthcatalyst.com/healthcare-big-data-realities, 2014, retrieved on: 2015-01-17.

19. K. Hung, Y. Zhang, and B. Tai, "Wearable medical devices for tele-home healthcare," in *26th Annual International Conference of the IEEE Engineering in Medicine and Biology Society (IEMBS'04)*, vol. 2, IEEE, 2004, pp. 5384–5387.

20. A. Pavlo, E. Paulson, A. Rasin, D. J. Abadi, D. J. DeWitt, S. Madden, and M. Stonebraker, "A comparison of approaches to large-scale data analysis," in *Proceedings of the 2009 ACM SIGMOD International Conference on Management of Data*, 2009, pp. 165–178.

21. J. C. Corbett, J. Dean, M. Epstein, A. Fikes, C. Frost, J. J. Furman, S. Ghemawat, A. Gubarev, C. Heiser, P. Hochschild, et al., "Spanner: googles globally distributed database," *ACM Transactions on Computer Systems (TOCS)*, vol. 31, no. 3, p. 8, 2013.

22. Techcrunch, "Healthcare's big data opportunity," http://techcrunch.com/2014/11/20/healthcares-big-data-opportunity/, 2014, retrieved at: 2015-02-20.

23. D. A. Reed and J. Dongarra, "Exascale computing and big data," *Communications of the ACM*, vol. 58, no. 7, pp. 56–68, 2015.

24. IBM, "Large gene interaction analytics at university at buffalo," http://www-03.ibm.com/software/businesscasestudies/no/no/corp? synkey=M947744T51514R54, 2012, retrieved at: 2015-02-23.

25. K. Priyanka and N. Kulennavar, "A survey on big data analytics in health care," *International Journal of Computer Science and Information Technologies*, vol. 5, no. 4, pp. 5865–5868, 2014.

26. N. T. Issa, S. W. Byers, and S. Dakshanamurthy, "Big data: the next frontier for innovation in therapeutics and healthcare," *Expert Review of Clinical Pharmacology*, vol. 7, no. 3, pp. 293–298, 2014.

27. A. Tippet, "Data capture and analytics in healthcare," http://blogs.zebra.com/data-capture-and-analytics-in-healthcare, 2014, retrieved at: 2015-02-08.

28. M. Zaharia, M. Chowdhury, M. J. Franklin, S. Shenker, and I. Stoica, "Spark: cluster computing with working sets," in *Proceedings of the 2nd USENIX Conference on Hot Topics in Cloud Computing*, 2010, pp. 10–10.

29. NetApp, "Netapp EHR solutions: efficient, high-availability EHR data storage and management," http://www.netapp.com/us/system/pdf-reader.aspx?cc=us&m=ds-3222.pdf&pdfUri=tcm:10-61401, 2011, retrieved at: 2015-02-20.

30. C. He, X. Fan, and Y. Li, "Toward ubiquitous healthcare services with a novel efficient cloud platform," *IEEE Transactions on Biomedical Engineering*, vol. 60, no. 1, pp. 230–234, 2013.

31. A. Madaan, W. Chu, Y. Daigo, and S. Bhalla, "Quasi-relational query language interface for persistent standardized EHRs: using NOSQL databases," in *Databases in Networked Information Systems*. Springer, 2013, pp. 182–196.

32. R. E. Cooke Jr, M. G. Gaeta, D. M. Kaufman, and J. G. Henrici, "Picture archiving and communication system," June 3, 2003, US Patent 6,574,629.

33. NetApp, http://www.netapp.com/us/solutions/industry/healthcare/, 2011, retrieved at: 2015-02-24.

34. Intel, "Distributed systems for clinical data analysis," http://www.intel.com/content/dam/www/public/us/en/documents/white-papers/big-data-hadoop-clinical-analysis-paper.pdf, 2011, retrieved at: 2015-02-24.

35. A. Crowne, "Preparing the healthcare industry to capture the full potential of big data," http://sparkblog.emc.com/2014/06/preparing-healthcare-industry-capture-full-potential-big-data/, 2014, retrieved at: 2015-02-25.

36. EMC, "Managing healthcare data within the ecosystem while reducing it costs and complexity," http://www.emc.com/collateral/emc-perspective/h8805-healthcare-costs-complexities-ep.pdf, 2011, retrieved at: 2015-02-25.

37. X. Dong, R. Li, H. He, W. Zhou, Z. Xue, and H. Wu, "Secure sensitive data sharing on a big data platform," *Tsinghua Science and Technology*, vol. 20, no. 1, pp. 72–80, 2015.

38. H. Banaee, M. U. Ahmed, and A. Loutfi, "Data mining for wearable sensors in health monitoring systems: a review of recent trends and challenges," *Sensors*, vol. 13, no. 12, pp. 17472–17500, 2013.

39. D. Sow, D. S. Turaga, and M. Schmidt, "Mining of sensor data in healthcare: a survey," in *Managing and Mining Sensor Data*. Springer, 2013, pp. 459–504.

40. R. R. Singh, S. Conjeti, and R. Banerjee, "An approach for real-time stress-trend detection using physiological signals in wearable computing systems for automotive drivers," in *14th International IEEE Conference on Intelligent Transportation Systems (ITSC)*, 2011, pp. 1477–1482.

41. Y. Mao, W. Chen, Y. Chen, C. Lu, M. Kollef, and T. Bailey, "An integrated data mining approach to real-time clinical monitoring and deterioration warning," in *Proceedings of the 18th ACM SIGKDD International Conference on Knowledge Discovery and Data Mining*, ACM, 2012, pp. 1140–1148.

42. A. S. Fauci, E. Braunwald, D. L. Kasper, S. L. Hauser, D. L. Longu, J. L. Jameson, and J. Loscalzo, *Harrison's Principles of Internal Medicine*, vol. 2. McGraw-Hill Medical: New York, 2008.

43. D. Apiletti, E. Baralis, G. Bruno, and T. Cerquitelli, "Real-time analysis of physiological data to support medical applications," *IEEE Transactions on Information Technology in Biomedicine*, vol. 13, no. 3, pp. 313–321, 2009.

44. F. Hu, M. Jiang, L. Celentano, and Y. Xiao, "Robust medical ad hoc sensor networks (MASN) with wavelet-based ECG data mining," *Ad Hoc Networks*, vol. 6, no. 7, pp. 986–1012, 2008.

45. C. A. Frantzidis, C. Bratsas, M. A. Klados, E. Konstantinidis, C. D. Lithari, A. B. Vivas, C. L. Papadelis, E. Kaldoudi, C. Pappas, and P. D. Bamidis, "On the classification of emotional biosignals evoked while viewing affective pictures: an integrated data-mining-based approach for healthcare applications," *IEEE Transactions on Information Technology in Biomedicine*, vol. 14, no. 2, pp. 309–318, 2010.

46. T. Meng, A. T. Soliman, M.-L. Shyu, Y. Yang, S.-C. Chen, S. Iyengar, J. S. Yordy, and P. Iyengar, "Wavelet analysis in current cancer genome research: a survey," *IEEE/ACM Transactions on Computational Biology and Bioinformatics*, vol. 10, no. 6, pp. 1442–14359, 2013.

47. A. A. Leema and M. Hemalatha, "An effective and adaptive data cleaning technique for colossal RFID data sets in healthcare," *WSEAS Transactions on Information Science and Applications*, vol. 8, no. 6, pp. 243–252, 2011.

48. A. S. Fialho, F. Cismondi, S. M. Vieira, S. R. Reti, J. M. Sousa, and S. N. Finkelstein, "Data mining using clinical physiology at discharge to predict icu readmissions," *Expert Systems with Applications*, vol. 39, no. 18, pp. 13158–13165, 2012.

49. R. Jané, H. Rix, P. Caminal, and P. Laguna, "Alignment methods for averaging of high-resolution cardiac signals: a comparative study of performance," *IEEE Transactions on Biomedical Engineering*, vol. 38, no. 6, pp. 571–579, 1991.

50. R. Martinez Orellana, B. Erem, and D. H. Brooks, "Time invariant multi electrode averaging for biomedical signals," in *IEEE International Conference on Acoustics, Speech and Signal Processing (ICASSP)*, IEEE, 2013, pp. 1242–1246.

51. A. G. Eapen, "Application of data mining in medical applications," Master's thesis, University of Waterloo, Ontario, Canada, 2004.

52. M. Hall, E. Frank, G. Holmes, B. Pfahringer, P. Reutemann, and I. H. Witten, "The WEKA data mining software: an update," *ACM SIGKDD Explorations Newsletter*, vol. 11, no. 1, pp. 10–18, 2009.

53. A. Thusoo, J. S. Sarma, N. Jain, Z. Shao, P. Chakka, S. Anthony, H. Liu, P. Wyckoff, and R. Murthy, "Hive: a warehousing solution over a map-reduce framework," *Proceedings of the VLDB Endowment*, vol. 2, no. 2, pp. 1626–1629, 2009.

54. A. T. Soliman, T. Meng, S.-C. Chen, S. Iyengar, P. Iyengar, J. Yordy, and M.-L. Shyu, "Driver missense mutation identification using feature selection and model fusion," *Journal of Computational Biology*, vol. 22, no. 12, pp. 1075–1085, 2015.

55. M. Pechenizkiy, A. Tsymbal, and S. Puuronen, "Pca-based feature transformation for classification: issues in medical diagnostics," in *Proceedings of the 17th IEEE Symposium on Computer-Based Medical Systems, (CBMS)*, IEEE, 2004, pp. 535–540.

56. A. Subasi and M. I. Gursoy, "Eeg signal classification using pca, ica, lda and support vector machines," *Expert Systems with Applications*, vol. 37, no. 12, pp. 8659–8666, 2010.

57. K. Mao, "Orthogonal forward selection and backward elimination algorithms for feature subset selection," *IEEE Transactions on Systems, Man, and Cybernetics, Part B: Cybernetics*, vol. 34, no. 1, pp. 629–634, 2004.

58. J. Sun, C. D. McNaughton, P. Zhang, A. Perer, A. Gkoulalas-Divanis, J. C. Denny, J. Kirby, T. Lasko, A. Saip, and B. A. Malin, "Predicting changes in hypertension control using electronic health records from a chronic disease management program," *Journal of the American Medical Informatics Association*, vol. 21, no. 2, pp. 337–344, 2014.

59. M. A. Hall, "Correlation-based feature selection for machine learning," Ph.D. dissertation, The University of Waikato, 1999.

60. J. S. Mathias, A. Agrawal, J. Feinglass, A. J. Cooper, D. W. Baker, and A. Choudhary, "Development of a 5 year life expectancy index in older adults using predictive mining of electronic health record data," *Journal of the American Medical Informatics Association*, vol. 20, no. e1, pp. e118–e124, 2013.

61. H. G. Li, X. Wu, Z. Li, and W. Ding, "Online group feature selection from feature streams," in *Twenty-Seventh AAAI Conference on Artificial Intelligence*, Citeseer, 2013, pp. 1627–1628.

62. K. Yu, X. Wu, W. Ding, and J. Pei, "Towards scalable and accurate online feature selection for big data," in *2014 IEEE International Conference on Data Mining (ICDM)*, 2014, pp. 660–669.

63. M. Tan, I. W. Tsang, and L. Wang, "Towards ultrahigh dimensional feature selection for big data," *The Journal of Machine Learning Research*, vol. 15, no. 1, pp. 1371–1429, 2014.

64. T. M. Mitchell, *Machine Learning*, vol. 45. Burr Ridge, IL: McGraw Hill, 1997.

65. P. C. Austin, J. V. Tu, J. E. Ho, D. Levy, and D. S. Lee, "Using methods from the data-mining and machine-learning literature for disease classification and prediction: a case study examining classification of heart failure subtypes," *Journal of Clinical Epidemiology*, vol. 66, no. 4, pp. 398–407, 2013.

66. D. J. Dittman, T. M. Khoshgoftaar, R. Wald, and A. Napolitano, "Simplifying the utilization of machine learning techniques for bioinformatics," in *12th International Conference on Machine Learning and Applications (ICMLA)*, vol. 2, IEEE, 2013, pp. 396–403.

67. Y. Zhang, S. Fong, J. Fiaidhi, and S. Mohammed, "Real-time clinical decision support system with data stream mining," *Biomedicine and Biotechnology*, vol. 2012, 580186, 2012.

68. P. Domingos and G. Hulten, "Mining high-speed data streams," in *Proceedings of the Sixth ACM SIGKDD International Conference on Knowledge Discovery and Data Mining*, 2000, pp. 71–80.

69. F. Estella, B. L. Delgado-Marquez, P. Rojas, O. Valenzuela, B. S. Roman, and I. Rojas, "Advanced system for automously classify brain MRI in neurodegenerative disease," in *International Conference on Multimedia Computing and Systems (ICMCS)*, IEEE, 2012, pp. 250–255.

70. D. M. Hawkins, "The problem of overfitting," *Journal of Chemical Information and Computer Sciences*, vol. 44, no. 1, pp. 1–12, 2004.

71. N. Cristianini and J. Shawe-Taylor, *An Introduction to Support Vector Machines and Other Kernel-Based Learning Methods*. Cambridge University Press, 2000.

72. S.-M. Zhou, R. A. Lyons, O. Bodger, J. C. Demmler, and M. D. Atkinson, "SVM with entropy regularization and particle swarm optimization for identifying children's health and socioeconomic determinants of education attainments using linked datasets," in *The 2010 International Joint Conference on Neural Networks (IJCNN)*, IEEE, 2010, pp. 1–8.

73. S. K. Pal and S. Mitra, "Multilayer perceptron, fuzzy sets, and classification," *IEEE Transactions on Neural Networks*, vol. 3, no. 5, pp. 683–697, 1992.

74. T. H. N. Vu, N. Park, Y. K. Lee, Y. Lee, J. Y. Lee, and K. H. Ryu, "Online discovery of heart rate variability patterns in mobile healthcare services," *Journal of Systems and Software*, vol. 83, no. 10, pp. 1930–1940, 2010.

75. B. Fritzke, "A growing neural gas network learns topologies," *Advances in Neural Information Processing Systems*, vol. 7, pp. 625–632, 1995.

76. T. Kohonen, "The self-organizing map," *Neurocomputing*, vol. 21, no. 1, pp. 1–6, 1998.

77. X.-W. Chen and X. Lin, "Big data deep learning: challenges and perspectives," *IEEE Access*, vol. 2, pp. 514–525, 2014.

78. M. Nielsen, *Neural networks and deep learning*. Determination Press, 2014.

79. G. E. Hinton, "Deep belief networks," *Scholarpedia*, vol. 4, no. 5, p. 5947, 2009.

80. H. Larochelle, Y. Bengio, J. Louradour, and P. Lamblin, "Exploring strategies for training deep neural networks," *The Journal of Machine Learning Research*, vol. 10, pp. 1–40, 2009.

81. R. Salakhutdinov and G. E. Hinton, "Deep Boltzmann machines," in *Proceedings of the 12th International Conference on Artificial Intelligence and Statistics (AISTATS)*, vol. 5, 2009, pp. 448–455.

82. A. Krizhevsky, I. Sutskever, and G. E. Hinton, "Imagenet classification with deep convolutional neural networks," in *Advances in Neural Information Processing Systems*, Curran Associates, Inc., 2012, pp. 1097–1105.

83. Z. Liang, G. Zhang, J. X. Huang, and Q. V. Hu, "Deep learning for healthcare decision making with emrs," in *2014 IEEE International Conference on Bioinformatics and Biomedicine (BIBM)*, 2014, pp. 556–559.

84. S. M. Plis, D. R. Hjelm, R. Salakhutdinov, E. A. Allen, H. J. Bockholt, J. D. Long, H. J. Johnson, J. S. Paulsen, J. A. Turner, and V. D. Calhoun, "Deep learning for neuroimaging: a validation study," *Frontiers in Neuroscience*, vol. 8, 229, 2014.

85. R. Li, W. Zhang, H.-I. Suk, L. Wang, J. Li, D. Shen, and S. Ji, "Deep learning based imaging data completion for improved brain disease diagnosis," in *Medical Image Computing and Computer-Assisted Intervention–MICCAI 2014*, Springer, 2014, pp. 305–312.

86. K. Huang and S. Aviyente, "Sparse representation for signal classification," in *Proceedings of the 19th International Conference on Neural Information Processing Systems*, 2006, pp. 609–616.

87. J. C. Ho, J. Ghosh, and J. Sun, "Marble: high-throughput phenotyping from electronic health records via sparse nonnegative tensor factorization," in *Proceedings of the 20th ACM SIGKDD International Conference on Knowledge Discovery and Data Mining*, 2014, pp. 115–124.

88. S. Li, H. Yin, and L. Fang, "Group-sparse representation with dictionary learning for medical image denoising and fusion," *IEEE Transactions on Biomedical Engineering*, vol. 59, no. 12, pp. 3450–3459, 2012.

89. Q. Xu, H. Yu, X. Mou, L. Zhang, J. Hsieh, and G. Wang, "Low-dose x-ray CT reconstruction via dictionary learning," *IEEE Transactions on Medical Imaging*, vol. 31, no. 9, pp. 1682–1697, 2012.

90. A. Gholipour, J. A. Estroff, and S. K. Warfield, "Robust super-resolution volume reconstruction from slice acquisitions: application to fetal brain mri," *IEEE Transactions on Medical Imaging*, vol. 29, no. 10, pp. 1739–1758, 2010.

91. R. Fang, T. Chen, and P. C. Sanelli, "Towards robust deconvolution of low-dose perfusion CT: sparse perfusion deconvolution using online dictionary learning," *Medical Image Analysis*, vol. 17, no. 4, pp. 417–428, 2013.

92. R. Fang, H. Jiang, and J. Huang, "Tissue-specific sparse deconvolution for brain CT perfusion," *Computerized Medical Imaging and Graphics*, 2015, online 21 May 2015, ISSN 0895-6111, doi:10.1016/j.compmedimag.2015.04.008.

93. R. Fang, K. Karlsson, T. Chen, and P. C. Sanelli, "Improving low-dose blood–brain barrier permeability quantification using sparse high-dose induced prior for patlak model," *Medical Image Analysis*, vol. 18, no. 6, pp. 866–880, 2014.

94. R. Fang, S. Zhang, T. Chen, and P. Sanelli, "Robust low-dose CT perfusion deconvolution via tensor total-variation regularization," *IEEE Transaction on Medical Imaging*, vol. 34, no. 7, pp. 1533–1548, 2015.

95. L. Breiman, "Random forests," *Machine Learning*, vol. 45, no. 1, pp. 5–32, 2001.

96. R. Díaz-Uriarte and S. A. De Andres, "Gene selection and classification of microarray data using random forest," *BMC Bioinformatics*, vol. 7, no. 1, p. 3, 2006.

97. J. J. Rodriguez, L. I. Kuncheva, and C. J. Alonso, "Rotation forest: a new classifier ensemble method," *IEEE Transactions on Pattern Analysis and Machine Intelligence*, vol. 28, no. 10, pp. 1619–1630, 2006.

98. Y. Freund and L. Mason, "The alternating decision tree learning algorithm," in *Proceedings of the 16th International Conference on Machine Learning*, vol. 99, 1999, pp. 124–133.

99. M. Liu, D. Zhang, and D. Shen, "Ensemble sparse classification of Alzheimer's disease," *NeuroImage*, vol. 60, no. 2, pp. 1106–1116, 2012.

100. Z. Ghahramani, "An introduction to hidden Markov models and Bayesian networks," *International Journal of Pattern Recognition and Artificial Intelligence*, vol. 15, no. 01, pp. 9–42, 2001.

101. B. Cooper and M. Lipsitch, "The analysis of hospital infection data using hidden Markov models," *Biostatistics*, vol. 5, no. 2, pp. 223–237, 2004.

102. Y. Zhu, "Automatic detection of anomalies in blood glucose using a machine learning approach," *Journal of Communications and Networks*, vol. 13, no. 2, pp. 125–131, 2011.

103. W. Wang, H. Wang, M. Hempel, D. Peng, H. Sharif, and H.-H. Chen, "Secure stochastic ECG signals based on Gaussian mixture model for e-healthcare systems," *Systems Journal, IEEE*, vol. 5, no. 4, pp. 564–573, 2011.

104. D. Giri, U. R. Acharya, R. J. Martis, S. V. Sree, T.-C. Lim, T. Ahamed, and J. S. Suri, "Automated diagnosis of coronary artery disease affected patients using LDA, PCA, ICA and discrete wavelet transform," *Knowledge-Based Systems*, vol. 37, pp. 274–282, 2013.

105. I. Ben-Gal, "Bayesian networks," *Encyclopedia of Statistics in Quality and Reliability.* John Wiley and Sons, 2007.

106. T. Meng, L. Lin, M.-L. Shyu, and S.-C. Chen, "Histology image classification using supervised classification and multimodal fusion," in *2010 IEEE International Symposium on Multimedia (ISM)*, 2010, pp. 145–152.

107. T. Meng and M.-L. Shyu, "Biological image temporal stage classification via multi-layer model collaboration," in *2013 IEEE International Symposium on Multimedia (ISM)*, 2013, pp. 30–37.

108. H. Yoshida, A. Kawaguchi, and K. Tsuruya, "Radial basis function-sparse partial least squares for application to brain imaging data," *Computational and Mathematical Methods in Medicine*, vol. 2013, 591032, 2013.

109. K.-A. Lê Cao, D. Rossouw, C. Robert-Granié, and P. Besse, "A sparse PLS for variable selection when integrating omics data," *Statistical Applications in Genetics and Molecular Biology*, vol. 7, no. 1, pp. 1544–6115, 2008.

110. B. Saha, K. Goebel, S. Poll, and J. Christophersen, "An integrated approach to battery health monitoring using Bayesian regression and state estimation," in *Autotestcon, 2007 IEEE*, 2007, pp. 646–653.

111. M. E. Tipping, "Sparse Bayesian learning and the relevance vector machine," *The Journal of Machine Learning Research*, vol. 1, pp. 211–244, 2001.

112. A. Criminisi, D. Robertson, E. Konukoglu, J. Shotton, S. Pathak, S. White, and K. Siddiqui, "Regression forests for efficient anatomy detection and localization in computed tomography scans," *Medical Image Analysis*, vol. 17, no. 8, pp. 1293–1303, 2013.

113. A. K. Jain, M. N. Murty, and P. J. Flynn, "Data clustering: a review," *ACM Computing Surveys (CSUR)*, vol. 31, no. 3, pp. 264–323, 1999.

114. D. Tomar and S. Agarwal, "A survey on data mining approaches for healthcare," *International Journal of Bio-Science and Bio-Technology*, vol. 5, no. 5, pp. 241–266, 2013.

115. P.-N. Tan, M. Steinbach, and V. Kumar, *Introduction to Data Mining*, vol. 1. Pearson Addison Wesley: Boston, 2006.

116. X. Yuan, N. Situ, and G. Zouridakis, "A narrow band graph partitioning method for skin lesion segmentation," *Pattern Recognition*, vol. 42, no. 6, pp. 1017–1028, 2009.

117. H.-P. Kriegel, P. Kröger, and A. Zimek, "Clustering high-dimensional data: a survey on subspace clustering, pattern-based clustering, and correlation clustering," *ACM Transactions on Knowledge Discovery from Data (TKDD)*, vol. 3, no. 1, pp. 1–58, 2009.

118. M. Hund, W. Sturm, T. Schreck, T. Ullrich, D. Keim, L. Majnaric, and A. Holzinger, "Analysis of patient groups and immunization results based on subspace clustering," in *Brain Informatics and Health.* Springer, 2015, pp. 358–368.

119. N. Nithya, K. Duraiswamy, and P. Gomathy, "A survey on clustering techniques in medical diagnosis," *International Journal of Computer Science Trends and Technology (IJCST)*, vol. 1, no. 2, pp. 17–23, 2013.

120. H. Chipman and R. Tibshirani, "Hybrid hierarchical clustering with applications to microarray data," *Biostatistics*, vol. 7, no. 2, pp. 286–301, 2006.

121. S. Belciug, "Patients length of stay grouping using the hierarchical clustering algorithm," *Annals of the University of Craiova- Mathematics and Computer Science Series*, vol. 36, no. 2, pp. 79–84, 2009.

122. M. Ester, H.-P. Kriegel, J. Sander, and X. Xu, "A density-based algorithm for discovering clusters in large spatial databases with noise," in *Proceedings of the 2nd International Conference on Knowledge Discovery and Data Mining*, vol. 96, no. 34, AAAI Press, 1996, pp. 226–231.

123. M. E. Celebi, Y. A. Aslandogan, and P. R. Bergstresser, "Mining biomedical images with density-based clustering," in *International Conference on Information Technology: Coding and Computing (ITCC)*, vol. 1, IEEE, 2005, pp. 163–168.

124. B. Andreopoulos, A. An, and X. Wang, "Hierarchical density-based clustering of categorical data and a simplification," in *Advances in Knowledge Discovery and Data Mining.* Springer LNCS, 2007, pp. 11–22.

125. G. Salton and D. Harman, "Information retrieval," in *Encyclopedia of Computer Science.* Chichester, UK: John Wiley and Sons Ltd., 2003, pp. 858–863.

126. F. Popowich, "Using text mining and natural language processing for health care claims processing," *ACM SIGKDD Explorations Newsletter*, vol. 7, no. 1, pp. 59–66, 2005.

127. A. Holzinger, J. Schantl, M. Schroettner, C. Seifert, and K. Verspoor, "Biomedical text mining: state-of-the-art, open problems and future challenges," in *Interactive Knowledge Discovery and Data Mining in Biomedical Informatics.* Springer, 2014, pp. 271–300.

128. K. Jung, P. LePendu, S. Iyer, A. Bauer-Mehren, B. Percha, and N. H. Shah, "Functional evaluation of out-of-the-box text-mining tools for data-mining tasks," *Journal of the American Medical Informatics Association*, vol. 22, no. 1, pp. 121–131, 2014.

129. R. Vijayakrishnan, S. R. Steinhubl, K. Ng, J. Sun, R. J. Byrd, Z. Daar, B. A. Williams, C. Defilippi, S. Ebadollahi, and W. F. Stewart, "Prevalence of heart failure signs and symptoms in a large primary care population identified through the use of text and data mining of the electronic health record," *Journal of Cardiac Failure*, vol. 20, no. 7, pp. 459–464, 2014.

130. P. Spyns, "Natural language processing in medicine: an overview," *Methods of Information in Medicine*, vol. 35, no. 4, pp. 285–301, 1996.

131. ADNI, "Alzheimer's disease neuroimaging initiative," http://adni.loni.usc.edu/about/, 2015, retrieved at: 2015-02-15.

132. J.-Y. Yeh, T.-H. Wu, and C.-W. Tsao, "Using data mining techniques to predict hospitalization of hemodialysis patients," *Decision Support Systems*, vol. 50, no. 2, pp. 439–448, 2011.

133. NCCHD, "Births: data from the national community child health database," http://gov.wale s/statistics-and-research/births-national-community-child-health-database/?lang=en, 2014, retrieved at: 2015-02-15.

134. K. U. Rani, "Analysis of heart diseases dataset using neural network approach," *International Journal of Data Mining & Knowledge Management Process,* vol. 1, no. 5, 2011.

135. I. Y. Khan, P. Zope, and S. Suralkar, "Importance of artificial neural network in medical diagnosis disease like acute nephritis disease and heart disease," *International Journal of Engineering Science and Innovative Technology*, vol. 2, pp. 210–217, 2013.

136. H. F. Huang, G. S. Hu, and L. Zhu, "Sparse representation-based heartbeat classification using independent component analysis," *Journal of Medical Systems*, vol. 36, no. 3, pp. 1235–1247, 2012.

137. PREDICT-HD, "Predict-hd project," https://www.predict-hd.net/, 2015, retrieved at: 2015-04-10.

138. A. Bate, M. Lindquist, I. Edwards, S. Olsson, R. Orre, A. Lansner, and R. M. De Freitas, "A Bayesian neural network method for adverse drug reaction signal generation," *European Journal of Clinical Pharmacology*, vol. 54, no. 4, pp. 315–321, 1998.

139. S. K. Singh, A. Malik, A. Firoz, and V. Jha, "CDKD: a clinical database of kidney diseases," *BMC Nephrology*, vol. 13, no. 1, p. 23, 2012.

140. HSCIC, "Hospital episode statistics," http://www.hscic.gov.uk/hes, 2012, retrieved at: 2015-02-20.

141. D. Comaniciu, P. Meer, and D. J. Foran, "Image-guided decision support system for pathology," *Machine Vision and Applications*, vol. 11, no. 4, pp. 213–224, 1999.

142. F. Schnorrenberg, C. Pattichis, C. Schizas, and K. Kyriacou, "Content-based retrieval of breast cancer biopsy slides," *Technology and Health Care*, vol. 8, no. 5, pp. 291–297, 2000.

143. H. C. Akakin and M. N. Gurcan, "Content-based microscopic image retrieval system for multi-image queries," *IEEE Transactions on Information Technology in Biomedicine*, vol. 16, no. 4, pp. 758–769, 2012.

144. X. S. Zhou, S. Zillner, M. Moeller, M. Sintek, Y. Zhan, A. Krishnan, and A. Gupta, "Semantics and CBIR: a medical imaging perspective," in *Proceedings of the 2008 International Conference on Content-Based Image and Video Retrieval*, ACM, 2008, pp. 571–580.

145. H. Müller, A. Geissbühler, and P. Ruch, "Imageclef 2004: combining image and multilingual search for medical image retrieval," in *Multilingual Information Access for Text, Speech and Images*, vol. 3491. Springer, 2005, pp. 718–727.

146. G. Langs, A. Hanbury, B. Menze, and H. Müller, "Visceral: towards large data in medical imaging—challenges and directions," in *Medical Content-Based Retrieval for Clinical Decision Support*, vol. 7723. Springer, 2013, pp. 92–98.

147. A. Hanbury, H. Müller, G. Langs, and B. H. Menze, "Cloud-based evaluation framework for big data," in *FIA book 2013*, ser. Springer LNCS, 2013.

148. A. Andoni and P. Indyk, "Near-optimal hashing algorithms for approximate nearest neighbor in high dimensions," in *47th Annual IEEE Symposium on Foundations of Computer Science (FOCS'06)*, 2006, pp. 459–468.

149. W. Liu, J. Wang, S. Kumar, and S.-F. Chang, "Hashing with graphs," in *Proceedings of the 28th International Conference on Machine Learning (ICML-11)*, 2011, pp. 1–8.

150. Y. Weiss, A. Torralba, and R. Fergus, "Spectral hashing," in *Advances in Neural Information Processing Systems 21*, D. Koller, D. Schuurmans, Y. Bengio, and L. Bottou, Eds. Curran Associates, Inc., 2009, pp. 1753–1760. [Online]. http://papers.nips.cc/paper/3383-spectral-hashing.pdf

151. W. Liu, J. Wang, R. Ji, Y.-G. Jiang, and S.-F. Chang, "Supervised hashing with kernels," in *IEEE Conference on Computer Vision and Pattern Recognition (CVPR)*, 2012, pp. 2074–2081.

152. X. Zhang, W. Liu, M. Dundar, S. Badve, and S. Zhang, "Towards large-scale histopathological image analysis: hashing-based image retrieval," *IEEE Transaction on Medical Imaging*, vol. 34, no. 2, pp. 496–506, 2014.

153. X. Zhang, F. Xing, H. Su, L. Yang, and S. Zhang, "High-throughput histopathological image analysis via robust cell segmentation and hashing," *Medical Image Analysis*, vol. 26, no. 1, pp. 306–315, 2015.

154. X. Zhang, H. Su, L. Yang, and S. Zhang, "Fine-grained histopathological image analysis via robust segmentation and large-scale retrieval," in *Proceedings of the IEEE Conference on Computer Vision and Pattern Recognition*, 2015, pp. 5361–5368.

155. X. Zhang, L. Yang, W. Liu, H. Su, and S. Zhang, "Mining histopathological images via composite hashing and online learning," in *Medical Image Computing and Computer-Assisted Intervention–MICCAI 2014*, vol. 8674, Springer, 2014, pp. 479–486.

156. D. Nister and H. Stewenius, "Scalable recognition with a vocabulary tree," in *IEEE Computer Society Conference on Computer Vision and Pattern Recognition*, vol. 2, IEEE, 2006, pp. 2161–2168.

157. M. Jiang, S. Zhang, H. Li, and D. N. Metaxas, "Computer-aided diagnosis of mammographic masses using scalable image retrieval," *IEEE Transactions on Biomedical Engineering*, vol. 62, no. 2, pp. 783–792, 2015.

158. P. D. Clayton and G. Hripcsak, "Decision support in healthcare," *International Journal of Bio-Medical Computing*, vol. 39, no. 1, pp. 59–66, 1995.

159. K. Kawamoto, C. A. Houlihan, E. A. Balas, and D. F. Lobach, "Improving clinical practice using clinical decision support systems: a systematic review of trials to identify features critical to success," *BMJ*, vol. 330, no. 7494, p. 765, 2005.

160. L. Xiao, J. Hanover, and S. Mukherjee, "Big data enables clinical decision support in hospital settings," http://www.idc.com/getdoc. jsp?containerId=CN245651, 2014, retrieved at: 2015-03-09.

161. IBM, "IBM patient care and insights," http://www-03.ibm.com/software/products/en/IBM-care-management, 2015, retrieved at: 2015- 03-05.

162. F. Wang, J. Sun, and S. Ebadollahi, "Composite distance metric integration by leveraging multiple experts' inputs and its application in patient similarity assessment," *Statistical Analysis and Data Mining: The ASA Data Science Journal*, vol. 5, no. 1, pp. 54–69, 2012.

163. M. Lichman, "UCI machine learning repository," http://archive.ics.uci.edu/ml, 2013, retrieved at: 2015-08-03.

164. B. Wang, A. M. Mezlini, F. Demir, M. Fiume, Z. Tu, M. Brudno, B. Haibe-Kains, and A. Goldenberg, "Similarity network fusion for aggregating data types on a genomic scale," *Nature Methods*, vol. 11, no. 3, pp. 333–337, 2014.

165. D. A. Keim, "Information visualization and visual data mining," *IEEE Transactions on Visualization and Computer Graphics*, vol. 8, no. 1, pp. 1–8, 2002.

166. A. Tsymbal, M. Huber, S. Zillner, T. Hauer, and S. K. Zhou, "Visualizing patient similarity in clinical decision support," in *LWA*, Martin-Luther-University Halle-Wittenberg, 2007, pp. 304–311.

167. B. Shneiderman, "Tree visualization with tree-maps: 2-D space-filling approach," *ACM Transactions on Graphics (TOG)*, vol. 11, no. 1, pp. 92–99, 1992.

168. G. T. Toussaint, "The relative neighbourhood graph of a finite planar set," *Pattern Recognition*, vol. 12, no. 4, pp. 261–268, 1980.

169. R. G. Verhaak, M. A. Sanders, M. A. Bijl, R. Delwel, S. Horsman, M. J. Moorhouse, P. J. van der Spek, B. Löwenberg, and P. J. Valk, "Heatmapper: powerful combined visualization of gene expression profile correlations, genotypes, phenotypes and sample characteristics," *BMC Bioinformatics*, vol. 7, no. 1, p. 337, 2006.

170. N. V. Chawla and D. A. Davis, "Bringing big data to personalized healthcare: a patient-centered framework," *Journal of General Internal Medicine*, vol. 28, no. 3, pp. 660–665, 2013.

171. M. E. Johnston, K. B. Langton, R. B. Haynes, and A. Mathieu, "Effects of computer-based clinical decision support systems on clinician performance and patient outcome: a critical appraisal of research," *Annals of Internal Medicine*, vol. 120, no. 2, pp. 135–142, 1994.

172. B. Menze, G. Langs, A. Montillo, M. Kelm, H. Müller, S. Zhang, W. T. Cai, and D. Metaxas, Medical computer vision: Algorithms for big data: International workshop, MCV 2014, held in conjunction with MICCAI 2014, Cambridge, MA, USA, September 18, 2014, revised selected papers. Springer, 2014, vol. 8848.

173. T. Schlegl, J. Ofner, and G. Langs, "Unsupervised pre-training across image domains improves lung tissue classification," in *Medical Computer Vision: Algorithms for Big Data*, vol. 8848. Springer, 2014, pp. 82–93.

174. O. A. J. del Toro and H. Müller, "Hierarchic multi–atlas based segmentation for anatomical structures: evaluation in the visceral anatomy benchmarks," in *Medical Computer Vision: Algorithms for Big Data*. Springer, 2014, pp. 189–200.

175. T. Neumuth, P. Jannin, G. Strauss, J. Meixensberger, and O. Burgert, "Validation of knowledge acquisition for surgical process models," *Journal of the American Medical Informatics Association*, vol. 16, no. 1, pp. 72–80, 2009.

176. P. Jannin and X. Morandi, "Surgical models for computer-assisted neurosurgery," *Neuroimage*, vol. 37, no. 3, pp. 783–791, 2007.

177. F. Lalys, L. Riffaud, D. Bouget, and P. Jannin, "A framework for the recognition of high-level surgical tasks from video images for cataract surgeries," *IEEE Transactions on Biomedical Engineering*, vol. 59, no. 4, pp. 966–976, 2012.

178. A. Nara, K. Izumi, H. Iseki, T. Suzuki, K. Nambu, and Y. Sakurai, "Surgical workflow monitoring based on trajectory data mining," in *New Frontiers in Artificial Intelligence*, vol. 6797. Springer, 2011, pp. 283–291.

179. B. Bhatia, T. Oates, Y. Xiao, and P. Hu, "Real-time identification of operating room state from video," *AAAI*, vol. 2, 2007, pp. 1761–1766.

180. H. Müller and J.-C. Freytag, "Problems, methods, and challenges in comprehensive data cleansing," Humboldt-Universitt zu Berlin, Technical Report, 2003, professoren des Inst. Für Informatik.

181. A. Holzinger and K.-M. Simonic, *Information Quality in e-Health*. Springer, 2011.

182. R. L. Richesson and J. Krischer, "Data standards in clinical research: gaps, overlaps, challenges and future directions," *Journal of the American Medical Informatics Association*, vol. 14, no. 6, pp. 687–696, 2007.

183. K. Kambatla, G. Kollias, V. Kumar, and A. Grama, "Trends in big data analytics," *Journal of Parallel and Distributed Computing*, vol. 74, no. 7, pp. 2561–2573, 2014.

184. A. Holzinger, M. Dehmer, and I. Jurisica, "Knowledge discovery and interactive data mining in bioinformatics-state-of-the-art, future challenges and research directions," *BMC Bioinformatics*, vol. 15, no. Suppl 6, p. I1, 2014.

185. G. Fitzmaurice, M. Davidian, G. Verbeke, and G. Molenberghs, *Longitudinal Data Analysis*. CRC Press, 2008, Handbooks of Modern Statistical Methods. New York: Chapman & Hall.

186. J. W. Twisk, "Longitudinal data analysis. a comparison between generalized estimating equations and random coefficient analysis," *European Journal of Epidemiology*, vol. 19, no. 8, pp. 769–776, 2004.

187. A. Labrinidis and H. Jagadish, "Challenges and opportunities with big data," *Proceedings of the VLDB Endowment*, vol. 5, no. 12, pp. 2032–2033, 2012.

188. P. Russom, "Big data analytics," TDWI Best Practices Report, Fourth Quarter, 2011.

189. E. Weippl, A. Holzinger, and A. M. Tjoa, "Security aspects of ubiquitous computing in health care," *e & i Elektrotechnik und Informationstechnik*, vol. 123, no. 4, pp. 156–161, 2006.

190. P. Kieseberg, J. Schantl, P. Frühwirt, E. Weippl, and A. Holzinger, "Witnesses for the doctor in the loop," in *Brain Informatics and Health*. Springer, 2015, pp. 369–378.

191. B. W. Wong, K. Xu, and A. Holzinger, "Interactive visualization for information analysis in medical diagnosis," in *Information Quality in e-Health*. Lecture Notes in Computer Science, vol. 7058. Springer, 2011, pp. 109–120.

192. F. Jeanquartier and A. Holzinger, "On visual analytics and evaluation in cell physiology: a case study," in *Availability, Reliability, and Security in Information Systems and HCI*. Springer, 2013, pp. 495–502.

193. M. Chen, S. Mao, and Y. Liu, "Big data: A survey," *Mobile Networks and Applications*, vol. 19, no. 2, pp. 171–209, 2014.

194. A. Holzinger, "Interactive machine learning for health informatics: when do we need the human-in-the-loop?" *Brain Informatics*, vol. 3, pp. 1–13, 2016.

Chapter 6

Fast Dual Optimization for Medical Image Segmentation

Jing Yuan, Ismail Ben Ayed, and Aaron Fenster

6.1 Introduction

During the last 20 years, convex optimization was successfully introduced as a powerful tool in image processing, computer vision and machine learning, which is mainly credited to the pioneering works from both theoretical and algorithmic studies [1–7], along with vast applications [8–13]. Typical applications include, for example, the total-variation-based image denoising [8, 12]

$$\min_{u} \int D(u - f)\, dx + a \int |\nabla u|\, dx,$$

where $D(\cdot)$ is a convex penalty function, e.g. L_1 or L_2 norm; the sparsity-based image reconstruction [3]

$$\min_{u} \int D(Au - f)\, dx + a \int |u|\, dx,$$

where A is some linear operator and its L_1 regularization term approximates the nonconvex sparsity penalty; and the min-cut image segmentation problem which is initially

143

formulated with a challenging combinatorial constraint $u(x) \in \{0,1\}$ but relaxed and solved by its convex version [6, 7]:

$$\min_{u(x) \in [0,1]} \int u(x) C(x) dx + a \int |\nabla u| dx, \tag{6.1}$$

where its binary constraint is relaxed as $u(x) \in [0,1]$, hence results in a convex optimization problem [6].

In this work, we consider the optimization problems related to medical image segmentation as the minimization of a finite sum of convex function terms:

$$\min_{u} \ f_1(u) + \cdots + f_n(u), \tag{6.2}$$

which actually includes the convex constrained optimization problem as one special case such that the convex constraint set C on the unknown function $u(x) \in C$ can be reformulated by adding its convex characteristic function

$$\chi_C(u) := \begin{cases} 0, & x \in C \\ +\infty, & x \in C \end{cases}.$$

into the energy function of Expression 6.2.

Given the very high dimension of the solution u, which is the usual case in medical image segmentation where the input image volume often includes over millions of pixels, the iterative first-order gradient-descent schemes play the central role in building up a practical algorithmic implementation, which typically has an affordable computational cost per iteration along with proved iteration complexity. In this perspective, the duality of each convex function term $f_i(u) = \langle u, p_i \rangle - f_i^*(p_i)$ provides one of the most powerful tools in both analyzing and developing such first-order iterative algorithms, where the introduced new dual variable p_i for each function term f_i just represents the first-order gradient of $f_i(u)$ implicitly; it brings two equivalent optimization models, aka the *primal-dual model*

$$\min_{u} \max_{p} \ \underbrace{\langle p_1 + \cdots + p_n, u \rangle - f_1^*(p_1) - \cdots - f_n^*(p_n)}_{\text{Lagrangian function } L(u,p)} \tag{6.3}$$

and the *dual model*

$$\begin{aligned} \max_{p} \quad & -f_1^*(p_1) - \cdots - f_n^*(p_n) \\ \text{s.t.} \quad & p_1 + \cdots + p_n = 0 \end{aligned} \tag{6.4}$$

to the studied convex minimization problem (Expression 6.2).

Comparing with the traditional first-order gradient-descent algorithms which directly evaluate the gradient of each function term at each iteration and improve the approximation of the optimum iteratively, the dual model (Expression 6.4) provides another expression to analyze the original convex optimization model (Expression 6.2) and delivers a novel point of view to design new first-order iterative algorithms, where the optimum u^* of Expression 6.2 just works as the optimal multiplier to the linear equality constraint as demonstrated in the Lagrangian function $L(u, p)$ of the primal-dual model (Expression 6.3) (see more details in Section 6.2). In practice, such a dual formulation-based approach enjoys great advantages in both mathematical analysis and algorithmic design: (a) each function term $f_i(p_i)$ of its energy function depends solely on an independent variable p_i, which naturally leads to an efficient splitting scheme to tackle the optimization problem in a simple separate-and-conquer way, or a stochastic descent scheme with low iteration-cost; (b) a unified algorithmic framework to compute the optimum multiplier u^* can be developed by the *augmented Lagrangian method* (ALM), which involves two sequential steps at each iteration:

$$p^{k+1} := \arg\max_{p} L(u^k, p) - \frac{c^k}{2} \left\| p_1 + \cdots + p_n \right\|^2 , \qquad (6.5)$$

$$u^{k+1} = u^k - c^k \left(p_1^{k+1} + \cdots + p_n^{k+1} \right), \qquad (6.6)$$

which is capable of setting up high-performance parallel implementations under the same numerical perspective; (c) the equivalent dual model in (6.4) additionally brings new insights to facilitate analyzing its original model (6.2) and discovers close connections from distinct optimization topics (see Section 6.3 for details).

Organization: The contents of this work are organized in three parts: In Section 6.2, we introduce the main theories and algorithmic scheme used in convex optimization under a unified dual optimization framework; the dual optimization approach sets up new equivalent optimization formulations to the studied convex optimization model and derives a new unified multiplier-based algorithmic framework; in addition, some applications of image processing are presented as examples. In Section 6.3, we study the clinically motivated applications of medical image segmentation and show how the introduced dual optimization approach can easily integrate various prior data, which largely improves the accuracy and robustness of optimization solutions, into the ALM-based optimization algorithms with much less effort than tackling the original convex optimization formulations directly.

6.2 Dual Optimization Method and Applications to Image Segmentation

In this section, we consider the convex optimization problem (Expression 6.2) of minimizing the sum of multiple convex function terms, which generalizes a big spectrum of convex optimization models including the convex constrained optimization problem for which some convex constraint set $\mathcal{C}$ on $u(x) \in \mathcal{C}$ can be well imposed by adding its characteristic function $\chi_{\mathcal{C}}(u)$ as one function term in Expression 6.2.

6.2.1 Primal and Dual Models

As one powerful tool to analyze convex functions, the duality of a convex function was developed [14] and largely exploited in designing fast convex optimization algorithms [4] to image processing recently, such that each convex function $f_i(u)$ of Expression 6.2 can be equally represented by

$$f_i(u) = \max_{p_i} \langle u, p_i \rangle - f_i^*(p_i) \qquad (6.7)$$

where $f_i^*(p_i)$ is the corresponding conjugate function of $f_i(u)$ and $\langle u, p_i \rangle$ is the inner product of u and the dual variable p_i. Therefore, by simple computation, we have the following result.

Proposition 1 *The convex optimization problem (Expression 6.2) is mathematically equivalent to the linear equality constrained convex optimization problem:*

$$\max_p - f_1^*(p_1) - \cdots - f_n^*(p_n), \quad s.t. \ p_1 + \cdots + p_n = 0; \tag{6.8}$$

i.e. the dual optimization model (Expression 6.4).

Its proof comes from the following facts: First, summing up each conjugate Expression 6.7 of the convex function $f_i(u)$, $i = 1 \ldots n$ results in a Lagrangian formulation

$$\max_p \min_u L(u, p) := \langle p_1 + \cdots + p_n, u \rangle - f_1^*(p_1) - \cdots - f_n^*(p_n). \tag{6.9}$$

Given the convexity of $L(u, p)$ on each variable of u and p, the minimization and maximization procedures of Expression 6.9 are actually interchangeable [15]. Then, the minimization of Expression 6.9 over u leads to the vanishing of the inner product term $\langle p_1 + \cdots + p_n, u \rangle$ and the corresponding linear equality constraint $p_1 + \cdots + p_n = 0$. This, hence, gives rise to the dual model (6.8).

Actually, for each convex function $f_i(u)$, the optimum of p_i for its dual Expression 6.7 is nothing but its corresponding gradient or subgradient at u; therefore, the linear equality constraint $p_1 + \cdots + p_n = 0$ for the dual model (Expression 6.8) exactly represents the first-order optimal condition to the studied convex optimization problem (Expression 6.2), i.e. the sum of all gradients or subgradients vanishes:

$$\partial f_1(u) + \cdots + \partial f_n(u) = 0.$$

Additionally, we can further conclude that:

Corollary 2 *For the dual optimization problem (Expression 6.8), the optimum multiplier u^* to its linear equality constraint $p_1 + \cdots + p_n = 0$ is just the minimum of the original convex optimization problem (6.2).*

This is clear from the above proof to Proposition 1. Especially, this result establishes the basis of a novel algorithmic framework for a wide spectrum of convex optimization problems using the classical ALM [15, 16] (see Section 6.2.2 for more details).

Especially, each function term $f_i^*(p_i)$, $i = 1 \ldots n$ in the energy function of the dual model (6.8) solely depends on an independent variable p_i which is loosely correlated to the other variables by the linear equality constraint $p_1 + \cdots + p_n = 0$. This is in contrast to its original optimization model (Expression 6.2) whose energy function terms are interacted with each other due to the common unknown variable u. This provides a big advantage in developing splitting optimization algorithms, as shown in Section 6.2.2, to tackle the underlying convex optimization problem, particularly at a large scale.

6.2.2 Dual Optimization Method

Now we consider the linearly constrained convex optimization problem (Expression 6.8), i.e. the dual model, and the Corollary 2, such that the energy function $L(u, p)$ of the primal-dual model (Expression 6.9) is exactly the Lagrangian function of the linearly constrained dual model (Expression 6.8). Hence, the classical ALM [15, 16] provides an optimization framework to develop the corresponding algorithmic scheme.

For this, we define the associated augmented Lagrangian function

$$L_c(u,p) := \underbrace{-f_1^*(p_1) - \cdots - f_n^*(p_n) + \langle p_1 + \cdots + p_n, u \rangle}_{L(u,p)} - \frac{c}{2} \| p_1 + \cdots + p_n \|^2, \tag{6.10}$$

where c is the positive parameter.

Algorithm 1: Augmented Lagrangian Method-Based Algorithm

Initialize u^0 and $\left(p_1^0, \ldots, p_n^0 \right)$; for each iteration k we explore the following two steps

- Fix u^k, compute p^{k+1}:

$$\left(p_1^{k+1}, \ldots, p_n^{k+1} \right) = \arg\max_p L_{c^k} \left(u^k, p \right); \tag{6.11}$$

- Fix p^{k+1}, then update u^{k+1} by

$$u^{k+1} := u^k - c^k \left(p_1^{k+1} + p_2^{k+1} + \cdots + p_n^{k+1} \right) \tag{6.12}$$

Then, an ALM-based algorithm can be developed as shown in Algorithm 1, which explores two consecutive optimization steps (Expressions 6.11 and 6.12) over p and u correspondingly. The convergence of such ALM-based algorithms can be proved to obtain a linear rate of $O(1/N)$ [17].

Clearly, at each iteration, the main computing load is from the first optimization step (Expression 6.11). In practice, the optimization sub-problem (Expression 6.11) is often solved by tackling each p_i, $i = 1 \ldots n$, separately. More specifically, this can be implemented either in parallel such that

$$p_1^{k+1} := \arg\max \left\{ -f_1^*(p_1) - \frac{c^k}{2} \left\| p_1 + p_2^k + \cdots + p_n^k - \frac{u^k}{c^k} \right\|^2 \right\} \tag{6.13}$$

$$\ldots\ldots$$

$$p_n^{k+1} := \arg\max \left\{ -f_n^*(p_n) - \frac{c^k}{2} \left\| p_1^k + p_2^k + \cdots + p_n - \frac{u^k}{c^k} \right\|^2 \right\}, \tag{6.14}$$

or in a sequential way such that

$$p_1^{k+1} := \arg\max \left\{ -f_1^*(p_1) - \frac{c^k}{2} \left\| p_1 + p_2^k + \cdots + p_n^k - \frac{u^k}{c^k} \right\|^2 \right\} \tag{6.15}$$

......

$$p_n^{k+1} := \arg\max\left\{-f_n^*(p_n) - \frac{c^k}{2}\left\|p_1^{k+1} + p_2^{k+1} + \cdots + p_n - \frac{u^k}{c^k}\right\|^2\right\}. \qquad (6.16)$$

Particularly, every optimization sub-problem of Expressions 6.13–6.14 or Expressions 6.15–6.16 is approximately solved by one step of gradient-descent in order to alleviate computational complexities of each iteration.

6.2.3 Application to Min-Cut-Based Image Segmentation

During the last number of decades, the min-cut model was developed to become one of the most successful models for image segmentation [18, 19], which has been well studied in the discrete graph setting and can be efficiently solved by the scheme of maximizing flows. In fact, such a min-cut model can be also formulated in a spatially continuous setting, i.e. the *spatially continuous min-cut problem* [6]:

$$\min_{u(x)\in\{0,1\}} \int_\Omega \{(1-u)C_t + u\,C_s\}(x)\,dx + a\int_\Omega |\nabla u|\,dx, \qquad (6.17)$$

where $C_s(x)$ and $C_t(x)$ are the cost functions such that, for each pixel $x \in \Omega$, $C_s(x)$ and $C_t(x)$ give the costs to label x as 'foreground' and 'background' respectively. The optimum $u^*(x)$ to the combinatorial optimization problem (Expression 6.17) defines the optimal foreground segmentation region S such that $u^*(x) = 1$ for any $x \in S$, and the background segmentation region $\Omega\backslash S$ otherwise.

Chan et al. [6] proved that the challenging non-convex combinatorial optimization problem (Expression 6.17) can be solved globally by computing its convex relaxation model, i.e. the *convex relaxed min-cut model*:

$$\min_{u(x)\in[0,1]} \int_\Omega \{(1-u)C_s + u\,C_t\}\,dx + a\int_\Omega |\nabla u|\,dx, \qquad (6.18)$$

while thresholding the optimum of (Expression 6.18) with any parameter $\beta \in (0,1)$. Hence, the difficult combinatorial optimization problem (Expression 6.17) can be exactly solved by a convex minimization problem (Expression 6.18) instead.

6.2.3.1 Primal and Dual Optimization Models

Yuan et al. [7, 20] proposed that the convex relaxed min-cut model (Expression 6.18) can be equivalently reformulated by its dual model, i.e. the *continuous max-flow model*:

$$\max_{p_s,p_t,p} \int_\Omega p_s(x)\,dx \qquad (6.19)$$

$$\text{s.t. } |p(x)| \le a, \quad p_s(x) \le C_s(x), \quad p_t(x) \le C_t(x); \qquad (6.20)$$

$$(\mathbb{D}\text{iv } p - p_t + p_s)(x) = 0. \qquad (6.21)$$

In addition, an efficient ALM-based algorithm to the linear equality, aka the flow conservation constraint (Expression 6.21), can be constructed with high numerical performance.

To see this, the key idea is to observe that

$$f(q) = \max_{p \leq C} p \cdot q = \begin{cases} q \cdot C & \text{if } q \geq 0 \\ \infty & \text{if } q < 0 \end{cases}; \qquad (6.22)$$

indeed, when $q < 0$, p can be chosen to be negative infinity in order to maximize the value $p \cdot q$, which results in $f(q) = +\infty$; furthermore, when $q \geq 0$, $p = C$ and $f(q)$ reaches maximum $q \cdot C$.

Clearly, the function $f(q)$ given by Expression 6.22 provides a prototype to see that, for each $x \in \Omega$,

$$\max_{p_s(x) \leq C_s(x)} (1 - u(x)) \cdot p_s(x) = \begin{cases} (1 - u(x)) \cdot C_s(x) & \text{if } (1 - u(x)) \geq 0 \\ \infty & \text{if } (1 - u(x)) < 0 \end{cases} \qquad (6.23)$$

and

$$\max_{p_t(x) \leq C_t(x)} u(x) \cdot p_t(x) = \begin{cases} u(x) \cdot C_t(x) & \text{if } u(x) \geq 0 \\ \infty & \text{if } u(x) < 0 \end{cases}. \qquad (6.24)$$

Therefore, the first energy function term of (Expression 6.18):

$$\int_\Omega \{(1 - u)C_s + u\, C_t\}\, dx, \quad \text{along with } 1 - u(x) \geq 0, u(x) \geq 0,$$

i.e. $u(x) \in [0,1]$ can be equally expressed by

$$\max_{p_s, p_t} \int_\Omega \{(1 - u)p_s + u\, p_t\}\, dx, \quad \text{s.t. } p_s(x) \leq C_s(x), \; p_t(x) \leq C_t(x).$$

Given the well-known dual formulation of the total-variation function, the convex relaxed min-cut model (6.18) can be identically reformulated as

$$\max_{p_s, p_t, p} \min_u \int_\Omega p_s\, dx + \int_\Omega u(\mathbb{D}\text{iv}\, p - p_s + p_t)\, dx\} \qquad (6.25)$$

$$\text{s.t.} \qquad p_s(x) \leq C_s(x), \quad p_t(x) \leq C_t(x), \quad |p(x)| \leq a.$$

We first minimize the energy function of Expression 6.25 over $u(x)$, which results in the continuous max-flow model (Expression 6.19). Then we can summarize the above discussions by the following proposition:

Proposition 3 *The continuous max-flow model (Expression 6.19), the primal-dual model (Expression 6.25) and the convex relaxed min-cut model (Expression 6.78) are equivalent to each other.*

Actually, the continuous max-flow model (Expression 6.19) is directly analogous to the graph-based max-flow for the *s-t* cut: Given the continuous image domain Ω, we assume there are two extra terminals, the source s and the sink t. We assume that for each image position $x \in \Omega$, there are three concerning flows: The source flow $p_s(x) \in \mathbb{R}$ directed from the source s to $x \in \Omega$, the sink flow $p_t(x) \in \mathbb{R}$ directed from x to the sink t

and the spatial flow field $p(x) \in \mathbb{R}^2$. The total source flow $\int_\Omega p_s(x)\,dx$ is maximized, provided that three flow fields are constrained by capacities (Expression 6.20) and conserved by the linear equality constraint (Expression 6.21) simultaneously.

In addition, one can also prove that:

Proposition 4 *Let p_s^*, p_t^*, p^* and $u^*(x)$ be a global optimum of the primal-dual model (Expression 6.25). Then each ℓ-upper level set $S^\ell := \{x \mid u^*(x) \geq \ell, \ell \in (0,1]\}$, $\ell \in (0,1]$, of $u^*(x)$ and the indicator function u^ℓ*

$$u^\ell(x) := \begin{cases} 1, & u^*(x) \geq \ell \\ 0, & u^*(x) < \ell \end{cases},$$

is a global binary solution of the combinatorial min-cut problem (Expression 6.17). Moreover, each cut energy given by S^ℓ has the same energy as its optimal max-flow energy, i.e.

$$P\left(p_s^*, p_t^*, p^*\right) = \int_\Omega p_s^*(x)\,dx.$$

See [7, 20] for its proof in details. In addition, it is easy to see the optimum $u^*(x)$ to the convex relaxed min-cut model (Expression 6.78) is exactly the optimum multiplier to the linear equality constraint (Expression 6.21).

6.2.3.2　Duality-Based Algorithm

Correspondingly, we define the augmented Lagrangian function

$$L_c(u, p_s, p_t, p) = \int_\Omega p_s(x)\,dx + \langle u, \mathbb{D}\mathrm{iv}\ p - p_s + p_t \rangle - \frac{c}{2} \| \mathbb{D}\mathrm{iv}\ p - p_s + p_t \|^2 \qquad (6.26)$$

and derive the ALM-based continuous max-flow algorithm, see Algorithm 2. An example of the foreground-background segmentation result of a 3D cardiac ultrasound image is shown by Figure 6.1 (a), which is computed by the proposed continuous max-flow model (Expression 6.19) based Algorithm 2 (code is available at [21]).

Algorithm 2: ALM-Based Continuous Max-Flow

Initialize u^0 and $\left(p_s^0, p_t^0, p^0\right)$; for each iteration k we explore the following two steps

- Fix u^k, compute $\left(p_s^{k+1}, p_t^{k+1}, p^{k+1}\right)$:

$$\left(p_s^{k+1}, p_t^{k+1}, p^{k+1}\right) = \arg \max_{p_s, p_t, p} L_{c^k}\left(u^k, p_s, p_t, p\right), \quad \text{s.t. (20) and (20) (20),} \qquad (6.27)$$

provided the augmented Lagrangian function $L_c(u, p_s, p_t, p)$ in Expression 6.26.

The maximization of $L_{c^k}\left(u^k, p_s, p_t, p\right)$ over (p_s, p_t, p) can be implemented by the following three sub-steps:

- Optimizing p by fixing the other variables

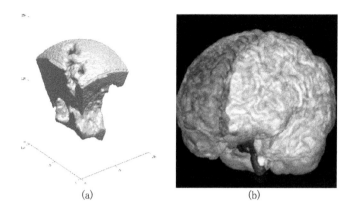

(a) (b)

FIGURE 6.1: (a) Foreground-background segmentation result of a 3D cardiac ultrasound image, computed by the continuous max-flow model (6.19) based Algorithm 2 (code is available at [21]). (b) Multiphase segmentation of a 3D brain CT image, computed by the continuous max-flow model (6.19) based Algorithm 2 (code is available at [23]).

$$p^{k+1} := \arg\max_{\|p\|_\infty \le a} L_{c^k}\left(u^k, p_s^k, p_t^k, p\right).$$

$$:= \arg\max_{\|p\|_\infty \le a} -\frac{c^k}{2}\left\|\mathrm{Div}\ p - F^k\right\|^2,$$

(6.28)

where $F^k(x)$ is a fixed variable. This problem can either be solved iteratively by Chambolle's projection algorithm [2] or approximately by one step of Expression 6.30.

- Optimizing p_s by fixing the other variables

$$p_s^{k+1} := \arg\max_{p_s(x) \le C_s(x)} L_{c^k}\left(u^k, p_s, p_t^k, p^{k+1}\right)$$

$$:= \arg\max_{p_s(x) \le C_s(x)} \int_\Omega p_s\ dx - \frac{c^k}{2}\left\|p_s - G^k\right\|^2$$

where $G^k(x)$ is a fixed variable and optimizing p_s can be easily computed at each $x \in \Omega$ pointwise;

- Optimizing p_t by fixing the other variables

$$p_t^{k+1} := \arg\max_{p_t(x) \le C_t(x)} L_{c^k}\left(u^k, p_s^{k+1}, p_t, p^{k+1}\right)$$

$$:= \arg\max_{p_t(x) \in C_t(x)} -\frac{c^k}{2}\left\|p_t - H^k\right\|^2,$$

where $H^k(x)$ is a fixed variable and optimizing p_t can be simply solved by

$$p_t(x) = \min\left(H^k(x), C_t(x)\right).$$

- Fix $\left(p_s^{k+1}, p_t^{k+1}, p^{k+1}\right)$, then update u^{k+1} by

$$u^{k+1} := u^k - c^k \left(\mathbb{D}\mathrm{iv}\ p^{k+1} - p_s^{k+1} + p_t^{k+1}\right). \tag{6.29}$$

Algorithm 2 is also based on the alternating direction method of multipliers. Convergence can be validated by optimization theories. In practice, the sub-optimization problem (Expression 6.28) can also be solved by one step of the following iterative procedure:

$$p^{k+1} = \Pi_a \left(p^k + c\nabla\left(\mathrm{div}\ p^k - F^k\right)\right) \tag{6.30}$$

where Π_a is the projection onto the convex set $C_a = \{q \mid \| q \|_\infty \leq a\}$. This requires much less computational effort.

6.2.4 Application to Potts-Model-Based Image Segmentation

For multiphase image segmentation, *Pott model* is used as the basis to formulate the associated mathematical model [18, 19] by minimizing the following energy function

$$\min_u \sum_{i=1}^n \int_\Omega u_i(x)\ \rho(l_i, x)\ dx + a \sum_{i=1}^n \int_\Omega |\nabla u_i| dx \tag{6.31}$$

subject to

$$\sum_{i=1}^n u_i(x) = 1, \quad u_i(x) \in \{0,1\}, i = 1 \ldots n, \quad \forall x \in \Omega, \tag{6.32}$$

where $\rho(l_i, x)$, $i = 1 \ldots n$, are the cost functions: For each pixel $x \in \Omega$, $\rho(l_i, x)$ gives the cost to label x as the segmentation region i. Potts model seeks the optimum labeling function $u_i^*(x)$, $i = 1 \ldots n$, to the combinatorial optimization problem (Expression 6.31), which defines the segmentation region Ω_i such that $u_i^*(x) = 1$ for any $x \in \Omega_i$. Clearly, the linear equality constraint $u_1(x) + \cdots + u_n(x) = 1$ states that each pixel x belongs to a single segmentation region.

Similar to the convex relaxed min-cut model (Expression 6.78), we can simply relax each binary constraint $u_i(x) \in \{0,1\}$ in Expression 6.32 to the convex set $u_i(x) \in [0,1]$, then formulate the convex relaxed optimization problem of the Potts model (Expression 6.31) as

$$\min_{u \in S} \sum_{i=1}^n \int_\Omega u_i(x)\ \rho(l_i, x)\ dx + a \sum_{i=1}^n \int_\Omega |\nabla u_i| dx \tag{6.33}$$

where

$$S = \left\{ u(x) \mid (u_1(x), \ldots, u_n(x)) \in \Delta_n^+, \ \forall x \in \Omega \right\}. \tag{6.34}$$

Δ_n^+ is the simplex set in the space $\mathbb{R}^n$.

6.2.4.1 Primal and Dual Models

By simple variational analysis [22], the convex relaxed Potts model is equivalent to the following primal-dual formulation

$$\max_{p_s, p, q} \min_u \underbrace{\int_\Omega p_s \, dx + \sum_{i=1}^n \int_\Omega u_i (p_i - p_s + \mathbb{D}\mathrm{iv}\, q_i) \, dx}_{L(u, p_s, p, q)} \tag{6.35}$$

$$\text{s.t.} \quad p_i(x) \le \rho(\ell_i, x), \quad |q_i(x)| \le \alpha, \quad i = 1 \ldots n$$

To see this, we apply Expression 6.22 and directly obtain

$$\max_{p_i(x) \le \rho(l_i, x)} u_i(x) \cdot p_i(x) = \begin{cases} u_i(x) \cdot \rho(l_i, x) & \text{if } u_i(x) \ge 0 \\ \infty & \text{if } u_i(x) < 0 \end{cases}. \tag{6.36}$$

Given the free $p_s(x)$, we have

$$\max_{p_s(x)} p_s(x) \cdot \left(1 - \sum_{i=1}^n u_i(x) \right) \Rightarrow 1 - \sum_{i=1}^n u_i(x) = 0.$$

Therefore, the energy function term of Expression 6.33:

$$\sum_{i=1}^n \int_\Omega u_i(x)\, \rho(l_i, x)\, dx, \quad \text{along with } u(x) \in S \text{ for (6.34)}$$

can be identically written as

$$\max_{p_s, p} \int_\Omega p_s \, dx + \sum_{i=1}^n \int_\Omega u_i (p_i - p_s) \, dx, \quad \text{s.t.} \quad p_i(x) \le \rho(\ell_i, x), \quad i = 1 \ldots n.$$

Given the dual expression for each total-variation term of Expression 6.33, we can then prove the equivalence between the primal-dual formulation (Expression 6.35) and the convex relaxed Potts model (Expression 6.33).

Minimizing the energy function of Expression 6.35 over the free variable $u_i(x)$, it is easy to obtain its equivalent dual formulation, i.e. the *continuous max-flow model*, for the convex relaxed Potts model (Expression 6.33) such that

$$\max_{p_s, p, q} \int_\Omega p_s \, dx \tag{6.37}$$

subject to

$$|q_i(x)| \le a, \quad p_i(x) \le \rho(\ell_i, x), \quad i = 1 \ldots n; \qquad (6.38)$$

$$(\mathbb{D}\mathrm{iv}\, q_i - p_s + p_i)(x) = 0, \quad i = 1 \ldots n. \qquad (6.39)$$

Summarizing the above facts, we end up with the following proposition:

Proposition 5 *The continuous max-flow model (Expression 6.37), the primal-dual model (Expression 6.35) and the convex relaxed Potts model (Expression 6.33) are equivalent to each other.*

Actually, the continuous max-flow model described by Expression 6.37 can be clearly explained in terms of "k-way" max-flows with one extra source s and n extra sinks such that:

1. n copies Ω_i, $i = 1 \ldots n$, of the image domain Ω are given in parallel.
2. For each position $x \in \Omega$, the source flow $p_s(x)$ tries to stream from the source s to x at each copy Ω_i, $i = 1 \ldots n$, of Ω. The source flow field is the same for each Ω_i, $i = 1 \ldots n$, i.e. $p_s(x)$ is unique.
3. For each position $x \in \Omega$, the sink flow $p_i(x)$, $i = 1 \ldots n$, is directed from x at the i-th copy Ω_i to the sink t. The n sink flow fields $p_i(x)$, $i = 1 \ldots n$, may be different.
4. The spatial flow fields $q_i(x)$, $i = 1 \ldots n$, are defined within each copy Ω_i, $i = 1 \ldots n$. They may also be different from each other.

One tries to maximize the total source flow $\int_\Omega p_s(x)\, dx$, provided that all sink flows and spatial flows are constrained by the capacities (Expression 6.38) and conserved by the linear equality constraints (Expression 6.39) simultaneously.

6.2.4.2 Duality-Based Algorithm

In view of the Lagrangian function $L(p_s, p, q, u)$ given in Expression 6.35, we define its corresponding augmented Lagrangian function

$$L_c(u, p_s, p, q) = \underbrace{\int_\Omega p_s\, dx + \sum_{i=1}^n \langle u_i, p_i - p_s + \mathbb{D}\mathrm{iv}\, q_i \rangle}_{L(p_s, p, q, u)} - \frac{c}{2} \sum_{i=1}^n \|p_i - p_s + \mathbb{D}\mathrm{iv}\, q_i\|^2 \qquad (6.40)$$

and derive the ALM-based continuous max-flow algorithm, see Algorithm 3. An example of the multiphase segmentation result of a 3D brain CT image is shown by Figure 6.1 (b), which is computed by the proposed continuous max-flow model (Expression 6.37)-based Algorithm 3 (code is available at [23]).

Algorithm 3: ALM-Based Continuous Max-Flow

Initialize u^0 and $\left(p_s^0, p_t^0, p^0\right)$; for each iteration k we explore the following two steps

- Fix u^k, compute $\left(p_s^{k+1}, p^{k+1}, q^{k+1}\right)$:

$$\left(p_s^{k+1}, p^{k+1}, q^{k+1}\right) = \arg\max_{p_s, p, q} L_{c^k}\left(u^k, p_s, p, q\right), \quad \text{s.t. (6.38)}, \tag{6.41}$$

provided the augmented Lagrangian function $L_c(u, p_s, p, q)$ in Expression 6.40.

The maximization of $L_{c^k}(u^k, p_s, p, q)$ over (p_s, p, q) can implemented by the following three sub-steps:

- Optimize spatial flows q_i, $i = 1 \dots n$, by fixing the other variables:

$$q_i^{k+1} := \arg\max_{|q_i|_\infty \leq a} -\frac{c^k}{2}\left\|\mathbb{D}\text{iv } q_i + \left(p_i^k - p_s^k - u_i^k / c^k\right)\right\|^2, \tag{6.42}$$

which can be solved by Chambolle's projection algorithm [2].

- Optimize sink flows p_i, $i = 1 \dots n$, by fixing the other variables

$$p_i^{k+1} := \arg\max_{p_i(x) \leq \rho(\ell_i, x)} -\frac{c^k}{2}\left\|p_i + \left(\mathbb{D}\text{iv } q_i^{k+1} - p_s^k - u_i^k / c^k\right)\right\|^2, \tag{6.43}$$

which can be computed at each $x \in \Omega$ in a closed form.

- Optimize the source flow p_s by fixing the other variables

$$p_s^{k+1} := \arg\max_{p_s} \int_\Omega p_s \, dx - \frac{c^k}{2}\sum_{i=1}^n\left\|p_s - \left(p_i^{k+1} + \mathbb{D}\text{iv } q_i^{k+1} - u_i^k / c^k\right)\right\|^2 \tag{6.44}$$

which can be obtained in a closed form.

- Fix $\left(p_s^{k+1}, p^{k+1}, q^{k+1}\right)$, then update u^{k+1} by

$$u_i^{k+1} := u_i^k - c^k\left(\mathbb{D}\text{iv } q_i^{k+1} - p_s^{k+1} + p_i^{k+1}\right), \quad i = 1 \dots n. \tag{6.45}$$

6.3 Medical Image Segmentation Applications

Medical image segmentation is often much more challenging than segmenting camera photos, since medical imaging data usually suffers from low image quality, loss of imaging information, high inhomogeneity of intensities and wrong imaging signals recorded, etc. Prior knowledge about target regions is thus incorporated into the related optimization models of medical image segmentation so as to improve the accuracy and robustness of segmentation results and reduce manual efforts and intra- and inter-observer variabilities. In addition, the input 3D or 4D medical images often have a big data volume; therefore, optimization algorithms with low iteration-complexity are appreciated in practice. With these respects, dual optimization approaches gained big successes in many applications of medical image segmentation [24–36], etc. In the following section, we will see a spectrum of priors can be easily integrated into the introduced dual optimization framework without adding big efforts in numerics.

6.3.1 Medical Image Segmentation with Volume-Preserving Prior

Medical image data often have low image quality, for example, the prostate transrectal ultrasound (TRUS) images (shown in Figure 6.2 (a) and (b)) usually with strong US speckles and shadowing due to calcifications, missing edges or texture similarities between the inner and outer regions of prostate. For segmenting such medical images, the volume information about the interesting object region provides a global description for the image segmentation task [35]; on the other hand, such knowledge can be easily obtained in most cases by learning the given training images or the other information sources.

To impose preserving the specified volume $\mathcal{V}$ in the continuous min-cut model (Expression 6.78), we penalize the difference between the volume of the target segmentation region and $\mathcal{V}$ such that

$$\min_{u(x)\in[0,1]} \int_\Omega \{(1-u)C_s + u\,C_t\}\,dx + a\int_\Omega |\nabla u|\,dx + \gamma\left|\mathcal{V} - \int_\Omega u\,dx\right|. \qquad (6.46)$$

By means of the conjugate expression of the absolute function such that $\gamma|v| = \max_{r\in[-\gamma,\gamma]} r \cdot v$, we have

$$\gamma\left|\mathcal{V} - \int_\Omega u\,dx\right| = \max_{r\in[-\gamma,\gamma]} r\left(\mathcal{V} - \int_\Omega u\,dx\right).$$

With variational analysis, it can be provided that the volume-preserving min-cut model (Expression 6.46) can be equally represented by [35]

$$\max_{p_s,p_t,p,\lambda} \int_\Omega p_s(x)\,dx + r\mathcal{V} \qquad (6.47)$$

$$s.t. \ \ |p(x)| \le a, \quad p_s(x) \le C_s(x), \quad p_t(x) \le C_t(x), \quad r \in [-\gamma,\gamma]; \qquad (6.48)$$

$$(\mathbb{D}\mathrm{iv}\ p - p_s + p_t)(x) - r = 0. \qquad (6.49)$$

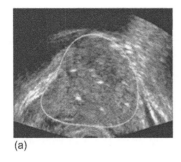
(a)
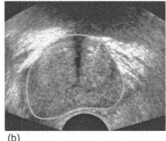
(b)
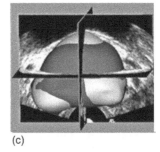
(c)

FIGURE 6.2: Example of 3D prostate TRUS segmentation: (a) sagittal view; (b) coronal view; (c) segmentation result (green surface) overlapped with manual segmentation (red surface).

In contrast to the linear equality constraint (Expression 6.21), i.e. the exact flow balance constraint in the maximal flow setting (Expression 6.19), such exact flow balance constraint is relaxed to be within a range of $r \in [-\gamma, \gamma]$ as shown in Expression 6.49. In addition, the value r is penalized in maximum flow configuration as shown in Expression 6.47.

Under such a dual optimization perspective, the optimum labeling $u^*(x)$ to Expression 6.46 works just as the optimal multiplier to the linear equality constraint (Expression 6.49), i.e. $(\mathbb{D}\mathrm{iv}\ p - p_s + p_t)(x) - r = 0$ and $r \in [-\gamma, \gamma]$.

As in Expression 6.26, we define the augmented Lagrangian function w.r.t. Expression 6.47

$$L_c(u, p_s, p_t, p, r) = \int_\Omega p_s(x)\ dx + \langle u, \mathbb{D}\mathrm{iv}\ p - p_s + p_t - r \rangle$$
$$- \frac{c}{2} \| \mathbb{D}\mathrm{iv}\ p - p_s + p_t - r \|^2, \tag{6.50}$$

and derive the related ALM-based algorithm, see Algorithm 1.

The example of segmenting 3D prostate TRUS images demonstrated, as shown in Figure 6.2, the segmentation results with the volume-preserving prior (Expression 6.46) are significantly better than the results without the volume prior from DSC $78.3 \pm 7.4\%$ to $89.5 \pm 2.4\%$ in DSC [35].

Algorithm 4: ALM-Based Segmentation Algorithm with Volume-Preserving Prior

Initialize u^0 and (p_s^0, p_t^0, p^0, r^0); for each iteration k we explore the following two steps

- Fix u^k, compute $(p_s^{k+1}, p_t^{k+1}, p^{k+1}, r^{k+1})$:

$$\left(p_s^{k+1}, p_t^{k+1}, p^{k+1}, r^{k+1}\right) = \arg \max_{p_s, p_t, p, r} L_{c^k}(u^k, p_s, p_t, p, r), \quad \text{s.t. (6.48)} \tag{6.51}$$

provided the augmented Lagrangian function $L_c(u, p_s, p_t, p, r)$ in (6.50).

- Fix $\left(p_s^{k+1}, p_t^{k+1}, p^{k+1}, r^{k+1}\right)$, then update u^{k+1} by

$$u^{k+1} := u^k - c^k \left(\mathbb{D}\mathrm{iv}\ p^{k+1} - p_s^{k+1} + p_t^{k+1} - r^{k+1}\right). \tag{6.52}$$

6.3.2 Medical Image Segmentation with Compactness Priors

The star-shape prior is a powerful description of region shapes, which forces the segmentation region to be compact, i.e. a single region without any cavity, namely the compactness prior. Usually, the star-shape prior is defined with respect to a center point O (see Figure 6.3 (b)): An object has a star-shape if for any pixel x inside the object, all points on the straight line between the center O and x also lie inside the object; in other words, the object boundary can only pass any radial line starting from the origin O one single time.

To formulate such a compactness prior, let $d_O(x)$ be the distance map with respect to the origin point O and $e(x) = \nabla d_O(x)$. Then the compactness prior can be defined as

$$(\nabla u \cdot e)(x) \geq 0. \tag{6.53}$$

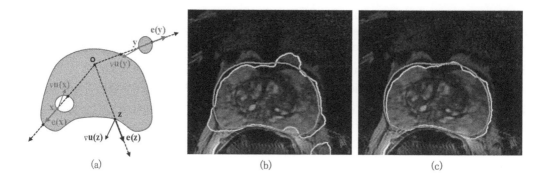

(a)   (b)   (c)

FIGURE 6.3: (a) Illustration of compactness (star-shape) prior; (b) the segmentation result of a 3D prostate MRI with compactness prior; (c) the segmentation result of a 3D prostate MRI without compactness prior.

Now we integrate the compactness prior (Expression 6.53) into the continuous min-cut model (Expression 6.78), which results in the following image segmentation model:

$$\min_{u(x)\in[0,1]} \int_\Omega \{(1-u)C_s + u\,C_t\}\,dx + a\int_\Omega |\nabla u|dx, \quad \text{s.t. } (\nabla u \cdot e)(x) \geq 0, \tag{6.54}$$

which is identical to

$$\max_{\lambda(x)\leq 0} \min_{u(x)\in[0,1]} \int_\Omega \{(1-u)C_s + u\,C_t\}\,dx + a\int_\Omega |\nabla u|dx + \int_\Omega \lambda(x)(\nabla u \cdot e)(x)\,dx. \tag{6.55}$$

Yuan et al. [32] showed that, with similar variational analysis as in [7, 20], the convex optimization model (Expression 6.54) is equivalent to the dual formulation below

$$\max_{p_s,p_t,p,\lambda} \int_\Omega p_s(x)\,dx \tag{6.56}$$

$$\text{s.t. } |p(x)| \leq a, \quad p_s(x) \leq C_s(x), \quad p_t(x) \leq C_t(x); \tag{6.57}$$

$$(\mathbb{D}\text{iv}\,(p - \lambda e) - p_s + p_t)(x) = 0, \quad \lambda(x) \leq 0. \tag{6.58}$$

With this perspective, the optimum labeling function $u^*(x)$ works just as the optimal multiplier to the linear equality constraint (Expression 6.58).

Clearly, the dual optimization model (Expression 6.56) is similar as the continuous max-flow model (Expression 6.19), with just an additional flow variable $\lambda(x)$ subject to the constraint $\lambda(x) \leq 0$. As in Expression 6.26, we define the augmented Lagrangian function w.r.t. Expression 6.56:

$$L_c(u, p_s, p_t, p, \lambda) = \int_\Omega p_s(x)\,dx + \langle u, \mathbb{D}\text{iv}\,(p - \lambda e) - p_s + p_t \rangle$$

$$-\frac{c}{2}\|\mathbb{D}\text{iv}\,(p - \lambda e) - p_s + p_t\|^2, \tag{6.59}$$

and derive the related ALM-based algorithm, see Algorithm 3.

An example of the segmentation of a 3D prostate MR image is shown by Figures 6.3b and 6.3c, with and without the compactness prior respectively. It is easy to see that some segmentation bias region is introduced in the result, which is in contrast to the segmentation result with the compactness prior.

Algorithm 5: ALM-Based Segmentation Algorithm with Compactness Prior

Initialize u^0 and $\left(p_s^0, p_t^0, p^0, \lambda^0\right)$; for each iteration k we explore the following two steps

- Fix u^k, compute $\left(p_s^{k+1}, p_t^{k+1}, p^{k+1}, \lambda^{k+1}\right)$:

$$\left(p_s^{k+1}, p_t^{k+1}, p^{k+1}, \lambda^{k+1}\right) = \arg \max_{p_s, p_t, p, \lambda} L_{c^k}(u^k, p_s, p_t, p, \lambda), \ \text{s.t. (6.57)}$$

provided the augmented Lagrangian function $L_c(u, p_s, p_t, p, \lambda)$ in (6.59).

- Fix $\left(p_s^{k+1}, p_t^{k+1}, p^{k+1}, \lambda^{k+1}\right)$, then update u^{k+1} by

$$u^{k+1} := u^k - c^k \left(\mathbb{D}\mathrm{iv}\left(p^{k+1} - \lambda^{k+1}e\right) - p_s^{k+1} + p_t^{k+1}\right).$$

6.3.3 Medical Image Segmentation with Region-Order Prior

In many practices of medical image segmentation, the target regions have exact inter-region relationships in geometry; for example, one region is contained in another region as its subregion. Such inclusion/overlay order between regions, aka region-order, appears quite often in medical image segmentation, such as the three regions of blood pool, myocardium and background are overlaid sequentially in 3D cardiac T2 MRI [29, 36], the central zone is well included inside the whole gland region of prostate in 3D T2w prostate MRIs [34], the region inside carotid artery adventitia boundary(AB) covers the region inside lumen-intima boundary(LIB) in input T1-weighted black-blood carotid magnetic resonance (MR) images [27, 28], etc. In practice, imposing such geometrical order for exacting target regions can significantly improve both accuracy and robustness of image segmentation.

Such overlaid regions, as well as linear-ordered regions, can be mathematically formulated as

$$\Omega_n(= \varnothing) \subseteq \Omega_{n-1} \subseteq \ldots \subseteq \Omega_1 \subseteq \Omega_0(:= \Omega). \tag{6.60}$$

In view of the Potts model-based multiphase image segmentation model (6.31), we can therefore encode the total segmentation cost and surface regularization terms as the following *coupled continuous min-cut model*:

$$\min_{u_i(x) \in [0,1]} \sum_{i=1}^{n} \int_\Omega (u_{i-1} - u_i) D_i \, dx + \sum_{i=1}^{n-1} a_i \int_\Omega |\nabla u_i| dx, \tag{6.61}$$

$$\text{s.t. } 0 \le u_{n-1}(x) \le \ldots \le u_1(x) \le 1; \tag{6.62}$$

where $u_i(x)$, $i = 1 \ldots n$, is the indicator function of the respective region Ω_i and $D_i(x)$ is the cost for the pixel x inside the region $\Omega_{i-1} \setminus \Omega_i$ which is labeled by $u_{i-1}(x) - u_i(x)$.

It can be proved that, with simple variational computation, the coupled continuous min-cut model (Expression 6.61) can be equivalently reformulated as the following dual optimization problem [37]

$$\max_{p_i, q_i} \int_\Omega p_1 \, dx \tag{6.63}$$

$$\text{s.t. } |q_i(x)| \leq a_i, \; p_i(x) \leq D_i(x), \tag{6.64}$$

$$(\mathbb{D}\text{iv } q_i - p_i + p_{i+1})(x) = 0, \; i = 1 \ldots n-1. \tag{6.65}$$

Also, for the dual optimization model (Expression 6.63), the optimum labeling function $u_i^*(x)$, $i = 1 \ldots n-1$, works as the optimal multiplier to the respective linear equality constraint (6.65), which can be seen from the corresponding Lagrangian function of Expression 6.63:

$$L(u, p, q) = \int_\Omega p_1 \, dx + \sum_{i=1}^{n-1} \langle u_i, \mathbb{D}\text{iv } q_i - p_i + p_{i+1} \rangle. \tag{6.66}$$

Similarly, we define the augmented Lagrangian function w.r.t. Expression 6.66

$$L_c(u, p, q) = \int_\Omega p_s(x) \, dx + \sum_{i=1}^{n-1} \langle u_i, \mathbb{D}\text{iv } q_i - p_i + p_{i+1} \rangle$$

$$- \frac{c}{2} \sum_{i=1}^{n-1} \left\| \mathbb{D}\text{iv } q_i - p_i + p_{i+1} \right\|^2, \tag{6.67}$$

and derive its related ALM-based algorithm, see Algorithm 2.

Two examples are given in Figure 6.4b demonstrate that the two regions of the prostate central zone (CZ) and whole gland (WG) are extracted the segmentation from the given 3D T2w prostate MR image [34] subject to the enforced linear region-order

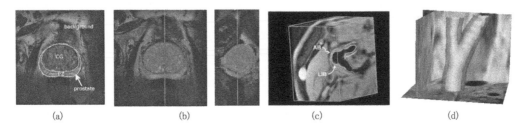

(a)   (b)   (c)   (d)

FIGURE 6.4: (a) Illustration of the segmented contours overlaid on a T2w prostate MRI slice, where the central zone (CZ) of prostate (inside the green contour) is included in the whole prostate region (inside the red contour); (b) the segmentation result in axial and sagittal views respectively [34]; (c) illustration of the overlaid surfaces by adventitia (AB) and lumen-intima (LIB) in 3D carotid MR images; (d) the result of segmented surfaces of AB and LIB in the input 3D carotid MR image [27, 28].

constraint $\Omega_{PZ} \subset \Omega_{WG}$; (d) show that the two regions of carotid lumen, i.e. contoured by AB and LIB, are segmented well by imposing the linear region-order constraint $\Omega_{LIB} \subset \Omega_{AB}$ [27, 28].

Algorithm 6: ALM-Based Segmentation Algorithm with Linear Region-Order Prior
   Initialize u^0 and $\left(p^0, q^0\right)$; for each iteration k we explore the following two steps

- Fix u^k, compute $\left(p^{k+1}, q^{k+1}\right)$:

$$(p^{k+1}, q^{k+1}) = \arg\max_{p,q} L_{c^k}(u^k, p, q), \quad \text{s.t. (6.64)} \tag{6.68}$$

provided the augmented Lagrangian function $L_c(u, p, q)$ in (6.67).

- Fix $\left(p^{k+1}, q^{k+1}\right)$, then update u^{k+1} by

$$u_i^{k+1} := u_i^k - c^k(\mathbb{D}\mathrm{iv}\, q_i^{k+1} - p_i^{k+1} + p_{i+1}^{k+1}). \tag{6.69}$$

6.3.4   Extension to Partially Ordered Regions

An extension to the linear-order of regions (Expression 6.60) is the partial-order of regions, for which the geometric inter-region relationship can be often formulated as

$$\Omega_k \supset \Omega_1 \cup \ldots \cup \Omega_{k-1}, \quad \text{or} \quad \Omega_k = \Omega_1 \cup \ldots \cup \Omega_{k-1}.$$

For example, the segmentation of 3D LE-MRIs [15, 30] targets to extract the thoracic background region R_B and its complementary region of the whole heart R_C in the input LE-MRIs

$$\Omega = R_C \cup R_B, \quad R_C \cap R_B = \varnothing,$$

and the cardiac region R_c contains three sub-regions of myocardium R_m, blood R_b and scar tissue R_s (see Figure 6.5 (a) for illustration):

$$R_C = (R_m \cup R_b \cup R_s) \tag{6.70}$$

where the three sub-regions R_m, R_b and R_s are mutually disjoint

$$R_m \cap R_b = \varnothing, \quad R_b \cap R_s = \varnothing, \quad R_s \cap R_m = \varnothing. \tag{6.71}$$

As in the Potts model (Expression 6.33), let $u_i(x) \in \{0,1\}$, $i \in \{m, b, s, C, B\}$, be the indicator function of the region R_i, then the region-order constraints (Expressions 6.70 and 6.71) can be expressed as

$$u_C(x) + u_B(x) = 1, \quad u_C(x) = u_m(x) + u_b(x) + u_s(x), \quad \forall x \in \Omega, \tag{6.72}$$

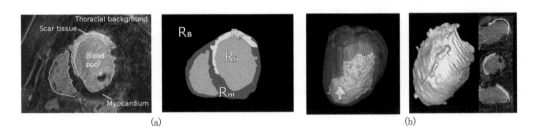

(a)   (b)

FIGURE 6.5: (a) Illustration of the anatomical spatial order of cardiac regions in a LE-MRI slice: The region R_C containing the heart is divided into three sub-regions including myocardium R_m, blood R_b and scar tissue R_s. R_B represents the thoracic background region. (b) 3D segmentation results of LE-MRI: Myocardium R_m (red) and scar tissue R_s (yellow).

the associated image segmentation model is formulated as

$$\min_{u(x)\in\{0,1\}} \sum_{i\in\{m,b,s,B\}} \int_\Omega u_i(x)\, \rho(l_i,x)\, dx + a \sum_{i\in\{m,b,s,B,C\}} \int_\Omega \left|\nabla u_i\right| dx \qquad (6.73)$$

where the first term sums up the costs of four disjoint segmentation regions $R_{m,b,s,B}$ and the second term regularizes the surfaces of all the regions including $R_{m,b,s,B,C}$.

Through variational computation [38, 30], we obtain the equivalent dual model to the convex relaxation of Expression 6.73

$$\max_{p_o,p,q} \int_\Omega p_o\, dx \qquad (6.74)$$

subject to

$$\left|q_i(x)\right| \le a, \quad i \in \{m,b,s,B,C\}; \quad p_i(x) \le \rho(\ell_i,x), \quad i \in \{m,b,s,B\}; \qquad (6.75)$$

$$(\mathbb{D}\text{iv}\, q_i - p_o + p_i)(x) = 0, \quad i \in \{B,C\}; \qquad (6.76)$$

$$(\mathbb{D}\text{iv}\, q_i - p_C + p_i)(x) = 0, \quad i \in \{m,b,s\}. \qquad (6.77)$$

The labeling function $u_i(x)$, $i \in \{m,b,s,B,C\}$, works as the multiplier to the linear equality constraint (Expressions 6.76–6.77) respectively. This hence gives the clue to define the related augmented Lagrangian function and build up the similar ALM-based optimization algorithm as Algorithm 3 which is omitted here (see [38, 30] for more details).

Actually, more complex priors of region-orders can be defined and employed in medical image segmentation; see [39] for references.

6.3.5 Medical Image Segmentation with Spatial Consistency Prior

For another kind of medical image segmentation task, the target regions usually appear with spatial similarities between two neighbor slices or multiple co-registered volumes, for example, 3D prostate ultrasound image segmentation [24–26], co-segmenting lung pulmonary ¹H and hyperpolarized ³He MRIs [33]. This greatly helps in making full use of image features in all related images and guides the simultaneous segmentation procedure to reach a higher accuracy in the result.

Now we consider jointly segmenting two input images [33], i.e. co-segmentation, for simplicities. Given two appropriately co-registered images (see Figure 6.6 (a)), we propose to simultaneously extract the same target region from both images, i.e., $\mathbb{R}^1$ and $\mathbb{R}^2$, and let $\mathbb{R}^1_B$ and $\mathbb{R}^2_B$ be the associate complementary background regions. Initially, the labeling function $u_i(x)$ for each target region $\mathbb{R}^i$, $i=1$, 2, can be computed through the continuous min-cut model (Expression 6.78), i.e.

$$\min_{u_i(x)\in[0,1]} \int_\Omega \left\{ (1-u_i)C_s^i + u_i C_t^i \right\} dx + a \int_\Omega |\nabla u_i| dx, \tag{6.78}$$

In addition, we impose the spatial similarity between the two target regions $\mathbb{R}^1$ and $\mathbb{R}^2$ by penalizing the total difference of $\mathbb{R}^1$ and $\mathbb{R}^2$, i.e. $\int_\Omega |u_1 - u_2| dx$. Hence, we have the convex optimization model for co-segmenting the two input images:

$$\min_{u_{1,2}(x)\in[0,1]} \sum_{i=1}^2 \int_\Omega \left\{ (1-u_i)C_s^i + u_i C_t^i \right\} dx + a \sum_{i=1}^2 \int_\Omega |\nabla u_i| dx$$
$$+ \beta \int_\Omega |u_1 - u_2| dx. \tag{6.79}$$

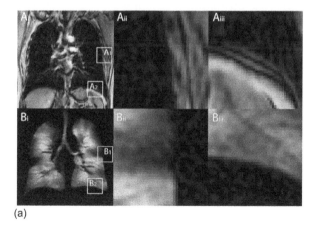
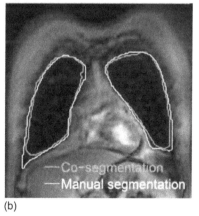

(a)   (b)

FIGURE 6.6: (a) Complementary edge information about ¹H and ³He 3D lung MRIs; (Ai) ¹H MRI coronal slice with inset boxes: A1 expanded in (Aii) and A2 expanded in (Aiii); (Bi) ³He MRI coronal slice with inset boxes: B1 expanded in (Bii) and B2 expanded in (Biii). (b) The co-segmentation result shown in a ¹H MRI coronal slice (pink contour) w.r.t. the manual segmentation result (yellow contour); see [33] for more details.

Observe that

$$\beta \int_\Omega |u_1 - u_2| \, dx = \max_{r(x) \in [-\beta, \beta]} \int_\Omega r \, (u_1 - u_2) \, dx,$$

and similar variational analysis as in [7, 20], we have the equivalent representation of the co-segmentation optimization model (Expression 6.79)

$$\min_{u_{1,2}} \max_{p_s, p_t, q, r} L(u, p_s, p_t, q, r) = \int_\Omega p_s^1 \, dx + \int_\Omega p_s^2 \, dx + \left\langle u_1, \mathbb{D}\text{iv } q_1 + p_t^1 - p_s^1 + r \right\rangle$$

$$+ \left\langle u_2, \mathbb{D}\text{iv } q_2 + p_t^2 - p_s^2 - r \right\rangle. \tag{6.80}$$

Minimizing the primal-dual optimization model (Expression 6.80) over $u_1(x)$ and $u_2(x)$ first, we then derive the dual optimization model to Expression 6.79 such that

$$\max_{p_s^{1,2}, p_t^{1,2}, q_{1,2}, r} \int_\Omega p_s^1 \, dx + \int_\Omega p_s^2 \, dx \tag{6.81}$$

subject to

$$p_s^i(x) \le C_s^i(x), \quad p_t^i(x) \le C_t^1(x), \quad |q_i(x)| \le a, \quad i = 1,2; \quad |r(x)| \le \beta; \tag{6.82}$$

$$(\mathbb{D}\text{iv } q_1 + p_t^1 - p_s^1 + r)(x) = 0, \quad (\mathbb{D}\text{iv } q_2 + p_t^2 - p_s^2 - r)(x) = 0. \tag{6.83}$$

The optimum $u_1^*(x)$ and $u_2^*(x)$ to the convex optimization model (Expression 6.79) are exactly the optimum multipliers to the linear equality constraints (Expression 6.83) respectively.

Correspondingly, we define the augmented Lagrangian function w.r.t. $L(u, p_s, p_t, q, r)$ in Expression 6.80

$$L_c(u, p_s, p_t, q, r) = L(u, p_s, p_t, q, r) - \frac{c}{2} \| \mathbb{D}\text{iv } q_1 - p_s^1 + p_t^1 + r \|^2$$

$$- \frac{c}{2} \| \mathbb{D}\text{iv } q_2 - p_s^2 + p_t^2 - r \|^2 . \tag{6.84}$$

and derive the ALM-based image co-segmentation algorithm, see Algorithm 7.

Algorithm 7: ALM-Based Image Co-Segmentation Algorithm

Initialize u^0 and $\left(p_s^0, p_t^0, q^0, r^0 \right)$; for each iteration k we explore the following two steps

- Fix u^k, compute $\left(p_s^{k+1}, p_t^{k+1}, q^{k+1}, r^{k+1} \right)$:

$$\left(p_s^{k+1}, p_t^{k+1}, q^{k+1}, r^{k+1} \right) = \arg \max_{p_s, p_t, q, r} L_{c^k}(u^k, p_s, p_t, q, r), \quad \text{s.t. (6.82)}$$

provided the augmented Lagrangian function $L_c(u, p_s, p_t, q, r)$ in (6.84).

- Fix $\left(p_s^{k+1}, p_t^{k+1}, q^{k+1}, r^{k+1} \right)$, then update u^{k+1} by

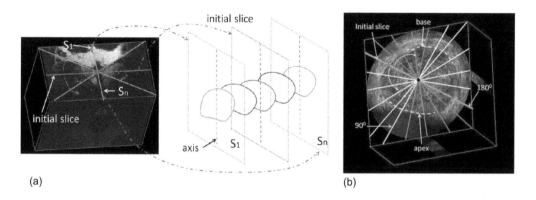

(a)   (b)

FIGURE 6.7: (a) Illustration of 3D prostate TRUS image slices; (b) the joint segmentation of all slices with the axial symmetry prior; see [24, 25] for more details.

$$u_1^{k+1} := u_1^k - c^k(\mathbb{D}\mathrm{iv}\, q_1^{k+1} - (p_s^1)^{k+1} + (p_t^1)^{k+1} + r),$$

$$u_2^{k+1} := u_2^k - c^k(\mathbb{D}\mathrm{iv}\, q_2^{k+1} - (p_s^2)^{k+1} + (p_t^2)^{k+1} - r).$$

Clearly, without $r(x)$, the dual model (Expression 6.81) of image co-segmentation can be viewed as two independent continuous max-flow models (Expression 6.19), and the associated ALM-based image co-segmentation algorithm (Algorithm 5) is just two separate continuous max-flossww algorithms (Algorithm 2) in combination. Observing this, we see that the ALM-based image co-segmentation algorithm (Algorithm 5) is exactly two joint continuous max-flow algorithms (Algorithm 2) along with optimizing the additional variable $r(x)$. This is much simpler than solving the convex optimization model (Expression 6.79) directly!

An example of image co-segmentation of ^{1}H and ^{3}He 3D lung MRIs is shown in Figure 6.6. The results showed that the introduced image co-segmentation approach (6.79) and (6.81) yields superior performance compared to the single-channel image segmentation in terms of precision, accuracy and robustness [33].

The introduced optimization method can be easily extended to simultaneously segmenting a series of images while enforcing their spatial consistencies between images [24–26]. For example, it can be used to enforce the axial symmetry prior between 3D prostate TRUS image slices [24, 25]; the prior can be formulated as

$$\sum_{i=1}^{n-1} \int_\Omega |u_{i+1} - u_i|\, dx + \int_\Omega |u_n(L-x) - u_1(x)|\, dx,$$

i.e. a sequence of spatial consistencies between the given image slices to be enforced. Similarly, it results in a series of joint continuous max-flow computations (see [24, 25]).

Acknowledgement

The authors would like to acknowledge the great support from the National Natural Science Foundation of China (Grant No. 61877047), the Canadian Institutes of Health

Research (CIHR), Ontario Institute for Cancer Research (OICR), and the Natural Sciences and Engineering Research Council of Canada (NSERC).

References

1. Yu Nesterov. Smooth minimization of non-smooth functions. *Mathematical Programming*, 103(1):127–152, 2005.
2. Antonin Chambolle. An algorithm for total variation minimization and applications. *Journal of Mathematical Imaging and Vision*, 20(1):89–97, 2004.
3. Amir Beck and Marc Teboulle. A fast iterative shrinkage-thresholding algorithm for linear inverse problems. *Siam Journal on Imaging Sciences*, 2(1):183–202, 2009.
4. Antonin Chambolle and Thomas Pock. A first-order primal-dual algorithm for convex problems with applications to imaging. *Journal of Mathematical Imaging and Vision*, 40(1):120–145, 2011.
5. Bingsheng He, Li Zhi Liao, Deren Han, and Hai Yang. A new inexact alternating directions method for monotone variational inequalities. *Mathematical Programming*, 92(1):103–118, 2002.
6. Tony F. Chan, Selim Esedoglu, and Mila Nikolova. Algorithms for finding global minimizers of image segmentation and denoising models. *SIAM Journal on Applied Mathematics*, 66(5):1632–1648 (electronic), 2006.
7. Jing Yuan, Egil Bae, and Xue-Cheng Tai. A study on continuous max-flow and min-cut approaches. In *IEEE Conference on Computer Vision and Pattern Recognition (CVPR), 2010*, 2010.
8. L. Rudin, S. Osher, and E. Fatemi. Nonlinear total variation based noise removal algorithms. *Physica D*, 60(1–4):259–268, 1992.
9. Jing Yuan, Christoph Schnorr, and Etienne Mémin. Discrete orthogonal decomposition and variational fluid flow estimation. *Journal of Mathematical Imaging and Vision*, 28(1):67–80, 2007.
10. Jan Lellmann, Jorg Kappes, Jing Yuan, Florian Becker, and Christoph Schnorr. Convex multi-class image labeling by simplex-constrained total variation. In *SSVM '09*, pages 150–162, 2009.
11. Tong Zhang. Analysis of multi-stage convex relaxation for sparse regularization. *The Journal of Machine Learning Research*, 11:1081–1107, 2010.
12. Tom Goldstein and Stanley Osher. The split Bregman method for l1 regularized problems. *Siam Journal on Imaging Sciences*, 2(2):323–343, 2009.
13. Kumaradevan Punithakumar, Jing Yuan, Ismail Ben Ayed, Shuo Li, and Yuri Boykov. A convex max-flow approach to distribution-based figure-ground separation. *SIAM Journal on Imaging Sciences*, 5(4):1333–1354, 2012.
14. R. Tyrrell Rockafellar. *Convex Analysis*. Princeton Mathematical Series, No. 28. Princeton University Press, Princeton, NJ, 1970.
15. D. P. Bertsekas. *Nonlinear Programming*. Athena Scientific, September 1999.
16. R. T. Rockafellar. Augmented Lagrangians and applications of the proximal point algorithm in convex programming. *Mathematics of Operations Research*, 1(2):97–116, 1976.
17. Bingsheng He and Xiaoming Yuan. On the o(1/n) convergence rate of the douglas-rachford alternating direction method. *Siam Journal on Numerical Analysis*, 50(2):700–709, 2012.
18. Yuri Boykov and Vladimir Kolmogorov. An experimental comparison of min-cut/max-flow algorithms for energy minimization in vision. *IEEE Transactions on Pattern Analysis and Machine Intelligence*, 26:359–374, 2001.
19. Yuri Boykov, Olga Veksler, and Ramin Zabih. Fast approximate energy minimization via graph cuts. *IEEE Transactions on Pattern Analysis and Machine Intelligence*, 23:1222–1239, 2001.

20. Jing Yuan, Egil Bae, Xue-Cheng Tai, and Yuri Boykov. A spatially continuous max-flow and min-cut framework for binary labeling problems. *Numerische Mathematik*, 126(3):559–587, 2014.

21. Jing Yuan. Fast continuous max-flow algorithm to 2d/3d image segmentation. 2011. http://www.mathworks.com/matlabcentral/fileexchange/34126.

22. Jing Yuan, Egil Bae, Xue-Cheng Tai, and Yuri Boykov. A continuous max-flow approach to potts model. In *ECCV*, 2010.

23. Jing Yuan. Fast continuous max-flow algorithm to 2d/3d multi-region image segmentation. 2011. http://www.mathworks.com/matlabcentral/fileexchange/34224.

24. Jing Yuan, Wu Qiu, Martin Rajchl, Eranga Ukwatta, Xue-Cheng Tai, and Aaron Fenster. Efficient 3d endfiring TRUS prostate segmentation with globally optimized rotational symmetry. In *IEEE Conference on Computer Vision and Pattern Recognition (CVPR), 2013*, pages 2211–2218, 2013.

25. Wu Qiu, Jing Yuan, Eranga Ukwatta, Yue Sun, Martin Rajchl, and Aaron Fenster. Prostate segmentation: An efficient convex optimization approach with axial symmetry using 3d TRUS and MR images. *IEEE Transactions on Medical Imaging*, 33(4):1–14, 2014.

26. Eranga Ukwatta, Jing Yuan, Wu Qiu, Martin Rajchl, Bernard Chiu, and Aaron Fenster. Joint segmentation of lumen and outer wall from femoral artery {MR} images: Towards 3d imaging measurements of peripheral arterial disease. *Medical Image Analysis*, 26(1):120–132, 2015.

27. Eranga Ukwatta, Jing Yuan, Martin Rajchl, and Aaron Fenster. Efficient global optimization based 3d carotid ab-lib MRI segmentation by simultaneously evolving coupled surfaces. In *MICCAI 2012*, volume 7512 of *Lecture Notes in Computer Science*, pages 377–384, 2012.

28. Eranga Ukwatta, Jing Yuan, Martin Rajchl, Wu Qiu, David Tessier, and Aaron Fenster. 3-d carotid multi-region MRI segmentation by globally optimal evolution of coupled surfaces. *IEEE Transactions on Medical Imaging*, 32(4):770–785, 2013.

29. M. Rajchl, J. Yuan, E. Ukwatta, and T. M. Peters. Fast interactive multi-region cardiac segmentation with linearly ordered labels. In *ISBI 2012*, pages 1409–1412, IEEE Conference Publications, 2012.

30. M. Rajchl, J. Yuan, J. White, E. Ukwatta, J. Stirrat, C. Nambakhsh, F. Li, and T. Peters. Interactive hierarchical max-flow segmentation of scar tissue from late-enhancement cardiac MR images. *IEEE Transactions on Medical Imaging*, 33(1):159–172, 2014.

31. Wu Qiu, Jing Yuan, Eranga Ukwatta, David Tessier, and Aaron Fenster. Rotational-slice-based prostate segmentation using level set with shape constraint for 3D end-firing TRUS guided biopsy. In *MICCAI(Part 1), LNCS 7510*, pages 536–543, 2012.

32. Jing Yuan, Wu Qiu, Eranga Ukwatta, Martin Rajchl, Yue Sun, and Aaron Fenster. An efficient convex optimization approach to 3d prostate MRI segmentation with generic star shape prior. In *Prostate MR Image Segmentation Challenge, MICCAI*, pages 13–24, 2012.

33. F. Guo, J. Yuan, M. Rajchl, S. Svenningsen, D. P. Capaldi, K. Sheikh, A. Fenster, and G. Parraga. Globally optimal co-segmentation of three-dimensional pulmonary 1h and hyper-polarized 3he MRI with spatial consistence prior. *Medical Image Analysis*, 23(1):43–55, 2015.

34. Jing Yuan, Eranga Ukwatta, Wu Qiu, Martin Rajchl, Yue Sun, Xue-Cheng Tai, and Aaron Fenster. Jointly segmenting prostate zones in 3d MRIs by globally optimized coupled level-sets. In *Energy Minimization Methods in Computer Vision and Pattern Recognition*, pages 12–25, Springer, Berlin Heidelberg, 2013.

35. Wu Qiu, Martin Rajchl, Fumin Guo, Yue Sun, Eranga Ukwatta, Aaron Fenster, and Jing Yuan. 3d prostate TRUS segmentation using globally optimized volume-preserving prior. In *MICCAI 2014*, pages 796–803, 2014.

36. Cyrus M. S. Nambakhsh, Jing Yuan, Kumaradevan Punithakumar, Aashish Goela, Martin Rajchl, Terry M. Peters, and Ismail Ben Ayed. Left ventricle segmentation in {MRI} via convex relaxed distribution matching. *Medical Image Analysis*, 17(8):1010–1024, 2013.

37. E. Bae, J. Yuan, X.-C. Tai, and Y. Boycov. A study on continuous max-flow and min-cut approaches. Part II: Multiple linearly ordered labels. Technical report CAM-10–62, UCLA, 2010.

38. Martin Rajchl, Yuan Jing, James A. White, Cyrus M. S. Nambakhsh, Eranga Ukwatta, Li Feng, John Stirrat, and Terry M. Peters. A fast convex optimization approach to segmenting 3D scar tissue from delayed-enhancement cardiac MR images. In *MICCAI 2012*, pages 659–666, 2012.

39. John S. H. Baxter, Martin Rajchl, A. Jonathan Mcleod, Jing Yuan, and Terry M. Peters. Directed acyclic graph continuous max-flow image segmentation for unconstrained label orderings. *International Journal of Computer Vision*, 123(3):1–20, 2017.

Chapter 7

Non-Parametric Bayesian Estimation of Rigid Registration for Multi-Contrast Data in Big Data Analysis

Stathis Hadjidemetriou, and Ismini Papageorgiou

7.1   Introduction

The accumulation of diverse data, namely, big data, in a variety of domains including the biomedical domain is an emerging reality [1, 2]. The latter involves large amounts of biomedical, diagnostic, and clinical data accumulating at an increasing rate in research centers and in hospital departments. The large data sizes create a variety of challenges such as for storage and retrieval [3], for the organization of heterogeneous data as a database [4], and for the standardization of the data for the purposes of an objective study. The normal operation of a clinical department requires the regular extraction of information from this data to evaluate specific as well as comprehensive effects of patient treatments. Moreover, statistical analysis of big data with techniques such as non-supervised clustering may provide novel directions towards disease classifications. Updated classifications create the potential to improve current therapeutic strategies towards personalized treatment approaches [5]. The techniques to process this data are under exploration [6]. It is not even clear whether a direct statistical analysis of this data is sufficient or whether more comprehensive analytical models are necessary [7].

One of these data types, the imaging data, is perhaps the most challenging one [7–9]. Many different imaging datasets can be available for a patient. They can be from different imaging contrasts and even from completely different modalities. There may even be differences between the anatomic domains depicted in different images of the same patient. The ability to analyze these data is not immediate, and hence appropriate methodology must be developed and applied in a sequence of phases [6]. The data must first be restored for a variety of artifacts. It must be restored and standardized with methodologies for denoising and normalization of the dynamic range of intensities. Different datasets of a patient must be standardized with registration to become spatially normalized [10]. These pre-processing steps are necessary to remove artifacts and systematic biases from the data. Thus, the pre-processing enables the extraction of valid conclusions from the analysis of the data and the avoidance of erroneous conclusions [11]. Statistical principles lead to the conclusion that data standardization also contributes to decreasing the number of datasets necessary to consider in a study and hence also decreases the storage requirements and the complexity of the analysis necessary to reach valid conclusions. Legal and ethical issues are also involved when analyzing big data that involve a balance between privacy and the extraction of beneficial conclusions [12–15].

One of the most significant issues to address in intra-patient imaging datasets is the spatial misregistration that needs to be compensated for to place different imaging data in a common anatomic space. The corrected data placed in the same anatomic space can subsequently be more directly used for semantic analysis. The spatial normalization can be a non-rigid spatial field [16]. This can be achieved by smoothing the spatial normalization with B-splines bases [17] or simply by smoothing with a Gaussian filter. In many cases, the non-rigid registration allows more degrees of freedom in spatial deformations than are necessary and a rigid registration is more meaningful. Some examples are the intra-patient anatomic registration or the pre-processing step for a more general non-rigid registration. A variety of methods have been developed for rigid registration [18]. The more widely used ones are those based on intensity [19–22]. The analytical rigid transformations have also been formulated with a variety of methods [23, 24]. A general and extensively studied methodology for rigid transformations is the Procrustes method [25–29]. The general solution of the Procrustes method has been expressed in terms of singular value decomposition [28, 29]. The standard case of the Procrustes method gives a rigid transformation [23, 24, 30].

Typically, spatial registration is preceded by intensity correction [31, 32]. This pre-processing step standardizes and normalizes the statistics and prepares the images for placement in a common anatomic space particularly with an intensity-based registration method. Traditionally, intensity-based registration involves the optimization of a cost function that is a functional of the joint statistics, most often the entropy. However, the entropy is a uni-modal convex functional, optimized for a uni-modal statistical distribution [33]. The joint statistics and their entropy are also used for intensity-based non-rigid registration of multi-contrast data [19]. In effect, its use makes the implicit assumption that the joint statistics of the imaging data to be registered are uni-modal. This assumption associated with the entropy is not representative of typical medical images. The statistics of typical medical images consist of multiple distributions resulting from various tissues. The result is that the entropy-based non-rigid registration methods distort some of the corresponding regions in the images, change their overall volumes, and have to rely on extensive spatial smoothing [34]. Similar misregistration effects result in rigid registration when it is based on the entropy.

This work presents a non-parametric method for rigid registration. The data is pre-processed to achieve intensity uniformity restoration that makes the methodology robust to spatial intensity non-uniformities that are very common in MRI data. A Bayesian non-parametric method for intensity restoration is used that is appropriate for the multi-contrast data used in this work [35]. The actual registration method uses directly and non-parametrically the joint statistics. Hence, it preserves and considers all the distributions in the statistics to estimate the rigid transformation. The Bayesian formulation for the registration gives the maximum posterior expectation. Non-parametric Bayesian methodologies have been used for various purposes in imaging and, in particular, in medical imaging [34, 36–39]. The non-parametric Bayesian formulation has been used for the non-rigid part of the NParBR image registration method that was used in combination with conventional rigid registration [34]. The high performance of this combination has been demonstrated extensively for MRI data, CT data, as well as for several anatomic regions [40, 41]. In this work the non-parametric formulation is developed for rigid registration as well.

The non-parametric Bayesian estimation provides a vector field in space. The spatial vector field is processed with the standard Procrustes method to provide a rigid transformation as the maximum likelihood estimate [23, 24, 30]. This study combines a Bayesian intensity based registration with a rigid model for the transformation. The Bayesian non-parametric methodology is effective for images of the same contrast as well as for images of different contrasts. The computation of the registration is over the global joint image statistics. These design choices improve the accuracy of the solution and enable the compensation of large misregistrations with a solution closer to the global maximum. The method was validated with datasets from three different databases. One database consists of images from the Brainweb simulator [42, 43], another database from the Human Connectome Project (HCP) [44–47], and the third database from clinical images of Parkinson's disease patients. The method successfully compensated for existing misregistrations and recovered extensive simulated transformations. The validation demonstrated the accuracy of the method for large rigid registrations.

7.2 Methods

7.2.1 Statistical and Spatial Representation

A moving image, $I_{Mov}(x)$, is assumed to have a geometric distortion compared to the reference image, $I_{Ref}(x)$, that can be represented as a rigid transformation. This is a rotation A in mean normalized coordinates. The transformation A^{-1} is computed so that $I_{Mov}(A^{-1}x)$ is in the same anatomic space as $I_{Ref}(x)$. The registered image becomes $I_{Ref}(x) = I_{Mov}(A^{-1}x)$. The initial transformation is the non-parametric Bayesian estimate computed as the posterior expectation. The statistical representation of the image is in terms of the 2D joint co-occurrence statistics of the reference image, $I_{Ref}(x)$, with the moving image, $I_{Mov}(x)$. The statistics are computed globally within a 3D image region. The co-occurrences use a spherical spatial neighborhood $\mathcal{N}$ of radius r around every voxel. It is assumed that the spatial misregistration between images results in a Point Spread Function (PSF), $P_{(NC)}$, of additive Gaussian noise to the joint statistics of the resulting images. It is also assumed that the misregistration results in a Gaussian PSF, $P_{(NS)}$, with standard deviation the size of $\mathcal{N}$ for the pixel displacement. The statistics are restored for denoising from $P_{(NC)}$ to provide the non-parametric prior. The posterior expectation is computed voxelwise to give the restoration back-projected over the entire image domain. It gives a rough initial spatial registration as an intermediate step.

This intermediate field provides a transformation that subsequently provides a rigid transformation as the maximum likelihood with the Procrustes method. As a preparation for the Procrustes method, the datasets are first mean normalized in space. This enables the application of the method to compute a rigid transformation as a rotation. An example of the original and of the transformed images as well as of the corresponding joint co-occurrence statistics is shown in Figure 7.1.

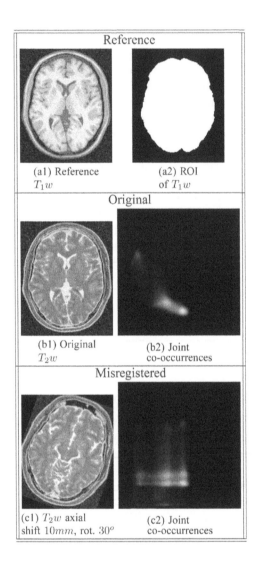

FIGURE 7.1: An example of a pair of a T_1w in (a1) and a T_2w image in (b1) from the BrainWeb simulator. The joint co-occurrence statistics between the original images in (a1) and in (b1) are shown in (b2). They are computed in the ROI shown in (a2). A rigidly misregistered T_2w image is in (c1). The joint co-occurrence statistics of the image in (a1) with the misregistered image in (c1) are shown in (c2). The spatial misregistration spreads the statistics in (c2) and also creates new distributions compared to those of the registered images in (b2).

7.2.2 Denoising of Statistics

The joint distribution of the intensity statistics of the reference image $I_R(x)$ with the moving image $I_M(x)$ are given by their joint co-occurrence statistics, $C(I_M(x), I_R(x))$. It is assumed that the spatial misregistration between these two images results in additive Gaussian noise superimposed upon ideal latent statistics $\hat{C}$. The PSF of the noise, $P_{(NC)}$, that is assumed to be a Gaussian distribution is

$$P_{(NC)}(j) = G(j; 0, \sigma_C^2), \tag{7.1}$$

where j is the noise level and σ_C is the standard deviation. It is assumed that the underlying registered images result in ideal statistics, $\hat{C}$, that can be obtained from the deconvolution of the PSF of the distortion, $P_{(NC)}$, from the actual image statistics, C. The deconvolution for denoising is obtained with the Van Cittert iterative, k, method given by [48, 49]:

$$\begin{aligned} C_0 &= C \\ C_{k+1} &= C_0 + \left(I - P_{(NC)}\right) * C_k, \quad k = 1, 2, 3, \end{aligned} \tag{7.2}$$

where $*$ is convolution and I is an image equal to unity everywhere. The restored statistics are given by $\hat{C} = C_4$.

7.2.3 Estimation of Vector Field in Space

The spatial displacement, d, of a voxel at location x is assumed to be given by the distribution of noise that is additive, independent, and identically distributed with x. The PSF of the noise, $P_{(NS)}$, is assumed to be Gaussian:

$$P_{(NS)}(d) = G(d; 0, \sigma_s^2), \tag{7.3}$$

where σ_s is the standard deviation.

Adjacent voxels are considered to be within a neighborhood $\mathcal{N}$, where a voxel in the moving image is at x_M and a voxel in the reference image is at x_R to give $u = x_M - x_R$. The method quantifies the probability that spatial location x_M in the moving image with intensity $I_M(x_M)$ corresponds to spatial location x_R in the reference image with intensity $I_R(x_R)$. This probability is assumed to be given by the density in the restored co-occurrence statistics indexed by the two intensities corresponding to these two spatial locations. That is,

$$\begin{aligned} P(u \mid x_R) &= \frac{P(u, x_R)}{P(x_R)} \\ &= \frac{P(x_R + u, x_R)}{P(x_R)} = \frac{P(x_M, x_R)}{P(x_R)} \\ &= \frac{\hat{C}(I_M(x_M), I_R(x_R))}{P(x_R)}. \end{aligned} \tag{7.4}$$

This relation gives the probability of displacement u at a certain location x_R. It corresponds to location $x_M = x_R + u$ in the moving image. The individual probabilities of all the pixels in the image are assumed equal, and hence the probability $P(x_R)$ in the denominator is ignored. The initial statistics are given by $P_v(v) = C$. The deconvolution in Equation 7.2 provides the restored statistics. These are used as a prior for the statistics of the restored image.

The assumption is that the initial displacement at voxel x is v and the final displacement is u. Then, the conditional probability $P(u|v,x)$ using Bayes's rule becomes:

$$P(u\,|\,v,x) = \frac{P_{(v|u,x)}(v\,|\,u,x)P_{(u|x)}(u\,|\,x)}{\int P_{(v|u,x)}(v\,|\,u,x)P_{(u|x)}(u\,|\,x)du}. \tag{7.5}$$

The likelihood term is the probability that at a voxel x the initial displacement is u and the updated displacement is v. The displacements are additive to the spatial coordinates. The dependence on displacement u and on v is only a dependence on the difference of the two displacements $d = v - u$ that is independent of x. The distribution $P_{(NS)}$ is a Gaussian distribution. Thus, overall,

$$P_{(v|u,x)}(v\,|\,u,x) = P_{(v-u|x)}(v-u\,|\,x)$$
$$= P_{(d|x)}(v-u\,|\,x)$$
$$= P_{(NS)}(v-u) \tag{7.6}$$
$$= G(d;0,\sigma_s^2)$$

The value of σ_s is set to the radius r of the spatial neighborhood $\mathcal{N}$, $\sigma_s = r$. The posterior expected value for the displacement is:

$$E(u\,|\,v,x) \overset{1}{=} \int P_{(u|v,x)}(u\,|\,v,x)u\,du$$

$$\overset{2}{=} \frac{\int P_{(v|u,x)}(v\,|\,u,x)P_{(u|x)}(u\,|\,x)u\,du}{\int P_{(v|u,x)}(v\,|\,u,x)P_{(u|x)}(u\,|\,x)\,du}$$

$$\overset{3}{=} \frac{\int P_{(v-u|x)}(v-u\,|\,x)P_{(u|x)}(u\,|\,x)u\,du}{\int P_{(v-u|x)}(v-u\,|\,x)P_{(u|x)}(u\,|\,x)\,du}$$

$$\overset{4}{=} \frac{\int P_{(d|x)}(v-u\,|\,x)P_{(u|x)}(u\,|\,x)u\,du}{\int P_{(d|x)}(v-u\,|\,x)P_{(u|x)}(u\,|\,x)\,du} \tag{7.7}$$

$$\overset{5}{=} \frac{\int P_{(NS)}(v-u)P_{(u|x)}(u\,|\,x)u\,du}{\int P_{(NS)}(v-u)P_{(u|x)}(u\,|\,x)du}$$

$$\overset{6}{=} \frac{\int G(v-u;0,\sigma_s^2)\hat{C}(I_{Mov}(x),I_{Ref}(u+x))u\,du}{\int G(v-u;0,\sigma_s^2)\hat{C}(I_{Mov}(x),I_{Ref}(u+x))du}.$$

The first relation ($\overset{1}{=}$) is the general expression for the conditional expectation. The second relation ($\overset{2}{=}$) makes use of Bayes's rule as is given in Equation 7.5. The third ($\overset{3}{=}$), fourth ($\overset{4}{=}$), and fifth ($\overset{5}{=}$) relations result from the sequence of relations developed in Equation 7.6 detailing the derivation for the expression of the likelihood. The last relation, sixth ($\overset{6}{=}$), results from Equation 7.4. Also in the last relation, sixth ($\overset{6}{=}$), the resulting probability, $\hat{C} = P_{(u)}(u)$, is obtained from the deconvolution of $P_{(NS)}(d)$ from $C = P_{(v)}(v)$ as described in Equation 7.2. The derivation of the Bayesian estimation for the posterior expected value thus far is summarized with the equations in Figure 7.2. The marginal likelihood in the denominator, that is, the total probability of the evidence, serves the purpose of normalization.

The overall expectation is the last relation from the series in Equations 7.7:

$$E(u \mid v, x) = \frac{\int G(v - u; 0, \sigma_s^2) \hat{C}(I_{Mov}(x), I_{Ref}(u + x)) u \, du}{\int G(v - u; 0, \sigma_s^2) \hat{C}(I_{Mov}(x), I_{Ref}(u + x)) \, du}. \tag{7.8}$$

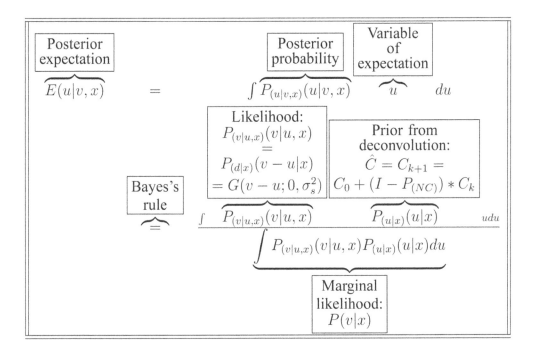

FIGURE 7.2: Overview of the non-parametric Bayesian derivation for the posterior conditional expectation. The conditional expectation is expanded with Bayes's rule. The expressions for the prior and the likelihood are then substituted. It is used to provide the intermediate vector field $\hat{U}$.

This relation is in continuous domain both for space as well as for intensity range. In this form it is useful for analytical purposes. However, in MRI imaging data both the image domain and the intensity range are discrete. Thus, the continuous Equation 7.8 is discretized to provide the relation,

$$\hat{u}(x) = E(u \mid v, x) = \frac{\sum_{u \in \mathcal{N}} G(v - u; 0, \sigma_s^2) \hat{C}(I_{Mov}(x), I_{Ref}(u + x)) u}{\sum_{u \in \mathcal{N}} G(v - u; 0, \sigma_s^2) \hat{C}(I_{Mov}(x), I_{Ref}(u + x))}. \tag{7.9}$$

This estimation provides a vector $\hat{u}(x)$ for every spatial location x. That is, the result of adding $\hat{u}_i$ to x_i in the reference image $I_{Ref}(x_i)$ gives the corresponding point in the moving image, $I_{Mov}(x_i + \hat{u}_i)$. The result for the entire image is a vector field, $\hat{U}$, that transforms the reference image $I_{Ref}(x)$ to the moving image, $I_{Mov}(x + \hat{u}_i)$. This discrete relation in Equation 7.9 is directly implemented.

7.2.4 Procrustes-Based Rigid Registration

The vector field $\hat{U}$ is the intermediate field used to estimate a rigid transformation A with maximum likelihood. The registration is in a 3D image space in which the points lie. An image of dimensions $n_1 \times n_2 \times n_3$ has $N = n_1.n_2.n_3$ pixels. They give two sets of points extracted from the reference image $x_{R,i}$ and from the moving image $x_{M,i}$, where $i = 0, ..., N-1$ corresponds to the index of the pixels. They are related with the vector field, $\hat{u}_i$, to give $x_{M,i} = x_{R,i} + \hat{u}_i$. The given N points form non-degenerate configurations $X_R = x_{R,i}$ and $X_M = x_{M,i}$. More specifically, these points $x_{R,i}$ are the coordinates of the points that give the rows of matrix $X_R = \left[x_{R,1}^T, ..., x_{R,N}^T \right]^T$. Similarly, the points $x_{M,i}$ are the coordinates of the points that give the rows of the matrix X_M, $X_M = \left[x_{M,1}^T, ..., x_{M,N}^T \right]^T$. Both matrices X_R and X_M are of size $N \times 3$.

The matrix A is computed with maximum likelihood to transform the points in X_R into AX_R. The likelihood function for the estimation of A is:

$$\mathcal{L}(A; X_M) = \prod_{i=0}^{N-1} P_{(x_M)}(x_{M,i} \mid A) = \prod_{i=0}^{N-1} \frac{1}{\sqrt{2\pi \mathrm{var}}} e^{-\frac{(x_{M,i} - A x_{R,i})^2}{2\mathrm{var}}}. \tag{7.10}$$

It enables the estimation of the parameters that are A and var by finding the combination of these parameters that maximize this function for the given data points $x_{R,i}$ and the corresponding data points $x_{M,i}$. These N data points are fixed with known values.

The objective is to estimate the unknown entries of matrix A. Here we derive the maximum likelihood estimate for A:

$$\hat{A} = \arg\max_A \mathcal{L}(A; x_M) = \arg\max_A \prod_{i=0}^{N-1} \frac{1}{\sqrt{2\pi\text{var}}} e^{-\frac{(x_{M,i} - Ax_{R,i})^2}{2\text{var}}}$$

$$= \arg\max_A \log\left(\prod_{i=0}^{N-1} \frac{1}{\sqrt{2\pi\text{var}}} e^{-\frac{(x_{M,i} - Ax_{R,i})^2}{2\text{var}}} \right)$$

$$= \arg\max_A \sum_{i=0}^{N-1} \left(\log\left(\frac{1}{\sqrt{2\pi\text{var}}} \right) + \log\left(e^{-\frac{(x_{M,i} - Ax_{R,i})^2}{2\text{var}}} \right) \right)$$

$$= \arg\max_A \sum_{i=0}^{N-1} \log\left(e^{-\frac{(x_{M,i} - Ax_{R,i})^2}{2\text{var}}} \right)$$

$$= \arg\max_A \sum_{i=0}^{N-1} -\frac{(x_{M,i} - Ax_{R,i})^2}{2\text{var}}$$

$$= \arg\min_A \sum_{i=0}^{N-1} (x_{M,i} - Ax_{R,i})^2 \tag{7.11}$$

The objective of the maximum likelihood in Equation 7.11 above is to obtain a transformation $\hat{A}$, which minimizes $\left(\sum_i \left\| Ax_{R,i} - x_{M,i} \right\|^2 \right)^{1/2}$. In matrix form this is equivalent to

$$\hat{A} = \arg\min_A \left\| AX_R - X_M \right\|^2, \tag{7.12}$$

where $\|...\|$ is the Euclidean matrix norm. In the standard case A is a rigid transformation [29, 50]. The solution to this is the standard Procrustes formulation, i.e. $A \in$ rigid transformations and has known solutions. A matrix representation of the rotational component can be computed with singular-value decomposition (SVD) [25–29].

The first step is to subtract the means from both X_R and X_M. This is because the optimal transformations are from centroid $\bar{x}_R$ to centroid $\bar{x}_M$. The demeaned datasets are obtained, respectively, with

$$\begin{aligned} x_{R,i} &\mapsto x_{R,i} - \bar{x}_R \\ x_{M,i} &\mapsto x_{M,i} - \bar{x}_M. \end{aligned} \tag{7.13}$$

The demeaned datasets reduce the problem to the standard orthogonal Procrustes formulation, which determines the orthogonal rotation A. Central to the problem is the 3×3 correlation matrix $K := X_R^T X_M$, as this matrix quantifies how much the points in X_M correspond to the points in X_R.

The matrices $X_R = \left[x_{R,1}^T, \ldots, x_{R,N}^T \right]^T$ and $X_M = \left[x_{M,1}^T, \ldots, x_{M,N}^T \right]^T$ give

$$K = \sum_i K_i, \tag{7.14}$$

where $K_i = x_{R,i} x_{M,i}^T$. Then, the SVD of K, $K = W_1 D W_2^T$, gives the rotation matrix A as

$$K = W_1 D W_2^T \Rightarrow A = W_2 W_1^T. \tag{7.15}$$

This formulation gives the transformation as a rotation A. Finally, the rotation A with the mean points in the moving image $\bar{x}_M$ and in the reference image $\bar{x}_R$ gives the translation b with $b = \bar{x}_M - A\bar{x}_R$. This is the standard case where A is a rigid body transformation as a rotation A and as a resulting translation b [29, 50]. The summary of the rigid transformation estimation method is given in Figure 7.3 that gives A and b.

7.3   Method Implementation

The images are pre-processed for artifact removal. The method considers only a Region of Interest (ROI) from the image domain. The restoration is iterative, t. The

$$x_{R,i} \mapsto x_{R,i} - \bar{x}_R, \qquad x_{M,i} \mapsto x_{M,i} - \bar{x}_M$$

$$X_M = X_R + \hat{U}$$

$$X_R = \begin{bmatrix} x_{R,0} \\ x_{R,1} \\ \cdots \\ x_{R,i} \\ \cdots \\ x_{R,N-1} \end{bmatrix}, X_M = \begin{bmatrix} x_{M,0} \\ x_{M,1} \\ \cdots \\ x_{M,i} \\ \cdots \\ x_{M,N-1} \end{bmatrix} = \begin{bmatrix} x_{R,0} \\ x_{R,1} \\ \cdots \\ x_{R,i} \\ \cdots \\ x_{R,N-1} \end{bmatrix} + \begin{bmatrix} \hat{u}_0 \\ \hat{u}_1 \\ \cdots \\ \hat{u}_i \\ \cdots \\ \hat{u}_{N-1} \end{bmatrix}$$

$$K = X_R^T X_M = W_1 D W_2^T$$

$$A = W_2 W_1^T$$

$$b = \bar{x}_M - A\bar{x}_R$$

FIGURE 7.3:   Overview of the processing of the intermediate vector field $\hat{U}$. The pixel coordinates are demeaned. Then, the covariance matrix between the point sets in the reference and in the moving images is decomposed with SVD. The result of the SVD gives the rigid transformation, A and the shift between the two point sets, b. This is the standard Procrustes method.

method has been implemented in *C++*. The matrix computations of the Procrustes method have been implemented with the numerics package VNL that is part of the VXL library [51, 52]. The VNL package is accessed from the ITK library [53]. Some of the parameters of the program are exported and are available for the user to set.

7.3.1 Pre-Processing and Region of Interest (ROI)

The images were pre-processed for noise removal. The images were denoised with a median filter of size $3 \times 3 \times 3$ $pixels^3$ along the axes. The dynamic range of an MRI image can be very large. It is compressed when its maximum value exceeds 750. The compression considers the intensity value that is the 95% of the cumulative histogram. The compression of the dynamic range is linear both before and after this value. However, above this value and up to the maximum intensity value in the image the factor of the linear compression is greater.

The method requires the specification of an ROI over the reference image $I_{R,ROI}$ as well as an ROI over the moving image $I_{M,ROI}$. These two regions can be different. The joint image statistics are computed over the ROI of both the reference image and of the moving image. The back-projection to space considers the co-occurrence statistics in $\mathcal{N}$ that are within the ROI. In the experiments the ROI of the reference image is explicitly specified. However, the ROI of the moving image is the entire image domain. This allows the registration of the moving image for extensive rigid transformations.

7.3.2 User-Defined Parameters

Two of the parameters of the program are exported and are set by the user. The first parameter is the standard deviation of the PSF of the noise in the statistics, that is set to $\sigma_C = 6\%$ of the length of the dynamic range along each of the two axes. The second parameter is the standard deviation of the PSF of the displacement in space that is equal to the radius of the spherical spatial neighborhood, $\mathcal{N}$, and is set to $\sigma_s = r = 18$. These parameters are kept the same for all the experiments. The remaining parameters and aspects of the program are preset in the code and only optionally set by the user.

7.3.3 Iterative Implementation

The registration is iterative, t. The registration is initialized with zero everywhere, $U(x)_{t=0} = 0, \forall x$. The transformation from iteration t, (A_t, b_t), aligns the two images for the computation of the co-occurrence statistics for iteration $t+1$. The posterior expectation at iteration $t+1$ gives the incremental image field as $\hat{U}_{t+1}^{inc}$. This is added to the complete rigid transformation up to iteration t, (A_t, b_t), to give the cumulative field up to iteration $t+1$, $\hat{U}_{t+1}^{cum}(x) = (Ax+b) + \hat{U}_{t+1}^{inc}(x)$. The field $\hat{U}_{t+1}^{cum}$ for iteration $t+1$ is provided to the Procrustes method to compute the rigid transformation up to iteration $t+1$, (A_{t+1}, b_{t+1}). The rigid transformation at iteration t, (A_t, b_t), is the field v_{t+1} for the next iteration $t+1$, $(A_t, b_t) = v_{t+1}$. Therefore, the conditional field estimation for iteration $t+1$ is $E(u_{t+1} | v_{t+1}) = E(u_{t+1} | A_t, b_t)$. The iterations also have explicit end conditions. They are terminated when the minimum of the variance of the registration field $\left(\sigma^2 \left(\hat{U}_{t,x}^{cum} \right) + \sigma^2 \left(\hat{U}_{t,y}^{cum} \right) + \sigma^2 \left(\hat{U}_{t,z}^{cum} \right) \right)$ is attained. A maximum number of iterations, $t_{max} = 50$, is also set.

7.4 Results

Three different databases of brain MRI images were used. The first consisted of images from the BrainWeb simulator [42, 43]. The second consisted of images from the Human Connectome Project (HCP) [44–47]. The last was a database of Parkinson's disease patients brain images. Two images were considered for each individual that were of the most commonly used brain anatomic MRI contrast mechanisms, a T_1w image and a T_2w image. The registration was intra-subject with reference to the original T_1w image. The moving was the T_2w image. The images were first restored for intensity uniformity [35].

The registration considers as ROI of the reference image the brain region from the entire head image. However, for the moving, T_2w image, the ROI was the entire image domain. The extensive ROI considered for the moving image allows registrations under extensive transformations. The brain region of the BrainWeb images is directly available through the union of the spatial tissue type classification images. The brain region of the real images was identified with the BET tool [54].

A further simulated misregistration was applied to the moving images considering selected planes. It consisted of a rotation around an axis normal to a selected plane as well as a translation of that same axis. Three planes of the brain were used: the axial, the sagittal, and the coronal. Two sets of transformations were applied to the moving T_2w images for each plane. The first was a shift of 10 mm and a rotation of 30°. The second was a shift of 15 mm and a rotation of 20°. They were applied to the 3D MRI datasets with the MIPAV tool [55]. These provided a total of six pairs of misregistered images. Eventually, the registration methodology was applied to give the corresponding registered moving images.

7.4.1 BrainWeb Simulator Data

The first database used images from the 1.5 Tesla BrainWeb brain MR simulator. The images were from the anatomic T_1 weighted, T_1w, and T_2 weighted, T_2w, MRI contrast mechanisms. Both images in a pair have a resolution of $1.0 \times 1.0 \times 1.0$ mm^3 in a matrix of size $181 \times 181 \times 217$. The pair of T_1w and T_2w images were corrupted with the highest level of noise available from the simulator, that is $N=5\%$. The highest level of intensity non-uniformity available, $B=100\%$, was also used. The use of these corrupted data demonstrated the robustness of the method to the restoration and the registration methodology.

The results of the most challenging registration examples are in Figure 7.4 and in Figure 7.5. In Figure 7.4 two misregistrations of the sagittal plane are shown. In Figure 7.5 a misregistration of the coronal plane and a misregistration of the axial plane are shown. The registered T_2w images are shown within the ROI of the original T_1w image. The registration is successful in all six cases. The registration decreases the number of distributions in the joint statistics and makes the remaining distributions sharper.

7.4.2 Human Connectome Project (HCP) Data

The second database consists of anatomic datasets from two studies of the Human Connectome Project (HCP) [44–47]. The first was anatomic data from the Lifespan (LS) pilot study, and the second was anatomic data from the Retest study. The data of the HCP LS study was from 27 volunteers from six age groups: four–six, eight–nine, 14–15, 25–35, 45–55, and 65–75 years old. The second was data from the HCP Retest study. It

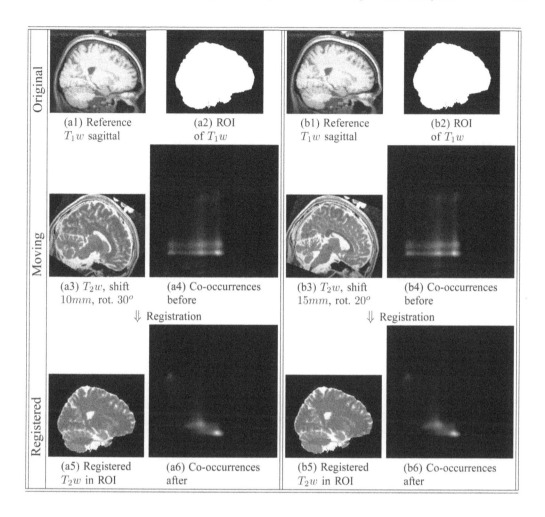

FIGURE 7.4: Two examples of misregistered images from the BrainWeb simulator under two different transformations together with the results of the registration. The images are shown contrast enhanced to improve visualization. The images are corrupted with a high level of noise as well as originally with a high level of intensity non-uniformity. The reference is the T_1w image, and the moving is the transformed T_2w image. The original joint statistics with the misregistered images are shown as well as the statistics after the registration. In both cases the misregistration is a rotation and a translation of the sagittal plane. A larger rotation is in (a3), and a larger translation is in (b3). In both cases the registered images are correctly placed in the reference image space. The joint statistics with the registered images have fewer and sharper distributions.

was data from 45 aged volunteers, 31 women and 14 men. The volunteers were imaged at 3.0 Tesla (Siemens, Connectom and Prisma). A T_1 weighted 3D structural MPRAGE sequence was acquired sagittally with $TR/TE/TI = 2400/2.14/1000$ ms. A T_2 weighted 3D structural SPACE sequence was acquired sagittally with $TR/TE = 3200/565$ ms. The voxel resolution of both the T_1w and the T_2w images is $0.8 \times 0.8 \times 0.8$ mm^3. The matrix size of both the T_1w and the T_2w images is $208 \times 300 \times 320$. The youngest group of age

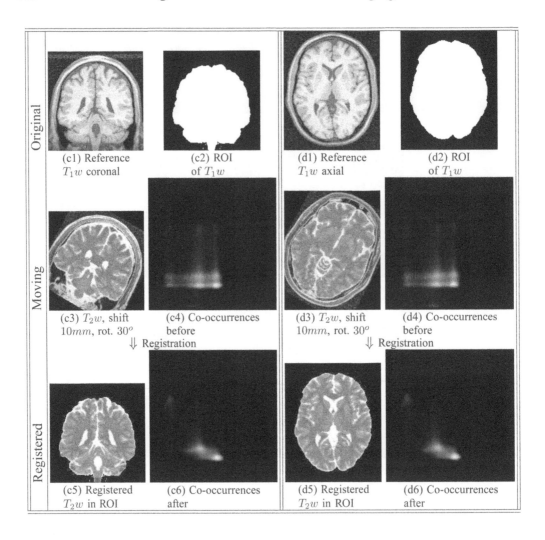

FIGURE 7.5: Two examples of misregistered images from the BrainWeb simulator under two different transformations together with the results of the registration. The images are shown contrast enhanced to improve visualization. The images are corrupted with a high level of noise as well as originally with a high level of intensity non-uniformity. The reference is the T_1w image, and the moving is the transformed T_2w image. The original joint statistics with the misregistered images are shown as well as the statistics after the registration. The misregistration is of the coronal plane in (c3) and of the axial plane in (d3). In both cases the registered images are correctly placed in the reference image space. The joint statistics with the registered images have fewer and sharper distributions.

range four–six years old of the HCP LS pilot project was scanned using a customized pediatric head coil and an optimized protocol for that age range. The original datasets are used without any pre-processing as they are available from the HCP.

The results of representative examples of registrations are in Figure 7.6 and in Figure 7.7. In Figure 7.6 two misregistrations of the sagittal planes are shown. In

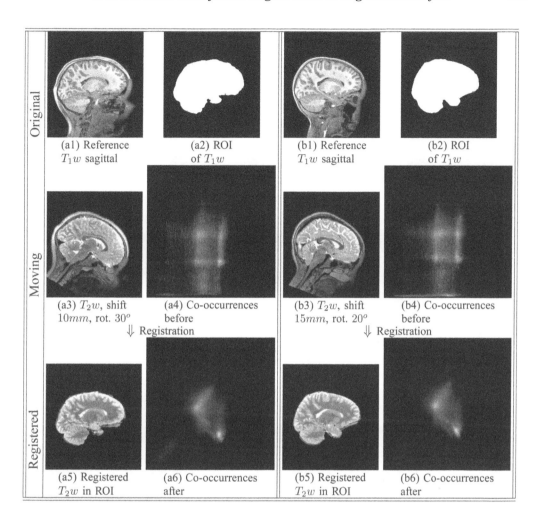

FIGURE 7.6: Two examples of misregistered images from the Human Connectome Project [44–47] together with the registration results. The images are shown contrast enhanced to improve visualization. The reference is the T_1w image, and the moving is the transformed T_2w image. The original statistics with the misregistered images are shown as well as the statistics after the registration. In both cases the misregistration is a rotation and a translation of the sagittal plane. A larger rotation is in (a3), and a larger translation is in (b3). In both cases the registered images are aligned with the reference image correctly. The joint statistics with the registered images have fewer and sharper distributions.

Figure 7.7 a misregistration of the coronal plane and a misregistration of the axial plane are shown. The registered images are shown to be within the ROI of the original T_1w image. The registration is successful in all cases. The registration decreases the number of distributions in the joint statistics and makes the remaining distributions sharper.

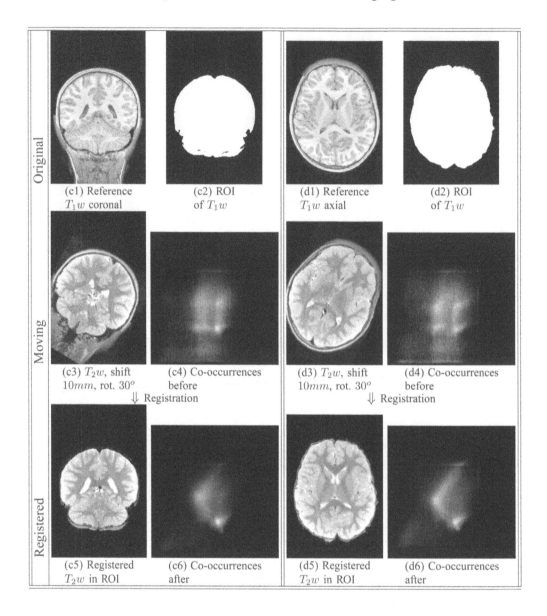

FIGURE 7.7: Two examples of misregistered images from the Human Connectome Project [44–47] together with the registration results. The images are shown contrast enhanced to improve visualization. The reference is the T_1w image, and the moving is the transformed T_2w image. The original statistics with the misregistered images are shown as well as the statistics after the registration. The misregistration is of the coronal plane in (c3) and of the axial plane in (d3). In all cases the registered images are aligned with the reference image correctly. The joint statistics with the registered images have fewer and sharper distributions.

7.4.3 Data of Parkinson's Disease Patients

The third database was of brain images of patients at an advanced stage of Parkinson's disease.* The datasets were retrieved retrospectively from the clinical database of University Hospital Goettingen, Germany. The use of the patient data was retrospective, fully anonymized, and according to the guidelines of the local ethical committee for clinical research. The patients gave informed consent for all imaging procedures. All pre-operative images were acquired under anesthesia to reduce motion artifacts and to increase anatomic precision. The quality of the images was evaluated by observation and images with artifacts were removed. The data were from 60 patients, 41 men of average age 59.91 ± 7.51 years and 19 women of average age 65.86 ± 7.40 years.

The Parkinson's disease patients were imaged at 3.0 Tesla (Siemens, TrioTim). A standard T_1 weighted 3D structural MPRAGE sequence was acquired sagittally with $TR / TE / TI = 2000 / 2.98 / 900$ ms, which gave an in-plane resolution of 1.0×1.0 mm^2 and a slice thickness of 1.1 mm. The matrix size of the $T_1 w$ images is sufficient to cover the entire head region and of size at least $240 \times 256 \times 160$. A high-resolution 3D SPACE structural T_2 weighted imaging that was axially planned was also acquired with $TR / TE = 1500 / 355$ ms that gave an in-plane resolution of 0.63×0.63 mm^2 and a slice thickness of 1.80 mm. The size of the $T_2 w$ images was $308 \times 384 \times 80$. The $T_1 w$ and the $T_2 w$ patient brain datasets were placed in the same resolution and in the same sampling grid. The reference grid was that of the $T_1 w$ image and the $T_2 w$ image was resampled with a closest neighbor filter [21]. In the high resolution $T_2 w$ image the caudal cerebellum is not always depicted, depending on the skull size and the "brain fitting" in a standard protocol for all Parkinson's disease patients. The primary target in Parkinson's disease diagnostic is the high-resolution imaging of the mesencephalon and of the basal ganglia. The in-plane imaging orientations of the two sequences are complementary to enable a more complete clinical evaluation.

The results of the registration of representative images are in Figure 7.8 and in Figure 7.9. In Figure 7.8 two misregistrations of the sagittal planes are shown. In Figure 7.9, a misregistration of the coronal plane and of the axial plane is shown. The registered images are shown to be within the ROI of the original $T_1 w$ image. The resulting registrations are successful. The joint statistics with the registered moving images make the distributions sharper and the distribution corresponding to the cerebrospinal fluid becomes apparent. The registration is successful in cases where the entire brain is imaged. It is also successful when there are no extensive artifacts in the data. One such type of artifact results when decreased time is used for imaging. The performance of the registration is lower when there are extensive lesions in the images.

7.5 Discussion

In many cases, rigid registrations are more appropriate than non-rigid registrations. One example is the intra-subject rigid registration of datasets that do not suffer from geometric distortions. Moreover, even a non-rigid registration is typically preceded by

---

* The MRI data of the Parkinson's disease patients was kindly provided by Marios Nikos Psychogios, PD Dr. med., Institute for Diagnostic and Interventional Neuroradiology, University Medical Center Goettingen, Robert Koch street 40, 37075 Göttingen, Germany.

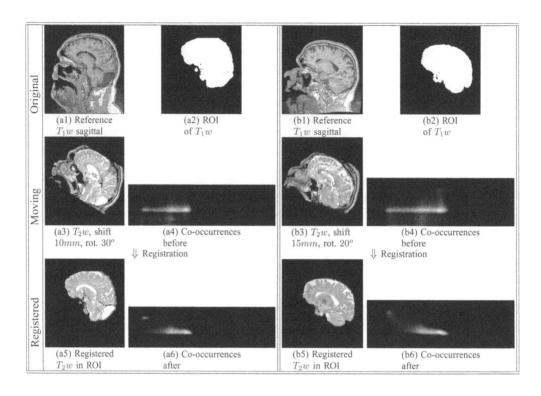

FIGURE 7.8: Two examples of misregistered images from the database of Parkinson's disease patients together with the registration results. The images are shown contrast enhanced to improve visualization. The reference is the T_1w image, and the moving is the transformed T_2w image. The original joint statistics with the misregistered images are shown as well as the joint statistics after the registration. In both cases the misregistration is a rotation and a translation of the sagittal plane. A larger rotation is in (a3), and a larger translation is in (b3). In both cases the registered images are placed in the space of the reference image correctly. The joint statistics with the registered images make the distribution of the cerebrospinal fluid apparent (in a computer monitor).

a rigid registration. This work describes a non-parametric Bayesian method to compute a rigid registration. The methodology developed performs a Bayesian statistical estimate for the linear transformation parameters. It has been validated with datasets from three different databases for which it has successfully performed registration. The resulting joint statistics of the registered images have fewer, sharper, and more apparent distributions. The performance of the registration is lower when there are artifacts in the data or when there are extensive lesions in the images.

The registration method uses the joint statistics of the two images. These are affected by spatial image intensity non-uniformity. Thus, the images are first pre-processed for intensity correction. The registration method assumes that the effect of the spatial misregistration on the joint statistics can be expressed as additive noise with unimodal Gaussian PSF. This PSF is deconvolved from the joint statistics to provide a prior for the statistics of the aligned images. The effective deconvolution to obtain the prior can significantly improve the performance of the registration method. The spatial misregistration is also assumed to be given by an additive displacement noise given by a unimodal

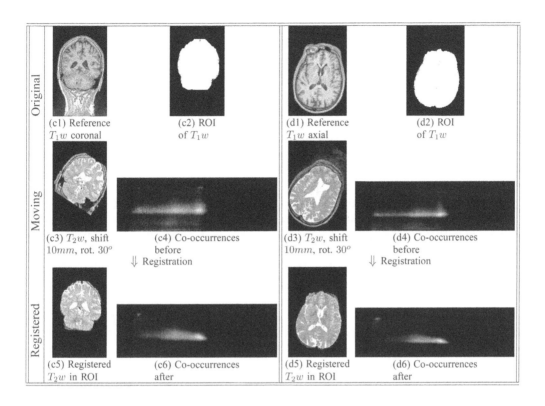

FIGURE 7.9: Two examples of misregistered images from the database of Parkinson's disease patients together with the registration results. The images are shown contrast enhanced to improve visualization. The reference is the T_1w image, and the moving is the transformed T_2w image. The original joint statistics with the misregistered images are shown as well as the joint statistics after the registration. The misregistration is of the coronal plane in (c3) and of the axial plane in (d3). In both cases the registered images are placed in the space of the reference image correctly. The joint statistics with the registered images make the distribution of the cerebrospinal fluid apparent (in a computer monitor).

Gaussian PSF. This PSF provides the likelihood. The prior and the likelihood compute a voxelwise non-parametric Bayesian posterior expectation. The voxelwise estimation for the whole image gives an intermediate vector field $\hat{U}$. This field processed with the simple and standard case of the Procrustes method computes a rigid body rotation with a translation as the maximum likelihood [29, 50]. The PSF of the noise in the statistics and the displacement in space are assumed to be unimodal distributions and mutually independent. However, as shown experimentally, this assumption is robust to datasets where the effective PSFs from the distortions can be more complicated. The setting of the two variance parameters considers the relation that exists between the two.

The use of the joint statistics together with the Bayesian methodology is effective both for the registration of datasets of the same contrast as well as for the registration of datasets of different contrasts. In same contrast registration, the statistics are constrained to lie along the diagonal in the domain of the 2D statistics. The methodology is also particularly effective for multi-contrast registration because of the potential for a larger distance between different tissues distributions in the domain of the joint statistics. The non-parametric methodology is able to more directly benefit from the greater distance between the distributions. In this work, this has been demonstrated extensively with the registration of multi-contrast MRI data. The method can also be extended for the co-registration of multi-modal images, such as for MRI data with CT data. The distance between tissues distributions cannot be used directly when optimizing with a uni-modal functional like entropy that strives to achieve uni-modality in the statistical domain. As a result, it decreases tissues contrast and tends to cause geometric misregistrations. The entropy can misregister the images depending on the magnitude, the spread, and the location of the distributions corresponding to different image regions.

The non-rigid registration for local distortions can have an ROI that is approximately the same for both the moving and for the reference image. Thus, the specification of a single ROI for the entire registration is sufficient. However, in registrations for large rotations the ROIs corresponding to the moving and to the reference images do not in general correspond and overlap. In fact, there might be a considerable difference between the two. In such cases, the use of a single ROI for both the moving and for the reference image in the registration is not representative. The ROI specification in such cases is not sufficiently addressed in the literature or by automated registration tools [19, 20, 22]. In this work, to accommodate this effect the ROI of the reference and of the moving images can be different. In fact, the ROI of the moving image is larger than that of the reference image and can even be the entire domain of the moving image. This property allows the method to perform registration for extensive misregistrations, as it is also shown experimentally. The use of the more extensive ROI in the moving image must be done with care, since it can affect the joint statistics and the overall registration results. One such example is when the region adjacent to the target ROI within the moving image is similar to the region in the actual target ROI in this image. The option for two different ROIs for the method enables the extension of the method to cases where the moving image in the registration has partially missing data [56]. The method has been demonstrated for a misregistration of translation together with a rotation of even an angle of 30°. This misregistration is quite large for practical clinical applications. However, it does demonstrate the robustness of the methodology.

This method is not only statistical Bayesian but also non-parametric. This formulation obviates the need for more conventional approaches based on differential calculus. The calculus-based approaches very often use gradient descent optimization. The latter optimization is part of more conventional implementations of registration in a variety of toolboxes [19, 21, 22]. However, the gradient descent is prone to local maxima both in terms of deformation in space and in terms of entropy values in the statistics. Also, the use of B-splines bases for spatial smoothing together with gradient descent optimization makes registration anisotropic along the axes. The estimation of the registration with the proposed method from global image statistics rather than from a scalar cost enables the identification of spatial transformations closer to the global optimum.

The method can be extended by combining the non-parametric Bayesian formulation that gives the intermediate vector field with additional larger classes of spatial transformations. The direct generalization of the Procrustes method to larger classes can be for scaling or for affine transformations. The Procrustes method is extended to include scaling with variance normalization [29]. The extension for the affine transformations case is the complete standard least squares [28]. The less constrained class of non-rigid registrations that has been implemented as the NParBR method can be formulated by estimating spatial smoothing with the maximum likelihood estimate whose likelihood is a Gaussian distribution over the gradient magnitude [34].

In general, big data are characterized in terms of their volume, variety, velocity, and veracity [1, 2, 12]. The method developed in this work involves the volume aspect of the data because of the large size of the 3D MRI datasets. It also involves the variety aspect because the MRI datasets are of different contrasts. Finally, it involves veracity due to the artifacts removed from the images as well as the validation of the restored images with a variety of databases.

The literature on big data currently still reports to a great extent the technical progress on data organization but not yet a sufficient number of studies on the development of the necessary analysis methodologies such as the non-parametric method described in this work [15]. The technical objectives of the storage and accessing of big data also involve legal and ethical issues beyond those traditionally considered by an institutional ethical review board. A clinical researcher must be aware of the legislation for data security protection and for the necessity for informed consent from the patients [14]. The ethical issues involve a balance between privacy and protection on one hand and beneficial conclusions as well as future prospects on the other [13, 14].

References

1. E. Filonenko and E. Seeram, "Big data: The next era of informatics and data science in medical imaging: A literature review," *Journal of Clinical & Experimental Radiology*, vol. 1, no. 1, pp. 1–6, 2018.
2. A. Kouanou, D. Tchiotsop, R. Kengne, D. Zephirin, N. Armele, and R. Tchinda, "An optimal big data workflow for biomedical image analysis," *Informatics in Medicine Unlocked*, vol. 11, pp. 68–74, 2018.
3. S. Bao, Y. Huo, P. Parvathaneni, A. Plassard, C. Bermudez, Y. Yao, I. Lyu, A. Gokhale, and B. Landman, "A data colocation grid framework for big data medical image processing: Backend design," in *Proceedings of SPIE International Society for Optical Engineering*, vol. 10597, 2018.
4. A. Belle, R. Thiagarajan, S. Soroushmehr, F. Navidi, D. Beard, and K. Najarian, "Big data analytics in healthcare," *BioMed Research International*, vol. 370194, pp. 1–16, 2015.
5. A. Kansagra, J. Yu, A. Chatterjee, L. Lenchik, D. Chow, A. Prater, J. Yeh, A. Doshi, C. Hawkins, M. Heilbrun, S. Smith, M. Oselkin, P. Gupta, and S. Ali, "Big data and the future of radiology informatics," *Academic Radiology*, vol. 23, no. 1, pp. 30–42, 2016.
6. A. Kharat and S. Singhal, "A peek into the future of radiology using big data applications," *Indian Journal of Radiology and Imaging*, vol. 27, no. 2, pp. 241–248, 2017.
7. M. Morris, B. Saboury, B. Burkett, J. Gao, and E. Siegel, "Reinventing radiology: Big data and the future of medical imaging," *Journal of Thoracic Imaging*, vol. 33, no. 1, pp. 4–16, 2018.
8. F. Maldonado, T. Moua, S. Rajagopalan, R. Karwoski, S. Raghunath, P. Decker, T. Hartman, B. Bartholmai, R. Robb, and J. Ryu, "Automated quantification of radiological patterns predicts survival in idiopathic pulmonary fibrosis," *European Respiratory Journal*, vol. 43, no. 1, pp. 204–212, 2014.
9. T. Simonite, "IBM's automated radiologist can read images and medical records," *MIT Technology Review*, February 4, 2016.
10. S. Toh and R. Platt, "Big data in epidemiology: Too big to fail?" *Epidemiology*, vol. 24, no. 6, p. 939, 2013.
11. M. Khoury and J. Ioannidis, "Big data meets public health: Human well-being could benefit from large-scale data if large-scale noise is minimized," *Science*, vol. 346, no. 6213, pp. 1054–1055, 2014.

12. C. Lee and H. Yoon, "Medical big data: Promise and challenges," *Kidney Research and Clinical Practice*, vol. 36, pp. 3–11, 2017.
13. D. Hand, "Aspects of data ethics in a changing world: Where are we now?" *Big Data*, vol. 6, no. 3, pp. 176–190, 2018.
14. E. Gray and J. Thorpe, "Comparative effectiveness research and big data: Balancing potential with legal and ethical considerations," *Journal of Comparative Effectiveness Research*, vol. 4, no. 1, pp. 61–74, 2015.
15. M. Ienca, A. Ferretti, S. Hurst, M. Puhan, C. Lovis, and E. Vayena, "Considerations for ethics review of big data health research: A scoping review," *PLoS One*, vol. 13, no. 10, pp. 1–15, 2018.
16. B. Avants, C. Epstein, M. Grossman, and J. Gee, "Symmetric diffeomorphic image registration with cross-correlation: Evaluating automated labeling of elderly and neurodegenerative brain," *Medical Image Analysis*, vol. 12, no. 1, pp. 26–41, 2008.
17. D. Rueckert, P. Aljabar, R. Heckemann, J. Hajnal, and A. Hammers, "Diffeomorphic registration using B-Splines," in *Proceedings of MICCAI*, vol. 4191, no. 2, 2006, pp. 702–709.
18. J. Maintz and M. Viergever, "A survey of medical image registration," *Medical Image Analysis*, vol. 2, no. 1, pp. 1–36, 1998.
19. J. Ashburner and K. Friston, "Chapter 4: Rigid body registration," in W. D. Penny, K. J. Friston, J. T. Ashburner, S. J. Kiebel, and T. E. Nichols, eds., *Statistical Parametric Mapping*, Academic Press, 2007.
20. S. Klein, M. Staring, K. Murphy, M. Viergever, and J. Pluim, "Elastix: A toolbox for intensity-based medical image registration," *IEEE Transactions on Medical Imaging*, vol. 29, no. 1, pp. 196–205, 2010.
21. A. Fedorov, R. Beichel, J. Kalpathy-Cramer, J. Finet, J. Fillion-Robin, S. Pujol, C. Bauer, D. Jennings, F. Fennessy, M. Sonka, J. Buatti, S. Aylward, J. Miller, S. Pieper, and R. Kikinis, "3D Slicer as an image computing platform for the quantitative imaging network," *Magnetic Resonance Imaging*, vol. 30, no. 9, pp. 1323–1341, July 2012.
22. R. Kikinis, S. Pieper, and K. Vosburgh, "3D Slicer: A platform for subject-specific image analysis, visualization, and clinical support," in F. Jolesz, ed., *Intraoperative Imaging and Image-Guided Therapy*, pp. 277–289. Springer, New York, 2013.
23. D. Eggert, A. Lorusso, and R. Fisher, "Estimating 3-D rigid body transformations: A comparison of four major algorithms," *Machine Vision and Applications*, vol. 9, pp. 272–290, 1997.
24. F. Pomerleau, "A review of point cloud registration algorithms for mobile robotics," *Foundations and Trends® in Robotics,* vol. 4, no. 1, pp. 1–104, 2015.
25. P. Schoenemann, "A generalized solution of the orthogonal procrustes problem," *Psychometrika*, vol. 31, no. 1, pp. 1–10, 1966.
26. S. Umeyama, "Least-squares estimation of transformation parameters between two point patterns," *IEEE Transactions on PAMI*, vol. 13, no. 4, pp. 376–380, 1991.
27. K. Kanatani, "Analysis of 3-D rotation fitting," *IEEE Transactions on PAMI*, vol. 16, no. 5, pp. 543–549, 1994.
28. G. Golub and C. Van Loan, *Matrix Computations*. Johns Hopkins University Press, Baltimore, MA, 1996.
29. I. Dryden and K. Mardia, *Statistical Shape Analysis*. Wiley, Chichester, 1998.
30. D. Hill, P. Batchelor, M. Holden, and D. Hawkes, "Medical image registration," *Physics in Medicine and Biology*, vol. 46, pp. R1–R45, 2001.
31. B. Likar, M. Viergever, and F. Pernus, "Retrospective correction of MR intensity inhomogeneity by information minimization," *IEEE Transactions on Medical Imaging*, vol. 20, no. 12, pp. 1398–1410, 2001.
32. U. Vovk, F. Pernus, and B. Likar, "Intensity inhomogeneity correction of multispectral MR images," *NeuroImage*, vol. 32, pp. 54–61, 2006.
33. T. Cover and J. Thomas, *Elements of Information Theory*. Wiley Series, 1991.
34. D. Pilutti, M. Strumia, M. Buechert, and S. Hadjidemetriou, "Non-parametric Bayesian registration (NParBR) of body tumors in DCE-MRI data," *IEEE Transactions on Medical Imaging*, vol. 35, no. 4, pp. 1025–1035, 2016.

35. S. Hadjidemetriou, M. Psychogios, P. Lingor, K. Eckardstein, and I. Papageorgiou, "Restoration of bi-contrast MRI data for intensity uniformity with Bayesian coring of co-occurrence statistics," *Journal of Imaging*, vol. 3, no. 67, pp. 1–23, 2017.

36. J. Rossi, "Digital techniques for reducing television noise," *Journal of SMPTE*, vol. 87, no. 3, pp. 134–140, 1978.

37. E. Simoncelli and E. Adelson, "Noise removal via Bayesian wavelet coring," in *Proceedings of 3rd IEEE ICIP*, vol. I, 1996, pp. 379–382.

38. J. Sled, A. Zijdenbos, and A. Evans, "A nonparametric method for automatic correction of intensity nonuniformity in MRI data," *IEEE Transactions on Medical Imaging*, vol. 17, no. 1, pp. 87–97, 1998.

39. A. Vidal-Pantaleoni and D. Marti, "Comparison of different speckle reduction techniques in SAR images using wavelet transform," *International Journal of Remote Sensing*, vol. 25, no. 22, pp. 4915–4932, 2004.

40. D. Pilutti, M. Strumia, and S. Hadjidemetriou, "Bimodal nonrigid registration of brain MRI data with deconvolution of joint statistics," *IEEE Transactions on Image Processing*, vol. 23, no. 9, pp. 3999–4009, 2014.

41. D. Pilutti, M. Strumia, and S. Hadjidemetriou, Non-Parametric Bayesian Registration (NParBR) on CT Lungs Data: EMPIRE10 Challenge, 2015.

42. D. Collins, A. Zijdenbos, V. Kollokian, J. Sled, N. Kabani, C. Holmes, and A. Evans, "Design and construction of a realistic digital brain phantom." *IEEE Transactions on Medical Imaging*, vol. 17, no. 3, pp. 463–468, 1998.

43. C. Cocosco, V. Kollokian, R.-S. Kwan, and A. Evans, "BrainWeb: Online interface to a 3D MRI simulated brain database," *NeuroImage,* vol. 5, no. 4-2/4, p. S425, 1997.

44. D. Van Essen, S. Smith, D. Barch, T. Behrens, E. Yacoub, and K. Ugurbil, "The WU-Minn human connectome project: An overview." *NeuroImage*, vol. 80, pp. 62–79, 2013.

45. M. Milchenko and D. Marcus, "Obscuring surface anatomy in volumetric imaging data," *Neuroinformatics*, vol. 11, no. 1, pp. 65–75, 2013.

46. C. Rorden, "MRIcron suite, dcm2nii utility," http://www.nitrc.org/projects/mricron/, 2008.

47. D. Marcus, J. Harwell, T. Olsen, M. Hodge, M. Glasser, F. Prior, M. Jenkinson, T. Laumann, S. Curtiss, and D. Van Essen, "Informatics and data mining: Tools and strategies for the human connectome project." *Frontiers in Neuroinformatics*, vol. 5, no. 4, 2011.

48. P. V. Cittert, "Zum einfluss der spaltbreite auf die intensitatswerteilung in spektrallinien II," *Zeitschrift für Physik*, vol. 69, pp. 298–308, 1931.

49. A. Katsaggelos, "Iterative image restoration algorithms," *Optical Engineering*, vol. 28, no. 7, pp. 735–748, 1989.

50. J. Fitzpatrick, J. West, and C. Maurer, "Predicting error in rigid-body point-based registration," *IEEE Transactions on Medical Imaging*, vol. 17, no. 5, pp. 694–702, 1998.

51. *The VXL Homepage*, https://vxl.github.io/.

52. *VXL Book*, http://paine.wiau.man.ac.uk/pub/doc_vxl/books/core/book.html, 2009.

53. The Insight Segmentation and Registration Toolkit, http://www.itk.org.

54. S. Smith, "Fast robust automated brain extraction," *Proceedings of Human Brain Mapping*, vol. 17, pp. 143–155, 2002.

55. *MIPAV (Medical Image Processing, Analysis, and Visualization)*, Center for Information Technology, National Institutes of Health, Bethesda, MA, 2007.

56. N. Chitphakdithai and J. Duncan, "Non-rigid registration with missing correspondences in preoperative and postresection brain images," in *Proceedings of MICCAI*, vol. 13, no. 1, 2011, pp. 367–374.

Chapter 8

Multimodal Analysis in Biomedicine

Mohammad-Parsa Hosseini, Aaron Lau,
Kost Elisevich, and Hamid Soltanian-Zadeh

8.1   Medical Big Data

As technology gets more advanced throughout all industries, there is an increasing interest in measuring and analyzing human behavior. Data mining in retail, finance, and medicine are among the industries where information about consumers will have a huge impact on the way businesses operate. Whether this data will be used to increase profits or to better our quality of life, it is certain that all of this data will lead to changes in our society in every aspect.

In particular, big data in medicine has greatly changed the way in which we analyze and manage information (Figure 8.1). The advancements in the medical device industry have allowed physicians to collect vast amounts of data about our personal well-being. The transformation and digitization of information drive a more proactive healthcare model creating more accessible information for both the patient and the physician. The generation of this data also comes from advances in the medical device industry as well as a shift in our culture and how we are choosing to manage our own health. Patient-generated health data is a large contributor to the amount of information that is being generated. This has been made possible by the growth in wearable technologies in recent years. This includes everyday lifestyle monitoring that tracks the number of steps taken in a day or monitors our blood pressure as we perform our normal daily functions. As the number of wearable medical devices increases, individuals will have growing access to objective information on their own well-being.

A potential issue in the field of medical big data analysis is the manner in which much of this data is stored [1]. Because of privacy concerns, a patient's information is distributed over several databases. As an example, as genetic sequencing becomes cheaper and more accessible, an increasing number of individuals will choose to sequence their DNA. This detailed genetic information, at the outset, would be excluded from the medical record. Connections between such datasets, however, would provide a more informed analysis for the individual and could be used in population studies across wider geographies and cultures opening new frontiers in the understanding of

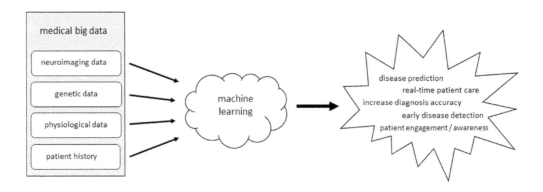

FIGURE 8.1:   A workflow representation of machine learning and medical big data.

variations in healthcare. Additionally, large amounts of data from medical equipment such as neuroimaging and electrodiagnostic analytic tasks provide a platform for deep learning techniques to be trained with a single model without the need for manual, labor-intensive screenings. Applications of big data in healthcare can increase our effectiveness in diagnosing disease and predicting outcomes. Current computer applications available to physicians only allow for basic functions such as context-sensitive warning messages, reminders, suggestions for economical prescribing, and results of quality improvement activities. The following five characteristics must be considered to further promote the use of big data: (1) standardization and categorization of patient groups; (2) automation of analytic methodologies; (3) optimization of data processing speed to facilitate new incoming data; (4) user-friendly data analytic methodologies; (5) translatability into a readable form for both the clinician and patient [1]. These five characteristics are important in applying analytics to healthcare datasets because of the wide variations in how a single disease may express itself in patients.

Medical imaging techniques such as computed tomography (CT), positron emission tomography (PET), and functional magnetic resonance imaging (fMRI) have generated vast amounts of data that provide information on different aspects of the body. Therefore, we need methods to analyze this big data which are captured and saved in servers. Machine learning, deep learning, and multimodal analysis are the methodologies that are used to analyze, manage, and capture information. Big data is very well suited for deep learning and multimodal analysis where a very large amount of data is needed for training and developing a high-performing algorithm. Also, these methods have been used widely for decision-making tasks in the form of computer-aided diagnosis systems. In the next part, we look at these methods.

8.2   Big Data Analysis with Machine Learning

Machine learning methods have been increasingly used in the medical imaging field for computer-aided analysis of diagnostics and prognostic models. Several of these methods such as supervised, unsupervised, and deep learning have more recently been employed to solve medical imaging related problems. Machine learning is a subset of artificial intelligence that allows machine algorithms to be programmed to an optimized performance criterion with the use of a known training dataset. The learning happens when an algorithm is trained to go through data and undertake an iteration process until it is able to correctly identify characteristic features. This machine learning is used mostly in cases where human expertise does not exist or is unable to quickly assimilate, process, and understand data. The unique advantage to machine learning is the ability to take in large amounts of data and be able to recognize patterns in a very short timeframe exceeding human ability. This has been made possible by the increase in access to large sources of clinical data and the increasing power of consumer computers.

As with every other field in science and healthcare, there will be ethical challenges that must be overcome as machine learning gains more access to data. A training algorithm may reflect human bias in decision-making. Likewise, inherent biases may exist within the available data. For example, an algorithm that is designed to predict certain outcomes may be biased against a certain population if less data exist for it. Large datasets are, therefore, a requirement to overcome queries that require subtle distinctions to be made in such studies. There is concern that, as machine learning advances in medical diagnostics, we will become too dependent upon computer applications without considering possible exceptions to standard presentations. This is a well-known circumstance in medicine. Quality in patient care must inevitably include physician consideration in the vetting of clinical decisions.

The two main classes of machine learning are supervised and unsupervised learning (see Figure 8.2). Supervised learning occurs when an algorithm or function is trained on a set of labeled data. The supervised learning algorithm analyzes a dataset and creates an inferred function from the generalized data. Classification and regression are two types of supervised learning. In classification, the algorithm attempts to separate the data into distinct groups. For example, classification is used in pattern and facial recognition – lighting, hair style, pose, and structure. In a medical context, certain symptoms may be grouped together and linked to a particular illness or to render definition for

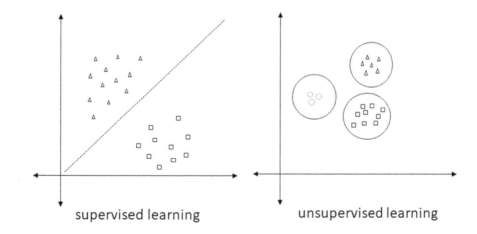

FIGURE 8.2: A representation of supervised learning compared to unsupervised learning. Supervised learning develops a model based on data from both input and output. Types of supervised learning include classification and regression. Unsupervised learning only depends on input data. Clustering is an example of unsupervised learning.

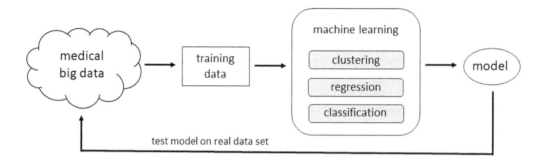

FIGURE 8.3: A workflow representation of applying machine learning algorithms in healthcare big data.

several possibilities. In regression, the relationship between two variables is modeled so that a continuous variable can be predicted from the other, e.g., the prediction of blood pressure from weight. Supervised learning is one of the basic learning methods that uses simple rules to predict an outcome from an input dataset. A slightly more complicated concept is that of unsupervised learning. This method forms inferences without labeled outcomes. An example of unsupervised learning would include anomaly detection and clustering. Objects that are similar to one another are grouped together. These types of methods are very common also for pattern recognition and for image analysis.

Machine learning has frequently been used to analyze and process large datasets. More recently, there is increasing research in using a multimodal method of machine learning where several machine learning algorithms are utilized to analyze imaging data from different modality sources (see Figure 8.3).

8.3 Multimodal Analysis

Multimodal imaging combines two or more imaging techniques in a single examination to allow for the integration of several analyses in order to have a better understanding of disease biology. A synchronous image acquisition is the best solution to achieve consistency in time and position of the scan. Common multimodal imaging techniques include single photon emission computed tomography (SPECT), PET-CT, and, more recently, PET-MR [2]. PET-CT is one of the most common combinations where the PET provides information regarding function through the distribution of radionuclide-labeled agents while the CT provides information about anatomical structure. This type of image acquisition allows physicians to better identify problems in the body. Additionally, electroencephalography (EEG), magnetoencephalography (MEG), and fMRI study neural activity and interactions. EEG and MEG provide high temporal resolution and fMRI uses blood oxygenation level dependent (BOLD) contrast to give high spatial resolution on an imaging platform that identifies signal location. A combination of these modalities increases the spatial resolution of electromagnetic source imaging to provide insight into an epileptogenic network within the brain [3].

Systems that are able to detect two or more modality signals at the same time take advantage of unique characteristics that each method possesses. Machine learning augments these systems with the use of a multimodal algorithm to extract additional desired features from each imaging modality. Multimodal integration allows high resolution classification using already existing methods [4] to achieve more informed results. A limitation of this multimodal data extraction is the increased complexity of the algorithm which creates some difficulty in determining true accuracy as existing methods may be individually weighted to provide more refined determinations. An example of this application in EEG is the diagnosis of multiple sclerosis patients. In this study, the T-test and Bhattacharyya criteria were used as a preprocessing step and were followed by a combination of KNN and SVM as the primary classification algorithm. This resulted in a total accuracy of 93% [5].

Multimodal imaging techniques are often differentiated into two categories: asymmetric and symmetric data analysis approaches (see Figure 8.4). In an asymmetric approach, the analysis uses one modality to bias the estimates of another [6]. Many asymmetric analyses are similar to the supervised learning method of regression where one modality is used to extract features of another. For example, the amplitude of an ERP component from EEG/MEG data can be extracted and correlated with fMRI data. The leveraging of one modality to bias another bears the potential of losing information from the second modality. Despite this, asymmetric analysis has the advantage of having one modality create a model of measurement noise. This noise model allows for subtle artifacts to be removed from the analysis.

In contrast, symmetric analysis processes both modalities simultaneously, and, in doing so, it identifies missing aspects of other modalities. Careful consideration is given to "preanalysis" steps such as feature selection in multimodal application. Attention may be given to employing unsupervised methods in order to identify critical features rather than relying on model assumptions or a manual feature selection. For example, unsupervised methods are able to learn features important for neurovascular coupling [7]. Unsupervised models are, therefore, used to find structure in the data when no stimulus variable may exist.

There are several unsupervised learning models that are being used in multimodal data analysis. These include principal component analysis (PCA), independent component

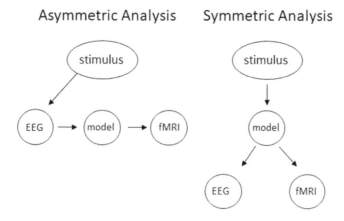

FIGURE 8.4: Multimodal methods usually have either an asymmetric or symmetric approach. In asymmetric analysis, features from one modality are used to improve on the features of another. Symmetric approaches analyze both modalities jointly [6].

analysis (ICA), functional connectivity analysis, and canonical correlation analysis (CCA) [6]. Clustering has also been used in many multimodal analyses where groups of objects similar to one another are grouped together. A well-known clustering analysis is called the k-means algorithm where each data point in the set is assigned a label (k) and then divided into groups where data points within each group have similarities. Clustering is commonly used for data exploration and to understand the structure of data.

Due to the vast amounts of imaging that are produced when analyzing the structure and function of the brain, neuroscience is an attractive field for the development of multimodal machine learning algorithms.

8.4 Multimodal Processing in Neuroscience

Functional changes in the brain may precede detectable structural changes [8, 9] and may be detected by existing noninvasive modalities. Functional connectivity analysis using EEG and resting-state fMRI (rs-fMRI) [10], complemented by diffusion tensor imaging (DTI), has provided such meaningful input in cases of temporal lobe epilepsy (TLE) [11]. To this end, the brain is modeled as a connected network of nodes in which connectivity matrices are estimated from EEG and rs-fMRI data. The nodes may be selected based on structural or functional parcellation of the brain using model-based or model-independent (data-driven) methods. The entire connectivity matrix or specific connections between any two groups of nodes may be compared between two groups of subjects. Whole brain connectivity analysis may reveal differences between groups but requires large samples and complicated statistical analysis [12].

Multimodal analysis of brain images to diagnose neurological disorders can also be paired with nonimaging factors to discover underlying correlations among illnesses. Schizophrenia and bipolar disorder are genetically related and are leading causes of disability worldwide [13]. Multimodal imaging along with clinical and behavioral variables must be analyzed in concert because MRI attributes can be influenced by lifestyle choices (smoking, substance abuse, etc.) [14]. Multivariate covariation between nonimaging and imaging variables must be analyzed to obtain measures of cortical thickness, subcortical volume, task-related brain activation, resting-state functional connectivity, and white matter fractional anisotropy (FA) [15]. Factors known to be associated with an MRI feature are included in the nonimaging dataset and may involve lifestyle factors, physical health, intelligence quotient (IQ), substance abuse, basal metabolic index (BMI), and medication. A sparse canonical correlation analysis (sCCA) is applied to take into account the varied sources of each dataset. Canonical correlation analysis is a multimodal analysis method that is very useful in determining correlations between multiple sets of variables. The sCCA analysis between imaging and nonimaging datasets typically demonstrates a substantial covariation among multiple variables. Prior study results highlight, for instance, the association between BMI and neuroimaging phenotypes. The relation of an individual's lifestyle to brain imaging features emphasizes the importance of such associations as possibly leading to early intervention in order to mitigate risk.

Multimodal analysis is also used in the diagnosis of Alzheimer's disease for treatment and possibly delaying illness. Deep polynomial network (DPN) is a deep learning algorithm that is able to perform well with large datasets and to learn effective feature representations on small datasets [16, 17, 18]. This is a new concept that improves

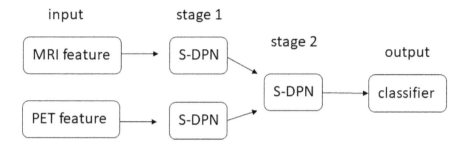

FIGURE 8.5: A representation of the multimodality framework using stacked Deep Polynomial Networks (S-DPN) for Alzheimer's disease classification with PET and MRI features. Image redrawn from Figure 2 of Zheng et al. [21].

performance on large datasets compared to deep belief networks and stacked autoencoder algorithms [19]. A multimodal stacked DPN is able to fuse and learn features from multimodal neuroimaging data for Alzheimer's disease diagnosis. The analysis first learns high level features of MRI and PET images. These features are extracted from only their corresponding imaging modality and therefore do not provide any correlation between PET and MRI. High-level features from PET and MRI are then fed into another DPN network to combine the two algorithms [20]. This creates a stacked, two-staged DPN network that is able to correlate information from features of both imaging modalities (see Figure 8.5). In the analysis done by Zheng et al. the stacked DPN network was able to demonstrate superior performance over the original DPN algorithm for both imaging modalities and achieved an accuracy of over 97% [21].

Before a deep learning algorithm is applied to a dataset, there are usually some preprocessing steps that must take place prior to the analysis. As an example, neuroimaging data were taken from the Alzheimer's Disease Neuroimaging Initiative (ADNI) [22]. Images from PET and MRI were fused to align the different modalities and improve the overall quality of the image. Noise reduction filters, such as the Wiener filter, were applied to reduce the amount of additive noise in the PET and MR images [22]. After filtering, the images went through a feature extraction step and then through the Elman Back Propagation Network for pattern classification [22]. Elman Back propagation Networks are Recurrent Neural Networks that undertake a supervised training algorithm [23]. Feature extraction was performed with both MR and PET images to classify the normal dataset as "No disease" and the subjects with Alzheimer's disease as "Alzheimer's disease." As there is no single imaging biomarker that can predict Alzheimer's disease with 100% certainty, image analysis from multiple modalities is used to increase the diagnostic accuracy for Alzheimer's disease.

8.5 Multimodal Analysis in Epilepsy, an Application

Multimodal analysis has potential to be used as a diagnosis tool to classify neurological diseases such as epilepsy. An example of this is for the detection of functional changes in the brain which may precede detectable structural changes [9] and may be detected by existing noninvasive modalities. Functional connectivity analysis through EEG and rs-fMRI [10], complemented by diffusion tensor imaging (DTI), has provided such meaningful input in cases of temporal lobe epilepsy (TLE) [11]. To this end, the brain is modeled as a connected network of nodes and connectivity matrices are estimated from EEG and rs-fMRI data. The nodes may be selected based on structural or functional parcellation of the brain using model-based or model-independent data-driven methods. The entire connectivity matrix or specific connections between any two groups of nodes may be compared between two groups of subjects. Whole brain connectivity analysis can reveal major differences between groups and requires large sample sizes and complicated statistical analysis [12]. As an example, in unilateral TLE patients, increased functional connectivity of the default mode network (DMN) with other brain regions has been identified in those cases with left TLE in contrast to a decreased connectivity in those with right TLE. Efficient handling and processing of medical big data can provide useful information about a patient and about diseases. To understand the task at hand, it is useful to review the current investigational aspects involved in elucidating the patient's epilepsy. In those patients declared to have an epileptogenicity that can be further investigated to establish its location in the brain, a number of standard neuroimaging, functional and electroencephalographic studies are undertaken. These include MRI, SPECT, PET, inpatient scalp EEG and video monitoring (phase I), sodium amobarbital study, and a neuropsychological profile. In select cases, a variety of further MRI postprocessing applications and MEG are applied. Several quantitative neuroimaging metrics have been applied to provide greater precision and reproducibility in defining putative sites of epileptogenicity particularly as it applies to the most common area of involvement, the mesial temporal lobe. These are correlated with the EEG data to render an initial assumption of the site of epileptogenicity, and these may be reported with varying degrees of certainty.

Based upon our previous studies [24–29], definitive therapy may be decided in the form of resective surgery or entirely discounted on the basis of multifocality suggesting greater than two sites of independent epileptogenicity. When uncertainty exists regarding the location of a particular focality or a need exists to establish the eloquence of cerebral function in the vicinity of a putative site, then intracranial electrographic investigation (i.e., phase II) is required in the form of extraoperative electrocorticography (eECoG). This requires the intracranial placement of surface and/or depth electrode arrays in specific locations of the brain to better understand the distribution of the epileptogenic network and a further admission to the Epilepsy Monitoring Unit (EMU). The results will often declare the approach to be taken therapeutically.

Beyond diagnosis, similar applications of this technology offer solutions for therapeutic interaction with an area of epileptogenicity that does not entail removal of a portion of the brain to achieve a beneficial outcome. Such an approach first requires adequate detection of ictal onset. The use of computers to help physicians in the acquisition, management, storage, and reporting of brain signals(i.e., EEG) is well established. To this end, there are computer-aided detection applications that use a brain–computer interface (BCI). In order for an autonomic computing system to work effectively, computational algorithms must reliably identify periods of increased probability of an impending ictal occurrence in order to abort its development. Such preictal periods may be of variable duration and may not afford suitable latency to provide current methodologies with sufficient time for signal deployment to achieve control in all circumstances. The development of an autonomic method for detection would optimize seizure control and bring about an improved quality of life.

The shape of the brain network using rs-fMRI and EEG data is shown in recent publications [30, 31]. The rs-fMRI data can determine the temporal dynamics of functional connectivity but is limited to the scanning time, usually less than 10 minutes. In contrast, both scalp and intracranial (i)EEG can be used for longer term analysis of dynamic changes in functional connectivity. Also, the temporal sampling rate of EEG is higher than that of rs-fMRI. The combination of rs-fMRI and EEG/iEEG in concert with one another may reveal more information about dynamic functional connectivity, although simultaneous fMRI imaging and EEG data acquisition does present challenges [32].

8.6   Conclusion

Multimodal analysis of big data in healthcare will continue to grow and provide greater efficiencies for diagnostics, categorization of disease, and therapeutic application. Neuroimaging is an example where machine learning analysis has made a lasting impact on the way we are able to diagnose disease. This is an area where vast amounts of data from different modalities have been generated. Machine learning techniques allow us to make connections between the modalities for a more robust diagnosis of patient data. Delayed diagnosis in neurological disease often leads to worsening consequences making early diagnosis even more important. The incorporation of multiple different imaging techniques into a comprehensive analytical scheme creates greater opportunity for recognizing features central to a particular disorder. With the addition of larger datasets into machine learning algorithms and rules for feature selection becoming better defined, greater sophistication will follow and this methodology will become indispensable for the many complex problems that exist in medicine. The applications of machine learning in healthcare are only in the beginning stages as there is much to validate, but the future is bright.

References

1. S. Schneeweiss. Learning from big health care data. *New England Journal of Medicine*, 370(23):2161–2163, 2014.
2. L. Martí-Bonmatí, R. Sopena, P. Bartumeus, and P. Sopena. Multimodality imaging techniques. *Contrast Media & Molecular Imaging*, 5(4):180–189, 2010.
3. Z. Liu, L. Ding, and B. He. Integration of EEG/MEG with MRI and fMRI. *IEEE Engineering in Medicine and Biology Magazine*, 25(4):46–53, 2006.
4. S. Dähne, F. Bießmann, F. C. Meinecke, J. Mehnert, S. Fazli, and K.-R. Mtüller. Multimodal integration of electrophysiological and hemodynamic signals. In *2014 International Winter Workshop on Brain-Computer Interface (BCI)*, pages 1–4. IEEE, 2014.
5. A. Torabi, M. R. Daliri, and S. H. Sabzposhan. Diagnosis of multiple sclerosis from EEG signals using nonlinear methods. *Australasian Physical & Engineering Sciences in Medicine*, 40(4):785–797, 2017.
6. F. Biessmann, S. Plis, F. C. Meinecke, T. Eichele, and K.-R. Muller. Analysis of multimodal neuroimaging data. *IEEE Reviews in Biomedical Engineering*, 4:26–58, 2011.
7. F. Bießmann, F. C. Meinecke, A. Gretton, A. Rauch, G. Rainer, N. K. Logothetis, and K.-R. Müller. Temporal kernel CCA and its application in multimodal neuronal data analysis. *Machine Learning*, 79(1–2):5–27, 2010.

8. M.-P. Hosseini, M.-R. Nazem-Zadeh, D. Pompili, K. Jafari-Khouzani, K. Elisevich, and H. Soltanian-Zadeh. Comparative performance evaluation of automated segmentation methods of hippocampus from magnetic resonance images of temporal lobe epilepsy patients. *Medical Physics*, 43(1):538–553, 2016.

9. M.-P. Hosseini, M. R. Nazem-Zadeh, D. Pompili, and H. Soltanian-Zadeh. Statistical validation of automatic methods for hippocampus segmentation in MR images of epileptic patients. In *Engineering in Medicine and Biology Society (EMBC), 36th Annual International Conference of the IEEE*, pages 4707–4710. IEEE, 2014.

10. J. Han, C. Chen, L. Shao, X. Hu, J. Han, and T. Liu. Learning computational models of video memorability from fMRI brain imaging. *IEEE Transactions on Cybernetics*, 45(8):1692–1703, 2015.

11. A. Iraji, V. D. Calhoun, N. M. Wiseman, E. Davoodi-Bojd, M. R. Avanaki, E. M. Haacke, and Z. Kou. The connectivity domain: analyzing resting state fMRI data using feature-based data-driven and model-based methods. *Neuroimage*, 134:494–507, 2016.

12. A. Zalesky, A. Fornito, and E. T. Bullmore. Network-based statistic: identifying differences in brain networks. *Neuroimage*, 53(4):1197–1207, 2010.

13. H. A. Whiteford, L. Degenhardt, J. Rehm, A. J. Baxter, A. J. Ferrari, H. E. Erskine, F. J. Charlson, R. E. Norman, A. D. Flaxman, N. Johns, et al. Global burden of disease attributable to mental and substance use disorders: findings from the global burden of disease study 2010. *The Lancet*, 382(9904):1575–1586, 2013.

14. K. L. Miller, F. Alfaro-Almagro, N. K. Bangerter, D. L. Thomas, E. Yacoub, J. Xu, A. J. Bartsch, S. Jbabdi, S. N. Sotiropoulos, J. L. Andersson, et al. Multimodal population brain imaging in the UK biobank prospective epidemiological study. *Nature Neuroscience*, 19(11):1523, 2016.

15. D. A. Moser, G. E. Doucet, W. H. Lee, A. Rasgon, H. Krinsky, E. Leibu, A. Ing, G. Schumann, N. Rasgon, and S. Frangou. Multivariate associations among behavioral, clinical, and multimodal imaging phenotypes in patients with psychosis. *JAMA Psychiatry*, 75(4):386–395, 2018.

16. M.-P. Hosseini. A cloud-based brain computer interface to analyze medical big data for epileptic seizure detection. In *The 3rd Annual New Jersey Big Data Alliance (NJBDA) Symposium*. Montclair State University, NJ, USA, 2016.

17. M.-P. Hosseini, T. X. Tran, D. Pompili, K. Elisevich, and H. Soltanian-Zadeh. Deep learning with edge computing for localization of epileptogenicity using multimodal rs-fMRI and EEG big data. In *2017 IEEE International Conference on Autonomic Computing (ICAC)*, pages 83–92. IEEE, 2017.

18. J. Shi, S. Zhou, X. Liu, Q. Zhang, M. Lu, and T. Wang. Stacked deep polynomial network based representation learning for tumor classification with small ultrasound image dataset. *Neurocomputing*, 194:87–94, 2016.

19. R. Livni, S. Shalev-Shwartz, and O. Shamir. An algorithm for training polynomial networks. arXiv preprint arXiv:1304.7045, 2013.

20. J. Shi, X. Zheng, Y. Li, Q. Zhang, and S. Ying. Multimodal neuroimaging feature learning with multimodal stacked deep polynomial networks for diagnosis of Alzheimer's disease. *IEEE Journal of Biomedical and Health Informatics*, 22(1):173–183, 2018.

21. X. Zheng, J. Shi, Y. Li, X. Liu, and Q. Zhang. Multi-modality stacked deep polynomial network based feature learning for Alzheimer's disease diagnosis. In *2016 IEEE 13th International Symposium on Biomedical Imaging (ISBI)*, pages 851–854. IEEE, 2016.

22. D. Akhila, S. Shobhana, A. L. Fred, and S. Kumar. Robust Alzheimer's disease classification based on multimodal neuroimaging. In *2016 IEEE International Conference on Engineering and Technology (ICETECH)*, pages 748–752. IEEE, 2016.

23. S. S. Nidhyananthan, and V. Shenbagalakshmi. Assessment of dysarthric speech using Elman back propagation network (recurrent network) for speech recognition. *International Journal of Speech Technology*, 19(3):577–583, 2016.

24. M. P. Hosseini. *Brain-computer interface for analyzing epileptic big data*. PhD thesis, Rutgers University-School of Graduate Studies, 2018.

25. M.-P. Hosseini, A. Hajisami, and D. Pompili. Real-time epileptic seizure detection from EEG signals via random subspace ensemble learning. In *2016 IEEE International Conference on Autonomic Computing (ICAC)*, pages 209–218. IEEE, 2016.

26. M.-P. Hosseini, M. R. Nazem-Zadeh, F. Mahmoudi, H. Ying, and H. Soltanian-Zadeh. Support vector machine with nonlinear-kernel optimization for lateralization of epileptogenic hippocampus in MR images. In *36th Annual International Conference of the IEEE Engineering in Medicine and Biology Society*, pages 1047–1050. IEEE, 2014.

27. M. P. Hosseini, S. Lu, K. Kamaraj, A. Slowikowski, H. C. Venkatesh. Deep Learning Architectures. *Deep Learning: Concepts and Architectures*, Springer, 2019.

28. M.-P. Hosseini, D. Pompili, K. Elisevich, and H. Soltanian-Zadeh. Optimized deep learning for EEG big data and seizure prediction BCI via Internet of Things. *IEEE Transactions on Big Data*, 3(4):392–404, 2017.

29. M.-P. Hosseini, D. Pompili, K. Elisevich, and H. Soltanian-Zadeh. Random ensemble learning for eeg classification. *Artificial Intelligence in Medicine*, 84:146–158, 2018.

30. J. Gong, X. Liu, T. Liu, J. Zhou, G. Sun, and J. Tian. Dual temporal and spatial sparse representation for inferring group-wise brain networks from resting-state fMRI dataset. *IEEE Transactions on Biomedical Engineering*, 65(5):1035–1048, 2018.

31. A. Coito, C. M. Michel, P. van Mierlo, S. Vulliémoz, and G. Plomp. Directed functional brain connectivity based on EEG source imaging: methodology and application to temporal lobe epilepsy. *IEEE Transactions on Biomedical Engineering*, 63(12):2619–2628, 2016.

32. H. Laufs. A personalized history of EEG–fMRI integration. *Neuroimage*, 62(2):1056–1067, 2012.

Chapter 9

Towards Big Data in Acute Renal Rejection

*Mohamed Shehata, Ahmed Shalaby, Ali Mahmoud,
Mohammed Ghazal, Hassan Hajjdiab, Mohammed A. Badawy,
Mohamed Abou El-Ghar, Ashraf M. Bakr, Amy C. Dwyer,
Robert Keynton, Adel Elmaghraby, and Ayman El-Baz*

9.1 Introduction

It is of great importance to accurately assess the function of renal transplant for graft survival [1] and because of the limited available transplantable kidneys – only 17,000 transplantations are performed annually [2–4] – no effort should be spared to prolong the renal allograft survival rate. There is no doubt that transplantation has severely improved patients' outcomes diagnosed with end-stage chronic kidney disease (ESKD). However, acute rejection (AR) [2, 5] remains one of the main barriers. Thus, kidney function routine clinical evaluation after transplantation is essentially required to prevent the graft loss. Therefore, many researchers have started to investigate the best markers for early detection of AR post-transplantation.

The glomerular filtration rate (GFR) is recommended by the National Kidney Foundation (NKF) to evaluate the overall function of kidneys. The GFR is estimated by quantifying the creatinine level. However, it is a late biomarker for renal dysfunction (i.e., 60% of renal function is lost before a serum creatinine level significant change is detected). Also, it has low sensitivity, and it cannot assess individual kidney function [6]. Needle biopsy is the current gold standard for diagnosing AR and is not preferable due to its invasiveness, difficulty to perform, cost, and time-consumption.

In addition to the GFR and the needle biopsy, several studies have investigated the ability of image-markers to evaluate renal transplants, such as scintigraphy or radionuclide which has been used by clinicians to evaluate the renal allograft function in qualitative and quantitative manners [7]. However, scintigraphy suffers from limited spatial resolution, which hinders the functional abnormalities inside different parts of the kidney (e.g., cortex and medulla) from being discriminated precisely [8]. Also,

205

radiation exposure is one of the radionuclide's main limitations [9], thus it is not preferably used [10]. Computed tomography (CT) is commonly used as an imaging modality for accurate evaluation of different renal transplant diseases [11]. However, CT uses contrast agents, which might lead to nephrogenic systemic fibrosis, and the information gathered by CT to detect acute rejection (AR) is unspecific, thus limiting the CT potential in diagnosing AR [12]. The aforementioned drawbacks have led to impractical clinical use of these imaging methods and have led researchers to investigate other alternatives such as ultrasound (US) and MRIs to evaluate renal transplants.

On the other hand, US is a safer imaging technique for diagnosing kidney diseases. However, shadowing artifacts, low signal-to-noise ratios (SNRs), and speckles notably decrease the quality of US image and diagnostic confidence. Additionally, renal graft dysfunction cannot be evaluated precisely using the conventional US parameters; instead, only a prognostic graft marker can be provided [13, 14] or even a similar indication like other diagnostic possibilities, such as acute tubular necrosis [15], [16]. Recently, these drawbacks have been overcome by using MRIs to evaluate kidney functions, which allows advanced analysis of renal function. Various MRI scan types are used to assess renal transplants. While some MRI modalities only provide anatomical information, others provide anatomical and functional kidney information together (e.g., diffusion-weighted (DW) and dynamic contrast-enhanced (DCE) MRI). DW-MRIs have several advantages over DCE-MRIs in that dynamic MRI imaging requires intravenous contrast agents be delivered to the patient, which may induce nephrogenic systemic fibrosis [17]. Therefore, DCE-MRI can only be used if the GFR > 30 ml/min; otherwise, nephrologists recommend the use of DW-MRI [18, 19]. On the other hand, DW-MRIs can be safely performed at any time on any patient (independent of GFR). Additionally, DW-MRI allows mapping of the diffusion process of water molecules, in biological tissues, in-vivo and non-invasively. These water molecule diffusion patterns are quantified by apparent diffusion coefficients (ADCs) and reveal microscopic details about tissue (e.g., kidney) architecture, either normal or in a diseased state [1], [20–23]. Moreover, quantitative assessment of the amount of rejection can be derived from the ADC, which can only be attained by DW-MRI. By using the ADC values, we remove patient variability and both perfusion and diffusion parameters can be obtained via DW-MRI while DCE-MRI will only yield perfusion parameters. Hence, DW-MRI can be used to evaluate changes in the amount or degree of rejection over time in response to treatment [24].

Therefore, several studies have used ADCs for functional renal assessment [1], [21–23]. However, most of them only performed a statistical analysis at certain b-values to investigate the significant difference between groups. For example, Abou-El-Ghar et al. [25] conducted a study to assess renal transplants function. Their study included 70 renal allograft patients. Using only two b-values of 0 and 800 s/mm^2, DW-MRI scans were conducted for 21 patients with acute graft impairment (group 1) and 49 normal renal allografts (group 2). A user-defined region of interest (ROI) was placed in the middle portion of the kidney in a selected cross-section, and pixie-wise ADCs were calculated. Their study revealed that group 1 had significantly higher ADC values than group 2. Using DW-MRIs, early identification of renal transplant dysfunction caused by AR was investigated by Liu et al. [1]. After a manual selection of ROIs, higher ADCs of the normal allografts than those of the AR ones were obtained. The feasibility of diagnosing acute rejection (AR) from DW-MRI was evaluated by Xu et al. [26] on 26 AR and 43 non-rejection (NR) patients. The NR patients showed higher ADCs than those of the R group and also demonstrated optimum sensitivity and specificity at b_{800} s/mm^2, as evidenced by the ROC curve. Katarzyna et al. [23] investigated possible relations between the diffusion parameters and selected laboratory results in the early stage after kidney transplantation. The measurements were conducted in kidneys over multiple user-defined ROIs at values of 600 and 1000 s/mm^2 only. According to the relative variability in the results and SNR, the optimum ADC value in the renal cortex was at b_{1000} s/mm^2 with a strong dependency between the ADC measured at the same b-value and the estimated GFR.

However, most of the aforementioned studies did not integrate laboratory biomarkers with image-markers to explore the correlations among renal transplant status and its biopsy diagnosis. In addition, most of these studies performed a statistical analysis using DW-MR image-markers alone at certain b-values (two at most) to investigate the significant difference between different groups. Moreover, image analysis in most of the previous studies only performed manual delineations of the kidney without accounting for motion effects, making their analysis imprecise. Furthermore, none of these studies utilized the results from the statistical analysis to build an automated computer-aided diagnostic (CAD) system for the final diagnosis. To account for these limitations, we present a comprehensive statistical analysis, which examines two categories of parameters for the assessment of transplant status: (*i*) laboratory biomarkers alone (creatinine clearance (CrCl) and serum creatinine (SCr)) and (*ii*) the average ADC (aADC) at 11 different b-values. Details of the comprehensive analysis are discussed below.

9.2 Methods and Materials

The ultimate goal of this study is to determine which of the aforementioned parameters are correlated with a more accurate diagnosis of AR in patients who have undergone kidney transplantation, using (3D + b-value) DW-MRI and can be integrated together to evaluate renal transplant status. To move towards this goal, all patients' kidneys were evaluated using both laboratory biomarkers (SCr and CrCl) and DW-MR image-markers (i.e., ADCs). Then, a comprehensive statistical analysis was conducted to find the possible correlations between the aforementioned markers and the AR and NR renal transplant biopsy diagnoses. The correlated parameters were then integrated/fused using stacked autoencoders (SAEs) towards the final diagnosis. Details of the proposed framework (shown in Figure 9.1) are outlined below.

9.2.1 Diffusion Weighted Imaging Markers

DW-MRI has the ability to interrelate local blood diffusion with the renal allograft status, which is a significant and important advantage, thanks to its ability to evaluate unique tissue (e.g., kidney) characteristics of inner spatial water behavior known as ADC [27, 28]. These ADCs can then be used to evaluate renal transplant status as AR or NR. Details about estimating ADCs are given below.

1) *Data Acquisition*: A total of 61 kidney-transplanted patients had provided consent for participation in this study. All DW-MRI scans and biopsies were performed in the period from July 2014 to February 2017. Patients were grouped into two groups. The first group (16 patients) included healthy graft function patients. Most of them only underwent DW-MRI scans as they have shown no symptoms of rejection. The second group (45 patients) included acute renal rejection patients, based on renal biopsy histology. All patients from the second group underwent DW-MRI and renal

biopsy. Before any biopsy procedure, coronal DW-MRIs were acquired using a 1.5 T scanner (SIGNA Horizon, General Electric Medical Systems, Milwaukee, WI) using a body coil and a gradient multi-shot spin-echo echo-planar sequence (TR/TE: 8000/61.2; bandwidth: 142 kHz; matrix: 1.25×1.25 mm^2; section thickness: 4 mm; intersection gap: 0 mm; FOV: 36 cm; signals acquired: 7; water signals acquired at multiple *b*-values of b_0, b_{50}, and b_{100}–b_{1000} s/mm^2 with 100 increment [see Figure 9.2]). Approximately 50 sections were obtained for a total acquisition time 60–120 s to cover the entire kidney. In the final analysis, both DW-MRIs and biopsies were included and examined by a board-certified radiologist and a nephrologist.

2) *3D Kidney Segmentation*: To obtain an accurate estimation of the ADCs, image inhomogeneities, inconsistencies, and noise effects were first reduced by employing an intensity histogram equalization technique on the nonparametric bias corrected DW-MRI data [29]. To account for the variability across different DW-MRI subjects and to handle kidney motion, a 3D B-splines based nonrigid registration [30] was performed. Then, a 3D level-set kidney segmentation approach, shown in Figure 9.3 [31–35], was performed. In order to more accurately segment the kidney, a guiding force that integrates regional statistics derived from the kidney and background regions is employed. Namely, it takes into account shape, regional appearance, and spatial DW-MRI features. Then, these features were combined using a joint

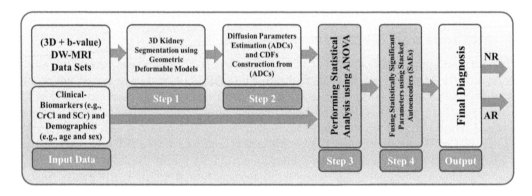

FIGURE 9.1: Illustrative block diagram of the presented framework to discriminate acute rejection (AR) from non-rejection (NR) renal transplants.

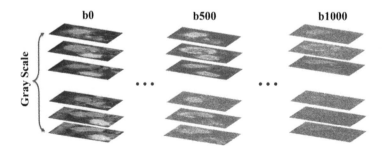

FIGURE 9.2: Illustration of the 3D DW-MRI data acquisition at different *b*-values (e.g., b_0, b_{500}, and b_{1000}).

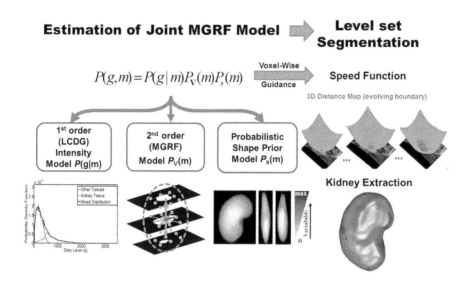

FIGURE 9.3: Illustration of the 3D kidney segmentation shape-based level-set framework from DW-MRIs. An adaptive shape prior information and current appearance features (i.e., first-order voxel-wise image intensity and their spatials) are integrated together into a joint MGRF model of the kidney and its background, which provides the voxel-wise guidance for the speed function (SF). This SF provides the 3D distance map, which controls the deformable model evolutions.

Markov–Gibbs random field (MGRF) [32–34] image model. For more details of each component of the segmentation approach, the reader is referred to [35].

3) *Diffusion Parameters Estimation:* After obtaining an accurate segmentation of the kidneys, DW-MR image-markers (i.e., ADCs) can be estimated precisely by using the simple equation defined by Le Bihan [36] as:

$$\text{ADC}_\text{v} = \frac{1}{b_0 - b} \ln\left(\frac{g_{b:\text{v}}}{g_{0:\text{v}}}\right)$$

where $\text{v} = (x, y, z)$ represents a voxel at a location with discrete Cartesian coordinates (x, y, z), while g_0 and g_b refer to the segmented DW-MR image acquired at the b_0 and at a given different b-value, respectively. The ADCs are estimated at a total of 11 different b-values to be used as discriminatory features for renal transplant status assessment.

9.2.2 Laboratory Markers and Demographics

The 61 participating patients were routinely assessed post-transplantation with both serum creatinine (SCr) and creatinine clearance (CrCl) laboratory values. Patient demographics included sex (50 males and 11 females) and mean age of 26.26 ± 9.87 years (range, 11–54 years). The NR group (16 patients) has a mean age of 25.125 ± 10.07 years, a mean SCr value of 1.54 ± 1.14 mg/dl, and a mean CrCl value of 73.8 ± 26.5 ml/min, while the AR group (45 patients) has a mean age of 26.76 ± 9.89 years, a mean SCr value of 1.78 ± 0.85 mg/dl, and a mean CrCl value of 59.4 ± 19.8 ml/min.

9.2.3 Statistical Analysis

After estimating the DW-MR image-markers (i.e., ADCs), we performed a comprehensive statistical analysis to investigate and determine all the correlations that might exist between renal transplant biomarkers and its biopsy diagnosis. Two categories of parameters were examined: (*i*) laboratory biomarkers alone (CrCl and SCr) and (*ii*) the average ADC (aADC) at the 11 individual *b*-values. R version 3.1.1 was used to perform the statistical calculations. The relationship of graft tolerance to ADC was tested using logistic regression of biopsy result (AR or NR) against demographics (e.g., age and sex), laboratory (CrCl and SCr), and imaging parameters (aADCs) at individual *b*-values. Statistical significance of each parameter was assessed using likelihood ratio (χ^2) tests.

9.3 Experimental Results

ANOVA showed that SCr ($\chi^2 = 10.1$, $p = 0.002$) and CrCl ($\chi^2 = 14.1$, $p = 0.0002$) had significant effects on the likelihood of AR, as did the aADC for b_{100} s/mm^2 ($\chi^2 = 3.79$, $p = 0.05$), b_{500} s/mm^2 ($\chi^2 = 3.98$, $p = 0.05$), b_{600} s/mm^2 ($\chi^2 = 5.81$, $p = 0.02$), b_{700} s/mm^2 ($\chi^2 = 5.65$, $p = 0.02$), and b_{900} s/mm^2 ($\chi^2 = 4.94$, $p = 0.03$). Patient age, sex, and aADC at the remaining *b*-values had no significance effect, as shown in Table 9.1.

Promising results from the statistical analysis encouraged us to investigate building an automated CAD system that might be able to differentiate the AR from the NR renal transplants using the resulting statistically significant parameters, namely, the DW-MR image-markers (voxel-wise ADCs at *b*-values of 100, 500, 600, 700, and 900 s/mm^2) and laboratory biomarkers (CrCl and SCr). However, a size mismatch challenge is added by dealing with the voxel-wise ADCs of the entire kidney volume due to varying input data size, which requires either zero padding for smaller volumes or data truncation for large ones. Therefore, we characterized these 3D ADCs at the *b*-values of 100, 500, 600, 700, and 900 s/mm^2 with their cumulative distribution coefficients (CDFs). The CDFs were constructed by dividing the range between the maximum and minimum ADC values into 100 steps, which helps to unify the input size for all subjects.

Our CAD system utilizes stacked autoencoders (SAEs) [37–40] based on a deep learning approach to evaluate renal transplants. First, the CDFs at the aforementioned *b*-values were fed as inputs to train an AE along with a leave-one-subject-out (LOSO) scenario yielding a probability of being a non-rejection (P_{NR}) and a probability of being an acute rejection (P_{AR}) for all the 61 subjects included in this study. Simultaneously, another AE was trained using the laboratory biomarkers (CrCl and SCr) along with the same LOSO scenario, yielding a probability of being a non-rejection (P_{NR}) and a probability of being an acute rejection (P_{AR}) as well. Then, the output probabilities from both AEs are fed as new inputs to train another AE, yielding the final probabilities of being AR or NR. Finally, a softmax classifier is used to give the final diagnosis of being AR or NR based on maximum probability voting; see Figure 9.4.

To highlight the advantage of integrating both the DW-MR image-markers represented by the CDFs and the laboratory biomarkers (CrCl and SCr), we first evaluated the proposed CAD system using the laboratory biomarkers alone, which had low accuracy because of the notable overlap between the AR and NR groups, as documented in Table 9.2. Then, the proposed CAD system was evaluated using the DW-MR

TABLE 9.1: Results of ANOVA on the Logistic Regression Model for AR. Note that CRCL: Creatine Clearance, and $aADC_n$: Average ADC at b-value $= n$

Variable	χ^2	*p-value*
$aADC_{50}$	2.62	0.11
$aADC_{100}$	3.79	**0.05**
$aADC_{200}$	0.61	0.44
$aADC_{300}$	1.24	0.27
$aADC_{400}$	0.02	0.90
$aADC_{500}$	3.98	**0.05**
$aADC_{600}$	5.81	**0.02**
$aADC_{700}$	5.65	**0.02**
$aADC_{800}$	1.45	0.23
$aADC_{900}$	4.94	**0.03**
$aADC_{1000}$	0.09	0.77
Age	0.85	0.36
Sex	2.29	0.13
SCr	14.07	**0.0002**
CrCl	10.07	**0.002**

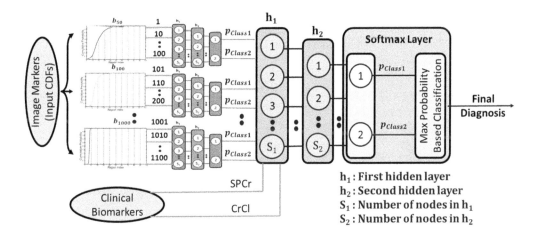

FIGURE 9.4: Stacked Autoencoders (SAE)-based deep learning classification after fusing both DW-MR image-markers (CDFs) and laboratory biomarkers (CrCl and SCr). Note that P_{NR} ($P_{Class}1$) and P_{AR} ($P_{Class}2$) represent the probabilities of non-rejection and acute rejection renal transplants, respectively.

image-markers alone (i.e., CDFs), which had a better performance in the terms of accuracy; see Table 9.2. Finally, we evaluated the CAD system using the integrated DW-MR image-markers with the laboratory biomarkers to assess the renal transplant status. As documented in Table 9.2, the integration of both markers has a positive effect on the overall diagnostic accuracy.

Results in Table 9.3 demonstrate a significant improvement of the final diagnostic accuracy after fusing/integrating all the CDFs of the ADCs calculated at the individual b-values. Furthermore, our CAD system describes the local voxel-wise diffusion of the segmented kidneys as color maps. These regional display mappings will help radiologists to investigate the kidney portions that need attention and follow-up with proper treatment. Figure 9.5 demonstrates the voxel-wise parametric diffusion maps of an NR and an AR renal transplant, respectively, which reveals the expected relation of the DW-MRI parameters for NR and AR status.

TABLE 9.2: Diagnostic Results of the Presented CAD Technique after Integrating/Fusing Laboratory Biomarkers and DW-MR Image-Markers Using the SAES. Let "Sens," "Spec," and "Acc" Stand for Sensitivity, Specificity, and Accuracy, Respectively

	Sens (%)	Spec (%)	Acc (%)
Image-markers	89	81	87
Laboratory biomarkers	78	56	72
Integrated biomarkers	**92**	**94**	**92**

TABLE 9.3: Renal Transplant Status Classification Accuracy Using the Input CDF for the Individual Statistically Significant b-Values and the Fused CDFs (F $5CDFs$) Model that Integrates/Fuses these Statistically Significant b-Values

Classification Accuracy%					
b_{100}	b_{500}	b_{600}	b_{700}	b_{900}	F $5CDFs$
74	71	71	72	72	87

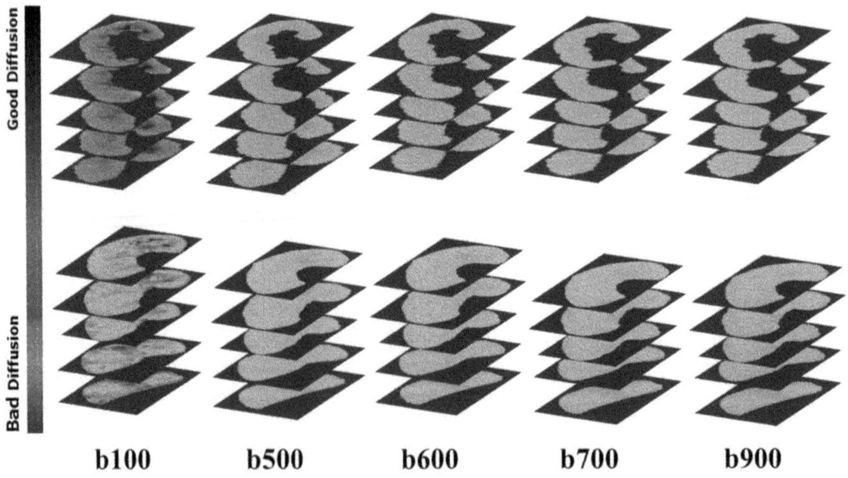

FIGURE 9.5: Voxel-wise color-maps reflecting the blood diffusion rate for DW-MRI data at b-values of 100, 500, 600, 700, and 900 s/mm^2 for a non-rejection kidney in the upper row and a rejection kidney in the lower row.

9.4 Conclusions

To sum up, our statistical analysis demonstrated the ADCs can be used as reliable discriminatory features to differentiate between accepted and rejected renal transplants, especially at the b-values of 100, 500, 600, 700, and 900 s/mm^2. Also, the laboratory biomarkers (CrCl and SCr) can be used as discriminatory features to assess renal transplants. Moreover, integrating both markers efficiently improved the final diagnostic accuracy and provided the best-quality discriminatory features. We were motivated by the promising results to build our CAD system utilizing the SAEs and statistically significant parameters to obtain the final diagnosis. The CAD system showed a 92% accuracy carrying promises that the presented technique might be useful as a non-invasive diagnostic tool.

Despite the promising results obtained by the proposed statistical analysis-based CAD system, this study had several limitations, including but not limited to lack of race diversity since all subjects were acquired from the same country (Mansoura University, Egypt); thus we are currently collecting more data from different locations in the USA. Additionally, the sex-imbalanced data due to the limited number of females was another issue, and this could be handled by including more female subjects in a bigger study. Moreover, this study did not include any patients who are below 11 or over 54 in terms of age, which might be another factor that could affect the results of the final statistical analysis. In the future, a larger and more diverse cohort of subjects will be tested using the proposed technique to confirm its accuracy and robustness. In addition, a new set of data will be acquired at lower b-values to investigate whether they can provide more help in finding more significant differences between different renal transplant groups. Furthermore, we plan to test the ability of the developed technique in diagnosing various renal rejection types (e.g., T-cell mediated rejection and antibody mediated rejection) for appropriate treatment purposes. This work could also be applied to various other applications in medical imaging, such as the heart, the prostate, the lung, and the retina, as well as several non-medical applications [41–44].

The heart is an important application for this work. The clinical assessment of myocardial perfusion plays a major role in the diagnosis, management, and prognosis of ischemic heart disease patients. Thus, there have been ongoing efforts to develop automated systems for accurate analysis of myocardial perfusion using first-pass images [45–61].

Moreover, the work could be applied for prostate cancer, which is the most common cancer in American men, and its related mortality rate is the second after lung cancer. Fortunately, the mortality rate can be reduced if prostate cancer is detected in its early stages. Early detection enables physicians to treat prostate cancer before it develops to a clinically significant disease [62–65].

Another application for this work could be the detection of retinal abnormalities. A majority of ophthalmologists depend on visual interpretation for the identification of disease types. However, inaccurate diagnosis will affect the treatment procedure which may lead to fatal results. Hence, there is a crucial need for computer automated diagnosis systems that yield highly accurate results. Optical coherence tomography (OCT) has become a powerful modality for the non-invasive diagnosis of various retinal abnormalities such as glaucoma, diabetic macular edema, and macular degeneration. The problem with diabetic retinopathy (DR) is that the patient is not aware of the disease until the changes in the retina have progressed to a level that treatment tends to be less effective. Therefore, automated early detection could limit the severity of the disease and assist ophthalmologists in investigating and treating it more efficiently [66–68].

Abnormalities of the lung could also be another promising area of research and a related application to this work. Radiation-induced lung injury is the main side effect of radiation therapy for lung cancer patients. Although higher radiation doses increase the radiation therapy effectiveness for tumor control, this can lead to lung injury as a greater quantity of normal lung tissues is included in the treated area. Almost one-third of patients who undergo radiation therapy develop lung injury following radiation treatment. The severity of radiation-induced lung injury ranges from ground-glass opacities and consolidation at the early phase to fibrosis and traction bronchiectasis in the late phase. Early detection of lung injury will thus help to improve management of the treatment [69–111].

This work can also be applied to other brain abnormalities, such as dyslexia in addition to autism. Dyslexia is one of the most complicated developmental brain disorders that affect children's learning abilities. Dyslexia leads to the failure to develop age-appropriate reading skills in spite of normal intelligence level and adequate reading instructions. Neuropathological studies have revealed an abnormal anatomy of some structures, such as the corpus callosum in dyslexic brains. There has been a lot of work in the literature that aims at developing CAD systems for diagnosing this disorder, along with other brain disorders [112–134].

For the vascular system [135], this work could also be applied for the extraction of blood vessels, e.g., from phase contrast (PC) magnetic resonance angiography (MRA). Accurate cerebrovascular segmentation using non-invasive MRA is crucial for the early diagnosis and timely treatment of intracranial vascular diseases [117, 118, 136–141].

9.5   Acknowledgment

This research is supported by National Institutes of Health Grant Number: 1R15AI135924-01A1.

References

1. G. Liu, F. Han, W. Xiao, Q. Wang, Y. Xu, and J. Chen, "Detection of renal allograft rejection using blood oxygen level-dependent and diffusion weighted magnetic resonance imaging: A retrospective study," *BMC Nephrology*, vol. 15, no. 1, p. 158, 2014.
2. E. Hollis, M. Shehata, F. Khalifa, M. A. El-Ghar, T. El-Diasty, and A. El-Baz, "Towards non-invasive diagnostic techniques for early detection of acute renal transplant rejection: A review," *The Egyptian Journal of Radiology and Nuclear Medicine*, vol. 48, no. 6, pp. 257–269, 2016.
3. K. org, "About chronic kidney disease," 2016. [Online]. Available: https://www.kidney.org/kidneydisease/aboutckd[cited2May2016].
4. N. K. Foundation, "Organ donation and transplantion statistics," 2016. [Online]. Available: https://www.kidney.org/news/newsroom/factsheets/Organ-Donation-and-Transplantation-Stats[cited25May2016].
5. W. Chon and D. Brennan, "Clinical manifestations and diagnosis of acute renal allograft rejection," *UpToDate Version*, vol. 21, 2014.

6. R. W. Katzberg, M. H. Buonocore, M. Ivanovic, C. Pellot-Barakat, J. M. Ryan, K. Whang, J. M. Brock, and C. D. Jones, "Functional, dynamic, and anatomic MR urography: Feasibility and preliminary findings," *Academic Radiology*, vol. 8, no. 11, pp. 1083–1099, 2001.

7. E. Brown, M. Chen, N. Wolfman, D. Ott, and N. Watson, "Complications of renal transplantation: Evaluation with US and radionuclide imaging," *Radiograph*, vol. 20, no. 3, pp. 607–622, 2000.

8. E. Giele, "Computer methods for semi-automatic MR renogram determination," Ph.D. dissertation, Eindhoven University of Technology, Eindhoven, 2002.

9. A. Taylor and J. V. Nally, "Clinical applications of renal scintigraphy," *American Journal of Roentgenology*, vol. 164, no. 1, pp. 31–41, 1995.

10. J. G. Heaf and J. Iversen, "Uses and limitations of renal scintigraphy in renal transplantation monitoring," *European Journal of Nuclear Medicine*, vol. 27, no. 7, pp. 871–879, 2000.

11. C. Sebastià, S. Quiroga, R. Boyé, C. Cantarell, M. Fernandez-Planas, and A. Alvarez, "Helical CT in renal transplantation: Normal findings and early and late complications," *Radiograph*, vol. 21, no. 5, pp. 1103–1117, 2001.

12. A. Grabner, D. Kentrup, M. Schäfers, S. Reuter, and U. Schnöckel, *Non-Invasive Diagnosis of Acute Renal Allograft Rejection-Special Focus on Gamma Scintigraphy and Positron Emission Tomography*. INTECH Open Access Publisher, 2013.

13. A. Kirkpantur, R. Yilmaz, D. E. Baydar, T. Aki, B. Cil, M. Arici, B. Altun, et al., "Utility of the doppler ultrasound parameter, resistive index, in renal transplant histopathology," *Transplantation Proceedings*, vol. 40, no. 1, pp. 104–106, 2008.

14. S. Seiler, S. M. Colbus, G. Lucisano, K. S. Rogacev, M. K. Gerhart, M. Ziegler, D. Fliser, et al., "Ultrasound renal resistive index is not an organ-specific predictor of allograft outcome," *Nephrology Dialysis Transplantation*, vol. 27, no. 8, pp. 3315–3320, 2012.

15. R. Kramann, D. Frank, V. M. Brandenburg, N. Heussen, J. Takahama, T. Krüger, J. Riehl, et al., "Prognostic impact of renal arterial resistance index upon renal allograft survival: The time point matters," *Nephrology Dialysis Transplantation*, vol. 27, no. 10, pp. 3958–3963, 2012.

16. D. O. Cosgrove and K. E. Chan, "Renal transplants: What ultrasound can and cannot do," *Ultrasound Quarterly*, vol. 24, no. 2, pp. 77–87, 2008.

17. E. Hodneland, Å. Kjørstad, E. Andersen, J. A. Monssen, A. Lundervold, J. Rørvik, and A. Munthe-Kaas, "In vivo estimation of glomerular filtration in the kidney using dce-mri," in *2011 7th International Symposium on Image and Signal Processing and Analysis (ISPA)*, IEEE, 2011, pp. 755–761.

18. A. C. of Radiology, "ACR manual on contrast media version 10.3," [Online]. Available: https://www.acr.org/-/media/ACR/Files/Clinical-Resources/Contrast Media.pdf.

19. B. Y. Cheong and R. Muthupillai, "Nephrogenic systemic fibrosis: A concise review for cardiologists," *Texas Heart Institute Journal*, vol. 37, no. 5, p. 508, 2010.

20. S. Y. Park, C. K. Kim, B. K. Park, S. J. Kim, S. Lee, and W. Huh, "Assessment of early renal allograft dysfunction with blood oxygenation level-dependent MRI and diffusion-weighted imaging," *European Journal of Radiology*, vol. 83, no. 12, pp. 2114–2121, 2014.

21. S. Palmucci, L. Mauro, G. Failla, P. Foti, P. Milone, N. Sinagra, N. Zerbo, et al., "Magnetic resonance with diffusion-weighted imaging in the evaluation of transplanted kidneys: Updating results in 35 patients," *Transplantation Proceedings*, vol. 44, no. 7, pp. 1884–1888, 2012.

22. P. Vermathen, T. Binser, C. Boesch, U. Eisenberger, and H. C. Thoeny, "Three-year follow-up of human transplanted kidneys by diffusion-weighted MRI and blood oxygenation level-dependent imaging," *Journal of Magnetic Resonance Imaging*, vol. 35, no. 5, pp. 1133–1138, 2012.

23. K. Wypych-Klunder, A. Adamowicz, A. Lemanowicz, W. Szczesny, Z. Włodarczyk, and Z. Serafin, "Diffusion-weighted MR imaging of transplanted kidneys: Preliminary report," *Polish Journal of Radiology*, vol. 79, p. 94, 2014.

24. V. N. Harry, S. I. Semple, D. E. Parkin, and F. J. Gilbert, "Use of new imaging techniques to predict tumour response to therapy," *The Lancet Oncology*, vol. 11, no. 1, pp. 92–102, 2010.

25. M. Abou-El-Ghar, T. El-Diasty, A. El-Assmy, H. Refaie, A. Refaie, and M. Ghoneim, "Role of diffusion-weighted MRI in diagnosis of acute renal allograft dysfunction: A prospective preliminary study," *The British Journal of Radiology*, vol. 85, no. 1014, pp. e206–e211, 2012.

26. J. Xu, W. Xiao, L. Zhang, and M. Zhang, "Value of diffusion-weighted MR imaging in diagnosis of acute rejection after renal transplantation," *Zhejiang da xue xue bao. Yi xue ban= Journal of Zhejiang University. Medical Sciences*, vol. 39, no. 2, pp. 163–167, 2010.

27. E. Hollis, M. Shehata, M. Abou El-Ghar, M. Ghazal, T. El-Diasty, M. Merchant, A. E. Switala, and A. El-Baz, "Statistical analysis of ADCs and clinical biomarkers in detecting acute renal transplant rejection," *The British Journal of Radiology*, vol. 90, no. 1080, p. 20170125, 2017.

28. F. Khalifa, M. Shehata, A. Soliman, M. A. El-Ghar, T. El-Diasty, A. C. Dwyer, M. El-Melegy, G. Gimel'farb, R. Keynton, and A. El-Baz, "A generalized MRI-based CAD system for functional assessment of renal transplant," in *2017 IEEE 14th International Symposium on Biomedical Imaging (ISBI 2017)*, IEEE, 2017, pp. 758–761.

29. N. J. Tustison, B. B. Avants, P. A. Cook, Y. Zheng, A. Egan, P. A. Yushkevich, and J. C. Gee, "N4ITK: Improved N3 bias correction," *IEEE Transactions on Medical Imaging*, vol. 29, no. 6, pp. 1310–1320, 2010.

30. B. Glocker, A. Sotiras, N. Komodakis, and N. Paragios, "Deformable medical image registration: Setting the state of the art with discrete methods," *Annual Review of Biomedical Engineering*, vol. 13, pp. 219–244, 2011.

31. M. Shehata, F. Khalifa, A. Soliman, R. Elrefai, M. A. El-Ghar, A. C. Dewyer, R. Ouseph, and A. El-Baz, "A novel framework for automatic segmentation of kidney from DW-MRI," in *IEEE International Symposium on Biomedical Imaging*, New York, 2015, pp. 951–954.

32. A. El-Baz, G. Gimelfarb, and J. S. Suri, *Stochastic Modeling for Medical Image Analysis*. CRC Press, 2015.

33. M. Shehata, F. Khalifa, A. Soliman, R. Elrefai, M. A. El-Ghar, A. C. Dewyer, R. Ouseph, and A. El-Baz, "A level set-based framework for 3D kidney segmentation from diffusion MR images," in *IEEE International Conference on Image Processing*, Québec City, QC, 2015, pp. 4441–4445.

34. M. Shehata, F. Khalifa, A. Soliman, A. T. Eldeen, M. A. El-Ghar, T. El-Diasty, A. El-Baz, and R. Keynton, "An appearance-guided deformable model for 4d kidney segmentation using diffusion MRI," In *Biomedical Image Segmentation: Advances and Trends*, pp. 271–285. CRC Press, 2016.

35. M. Shehata, A. Mahmoud, A. Soliman, F. Khalifa, M. Ghazal, M. A. El-Ghar, M. El-Melegy, and A. El-Baz, "3D kidney segmentation from abdominal diffusion MRI using an appearance-guided deformable boundary," *PloS One*, vol. 13, no. 7, p. e0200082, 2018.

36. D. Le Bihan and E. Breton, "Imagerie de diffusion in-vivo par résonance magnétique nucléaire," *Comptes-Rendus de l'Académie des Sciences*, vol. 93, no. 5, pp. 27–34, 1985.

37. P. Vincent, H. Larochelle, I. Lajoie, Y. Bengio, and P.-A. Manzagol, "Stacked denoising autoencoders: Learning useful representations in a deep network with a local denoising criterion," *Journal of Machine Learning Research*, vol. 11, no. December, pp. 3371–3408, 2010.

38. M. Shehata, F. Khalifa, E. Hollis, A. Soliman, E. Hosseini-Asl, M. A. El-Ghar, M. El-Baz, A. C. Dwyer, A. El-Baz, and R. Keynton, "A new non-invasive approach for early classification of renal rejection types using diffusion-weighted MRI," in *2016 IEEE International Conference on Image Processing (ICIP)*, IEEE, 2016, pp. 136–140.

39. M. Shehata, F. Khalifa, A. Soliman, M. A. El-Ghar, A. Dwyer, G. Gimelfarb, R. Keynton, and A. El-Baz, "A promising non-invasive CAD system for kidney function assessment," in *International Conference on Medical Image Computing and Computer-Assisted Intervention*, Springer, 2016, pp. 613–621.

40. M. Shehata, F. Khalifa, A. Soliman, M. Ghazal, F. Taher, M. A. El-Ghar, A. C. Dwyer, G. Gimelfarb, R. S. Keynton, and A. El-Baz, "Computer-aided diagnostic system for early detection of acute renal transplant rejection using diffusion-weighted mri," *IEEE Transactions on Biomedical Engineering*, vol. 66, no. 2, pp. 539–552, 2019.

41. A. H. Mahmoud, "Utilizing radiation for smart robotic applications using visible, thermal, and polarization images," Ph.D. dissertation, University of Louisville, 2014.

42. A. Mahmoud, A. El-Barkouky, J. Graham, and A. Farag, "Pedestrian detection using mixed partial derivative based his togram of oriented gradients," in *2014 IEEE International Conference on Image Processing (ICIP)*, IEEE, 2014, pp. 2334–2337.

43. A. El-Barkouky, A. Mahmoud, J. Graham, and A. Farag, "An interactive educational drawing system using a humanoid robot and light polarization," in *2013 IEEE International Conference on Image Processing*, IEEE, 2013, pp. 3407–3411.

44. A. H. Mahmoud, M. T. El-Melegy, and A. A. Farag, "Direct method for shape recovery from polarization and shading," in *2012 19th IEEE International Conference on Image Processing*, IEEE, 2012, pp. 1769–1772.

45. F. Khalifa, G. Beache, A. El-Baz, and G. Gimel'farb, "Deformable model guided by stochastic speed with application in cine images segmentation," in *Proceedings of IEEE International Conference on Image Processing, (ICIP'10)*, Hong Kong, September 26–29, 2010, pp. 1725–1728.

46. F. Khalifa, G. M. Beache, A. Elnakib, H. Sliman, G. Gimel'farb, K. C. Welch, and A. El-Baz, "A new shape-based framework for the left ventricle wall segmentation from cardiac first-pass perfusion MRI," in *Proceedings of IEEE International Symposium on Biomedical Imaging: From Nano to Macro, (ISBI'13)*, San Francisco, CA, April 7–11, 2013, pp. 41–44.

47. F. Khalifa, G. M. Beache, A. Elnakib, H. Sliman, G. Gimel'farb, K. C. Welch, and A. El-Baz, "A new nonrigid registration framework for improved visualization of transmural perfusion gradients on cardiac first–pass perfusion MRI," in *Proceedings of IEEE International Symposium on Biomedical Imaging: From Nano to Macro, (ISBI'12)*, Barcelona, Spain, May 2–5, 2012, pp. 828–831.

48. F. Khalifa, G. M. Beache, A. Firjani, K. C. Welch, G. Gimel'farb, and A. El-Baz, "A new nonrigid registration approach for motion correction of cardiac first-pass perfusion MRI," in *Proceedings of IEEE International Conference on Image Processing, (ICIP'12)*, Lake Buena Vista, Florida, September 30–October 3, 2012, pp. 1665–1668.

49. F. Khalifa, G. M. Beache, G. Gimel'farb, and A. El-Baz, "A novel CAD system for analyzing cardiac first-pass MR images," in *Proceedings of IAPR International Conference on Pattern Recognition (ICPR'12)*, Tsukuba Science City, Japan, November 11–15, 2012, pp. 77–80.

50. F. Khalifa, G. M. Beache, G. Gimel'farb, and A. El-Baz, "A novel approach for accurate estimation of left ventricle global indexes from short-axis cine MRI," in *Proceedings of IEEE International Conference on Image Processing, (ICIP'11)*, Brussels, Belgium, September 11–14, 2011, pp. 2645–2649.

51. F. Khalifa, G. M. Beache, G. Gimel'farb, G. A. Giridharan, and A. El-Baz, "A new image-based framework for analyzing cine images," in *Handbook of Multi Modality State-of-the-Art Medical Image Segmentation and Registration Methodologies*, A. El-Baz, U. R. Acharya, M. Mirmedhdi, and J. S. Suri, Eds. Springer, New York, 2011, vol. 2, ch. 3, pp. 69–98.

52. F. Khalifa, G. M. Beache, G. Gimel'farb, G. A. Giridharan, and A. El-Baz, "Accurate automatic analysis of cardiac cine images," *IEEE TBME*, vol. 59, no. 2, pp. 445–455, 2012.

53. F. Khalifa, G. M. Beache, M. Nitzken, G. Gimel'farb, G. A. Giridharan, and A. El-Baz, "Automatic analysis of left ventricle wall thickness using short-axis cine CMR images," in *Proceedings of IEEE International Symposium on Biomedical Imaging: From Nano to Macro, (ISBI'11)*, Chicago, Illinois, March 30–April 2, 2011, pp. 1306–1309.

54. M. Nitzken, G. Beache, A. Elnakib, F. Khalifa, G. Gimel'farb, and A. El-Baz, "Accurate modeling of tagged cmr 3D image appearance characteristics to improve cardiac cycle strain estimation," in *2012 19th IEEE International Conference on Image Processing (ICIP)*, Orlando, Florida, USA, IEEE, September 2012, pp. 521–524.

55. M. Nitzken, G. Beache, A. Elnakib, F. Khalifa, G. Gimel'farb, and A. El-Baz, "Improving full-cardiac cycle strain estimation from tagged cmr by accurate modeling of 3D image appearance characteristics," in *2012 9th IEEE International Symposium on Biomedical Imaging (ISBI)*, Barcelona, Spain, IEEE, May 2012, pp. 462–465, (Selected for oral presentation).

56. M. J. Nitzken, A. S. El-Baz, and G. M. Beache, "Markov–Gibbs random field model for improved full-cardiac cycle strain estimation from tagged cmr," *Journal of Cardiovascular Magnetic Resonance*, vol. 14, no. 1, pp. 1–2, 2012.

57. H. Sliman, A. Elnakib, G. Beache, A. Elmaghraby, and A. El-Baz, "Assessment of myocardial function from cine cardiac MRI using a novel 4D tracking approach," *Journal of Computer Science and System Biology*, vol. 7, pp. 169–173, 2014.

58. H. Sliman, A. Elnakib, G. M. Beache, A. Soliman, F. Khalifa, G. Gimel'farb, A. Elmaghraby, and A. El-Baz, "A novel 4D PDE-based approach for accurate assessment of myocardium function using cine cardiac magnetic resonance images," in *Proceedings of IEEE International Conference on Image Processing (ICIP'14)*, Paris, France, October 27–30, 2014, pp. 3537–3541.

59. H. Sliman, F. Khalifa, A. Elnakib, G. M. Beache, A. Elmaghraby, and A. El-Baz, "A new segmentation-based tracking framework for extracting the left ventricle cavity from cine cardiac MRI," in *Proceedings of IEEE International Conference on Image Processing, (ICIP'13)*, Melbourne, Australia, September 15–18, 2013, pp. 685–689.

60. H. Sliman, F. Khalifa, A. Elnakib, A. Soliman, G. M. Beache, A. Elmaghraby, G. Gimel'farb, and A. El-Baz, "Myocardial borders segmentation from cine MR images using bi-directional coupled parametric deformable models," *Medical Physics*, vol. 40, no. 9, pp. 1–13, 2013.

61. H. Sliman, F. Khalifa, A. Elnakib, A. Soliman, G. M. Beache, G. Gimel'farb, A. Emam, A. Elmaghraby, and A. El-Baz, "Accurate segmentation framework for the left ventricle wall from cardiac cine MRI," in *Proceedings of International Symposium on Computational Models for Life Science, (CMLS'13)*, vol. 1559, Sydney, Australia, November 27–29, 2013, pp. 287–296.

62. I. Reda, M. Ghazal, A. Shalaby, M. Elmogy, A. AbouEl-Fetouh, B. O. Ayinde, M. AbouEl-Ghar, A. Elmaghraby, R. Keynton, and A. El-Baz, "A novel ADCS-based CNN classification system for precise diagnosis of prostate cancer," in *2018 24th International Conference on Pattern Recognition (ICPR)*, IEEE, 2018, pp. 3923–3928.

63. I. Reda, A. Khalil, M. Elmogy, A. Abou El-Fetouh, A. Shalaby, M. Abou El-Ghar, A. Elmaghraby, M. Ghazal, and A. El-Baz, "Deep learning role in early diagnosis of prostate cancer," *Technology in Cancer Research & Treatment*, vol. 17, p. 1533034618775530, 2018.

64. I. Reda, B. O. Ayinde, M. Elmogy, A. Shalaby, M. El-Melegy, M. A. El-Ghar, A. A. El-fetouh, M. Ghazal, and A. El-Baz, "A new CNN-based system for early diagnosis of prostate cancer," in *2018 IEEE 15th International Symposium on Biomedical Imaging (ISBI 2018)*, IEEE, 2018, pp. 207–210.

65. I. Reda, A. Shalaby, M. Elmogy, A. A. Elfotouh, F. Khalifa, M. A. El-Ghar, E. Hosseini-Asl, G. Gimel'farb, N. Werghi, and A. El-Baz, "A comprehensive non-invasive framework for diagnosing prostate cancer," *Computers in Biology and Medicine*, vol. 81, pp. 148–158, 2017.

66. N. Eladawi, M. Elmogy, M. Ghazal, O. Helmy, A. Aboelfetouh, A. Riad, S. Schaal, and A. El-Baz, "Classification of retinal diseases based on OCT images," *Front Biosci (Landmark Ed)*, vol. 23, pp. 247–264, 2018.

67. A. ElTanboly, M. Ismail, A. Shalaby, A. Switala, A. El-Baz, S. Schaal, G. Gimelfarb, and M. El-Azab, "A computer-aided diagnostic system for detecting diabetic retinopathy in optical coherence tomography images," *Medical Physics*, vol. 44, no. 3, pp. 914–923, 2017.

68. H. S. Sandhu, A. El-Baz, and J. M. Seddon, "Progress in automated deep learning for macular degeneration," *JAMA Ophthalmology*, vol. 136, no. 12, pp. 1366–1367, 2018.

69. B. Abdollahi, A. C. Civelek, X.-F. Li, J. Suri, and A. El-Baz, "PET/CT nodule segmentation and diagnosis: A survey," in *Multi Detector CT Imaging*, L. Saba and J. S. Suri, Eds. Taylor & Francis, 2014, ch. 30, pp. 639–651.

70. B. Abdollahi, A. El-Baz, and A. A. Amini, "A multi-scale non-linear vessel enhancement technique," in *Engineering in Medicine and Biology Society, EMBC, 2011 Annual International Conference of the IEEE*, IEEE, 2011, pp. 3925–3929.

71. B. Abdollahi, A. Soliman, A. Civelek, X.-F. Li, G. Gimel'farb, and A. El-Baz, "A novel gaussian scale space-based joint MGRF framework for precise lung segmentation," in *Proceedings of IEEE International Conference on Image Processing, (ICIP'12)*, IEEE, 2012, pp. 2029–2032.

72. B. Abdollahi, A. Soliman, A. Civelek, X.-F. Li, G. Gimelfarb, and A. El-Baz, "A novel 3D joint MGRF framework for precise lung segmentation," in *Machine Learning in Medical Imaging*, Springer, 2012, pp. 86–93.

73. A. M. Ali, A. S. El-Baz, and A. A. Farag, "A novel framework for accurate lung segmentation using graph cuts," in *Proceedings of IEEE International Symposium on Biomedical Imaging: From Nano to Macro, (ISBI'07)*, IEEE, 2007, pp. 908–911.

74. A. El-Baz, G. M. Beache, G. Gimel'farb, K. Suzuki, and K. Okada, "Lung imaging data analysis," *International Journal of Biomedical Imaging*, vol. 2013, pp. 1–2, 2013.

75. A. El-Baz, G. M. Beache, G. Gimel'farb, K. Suzuki, K. Okada, A. Elnakib, A. Soliman, and B. Abdollahi, "Computer-aided diagnosis systems for lung cancer: Challenges and methodologies," *International Journal of Biomedical Imaging*, vol. 2013, pp. 1–46, 2013.

76. A. El-Baz, A. Elnakib, M. Abou El-Ghar, G. Gimel'farb, R. Falk, and A. Farag, "Automatic detection of 2D and 3D lung nodules in chest spiral CT scans," *International Journal of Biomedical Imaging*, vol. 2013, pp. 1–11, 2013.

77. A. El-Baz, A. A. Farag, R. Falk, and R. La Rocca, "A unified approach for detection, visualization, and identification of lung abnormalities in chest spiral CT scans," in *International Congress Series*, vol. 1256, Elsevier, 2003, pp. 998–1004.

78. A. El-Baz, A. A. Farag, R. Falk, and R. La Rocca, "Detection, visualization and identification of lung abnormalities in chest spiral CT scan: Phase-I," in *Proceedings of International Conference on Biomedical Engineering*, Cairo, Egypt, vol. 12, no. 1, 2002.

79. A. El-Baz, A. Farag, G. Gimel'farb, R. Falk, M. A. El-Ghar, and T. Eldiasty, "A framework for automatic segmentation of lung nodules from low dose chest CT scans," in *Proceedings of International Conference on Pattern Recognition, (ICPR'06)*, vol. 3, IEEE, 2006, pp. 611–614.

80. A. El-Baz, A. Farag, G. Gimelfarb, R. Falk, and M. A. El-Ghar, "A novel level set-based computer-aided detection system for automatic detection of lung nodules in low dose chest computed tomography scans," *Lung Imaging and Computer Aided Diagnosis*, vol. 10, pp. 221–238, 2011.

81. A. El-Baz, G. Gimel'farb, M. Abou El-Ghar, and R. Falk, "Appearance-based diagnostic system for early assessment of malignant lung nodules," in *Proceedings of IEEE International Conference on Image Processing, (ICIP'12)*, IEEE, 2012, pp. 533–536.

82. A. El-Baz, G. Gimel'farb, and R. Falk, "A novel 3D framework for automatic lung segmentation from low dose CT images," in *Lung Imaging and Computer Aided Diagnosis*, A. El-Baz and J. S. Suri, Eds. Taylor, Francis, 2011, ch. 1, pp. 1–16.

83. A. El-Baz, G. Gimel'farb, R. Falk, and M. El-Ghar, "Appearance analysis for diagnosing malignant lung nodules," in *Proceedings of IEEE International Symposium on Biomedical Imaging: From Nano to Macro (ISBI'10)*, IEEE, 2010, pp. 193–196.

84. A. El-Baz, G. Gimel'farb, R. Falk, and M. A. El-Ghar, "A novel level set-based CAD system for automatic detection of lung nodules in low dose chest CT scans," in *Lung Imaging and Computer Aided Diagnosis*, A. El-Baz and J. S. Suri, Eds. Taylor, Francis, 2011, vol. 1, ch. 10, pp. 221–238.

85. A. El-Baz, G. Gimel'farb, R. Falk, and M. A. El-Ghar, "A new approach for automatic analysis of 3D low dose CT images for accurate monitoring the detected lung nodules," in *Proceedings of International Conference on Pattern Recognition, (ICPR'08)*, IEEE, 2008, pp. 1–4.

86. A. El-Baz, G. Gimel'farb, R. Falk, and M. A. El-Ghar, "A novel approach for automatic follow-up of detected lung nodules," in *Proceedings of IEEE International Conference on Image Processing, (ICIP'07)*, vol. 5, IEEE, 2007, pp. V–501.

87. A. El-Baz, G. Gimel'farb, R. Falk, and M. A. El-Ghar, "A new CAD system for early diagnosis of detected lung nodules," in *ICIP 2007. IEEE International Conference on Image Processing, 2007*, vol. 2, IEEE, 2007, pp. II–461.

88. A. El-Baz, G. Gimel'farb, R. Falk, M. A. El-Ghar, and H. Refaie, "Promising results for early diagnosis of lung cancer," in *Proceedings of IEEE International Symposium on Biomedical Imaging: From Nano to Macro, (ISBI'08)*, IEEE, 2008, pp. 1151–1154.

89. A. El-Baz, G. L. Gimel'farb, R. Falk, M. Abou El-Ghar, T. Holland, and T. Shaffer, "A new stochastic framework for accurate lung segmentation," in *Proceedings of Medical Image Computing and Computer-Assisted Intervention, (MICCAI'08)*, New York, 2008, pp. 322–330.

90. A. El-Baz, G. L. Gimel'farb, R. Falk, D. Heredis, and M. Abou El-Ghar, "A novel approach for accurate estimation of the growth rate of the detected lung nodules," in *Proceedings of International Workshop on Pulmonary Image Analysis*, 2008, pp. 33–42.

91. A. El-Baz, G. L. Gimel'farb, R. Falk, T. Holland, and T. Shaffer, "A framework for unsupervised segmentation of lung tissues from low dose computed tomography images," in *Proceedings of British Machine Vision, (BMVC'08)*, 2008, pp. 1–10.

92. A. El-Baz, G. Gimelfarb, R. Falk, and M. A. El-Ghar, "3D MGRF-based appearance modeling for robust segmentation of pulmonary nodules in 3D LDCT chest images," in A. El-Baz and J. Suri, Eds., *Lung Imaging and Computer Aided Diagnosis*, CRC Press, 2011, ch. 3, pp. 51–63.

93. A. El-Baz, G. Gimelfarb, R. Falk, and M. A. El-Ghar, "Automatic analysis of 3D low dose CT images for early diagnosis of lung cancer," *Pattern Recognition*, vol. 42, no. 6, pp. 1041–1051, 2009.

94. A. El-Baz, G. Gimelfarb, R. Falk, M. A. El-Ghar, S. Rainey, D. Heredia, and T. Shaffer, "Toward early diagnosis of lung cancer," in *Proceedings of Medical Image Computing and Computer-Assisted Intervention, (MICCAI'09)*, Springer, 2009, pp. 682–689.

95. A. El-Baz, G. Gimelfarb, R. Falk, M. A. El-Ghar, and J. Suri, "Appearance analysis for the early assessment of detected lung nodules," in A. El-Baz and J. Suri, Eds., *Lung Imaging and Computer Aided Diagnosis*, CRC Press, 2011, ch. 17, pp. 395–404.

96. A. El-Baz, F. Khalifa, A. Elnakib, M. Nitkzen, A. Soliman, P. McClure, G. Gimel'farb, and M. A. El-Ghar, "A novel approach for global lung registration using 3D Markov Gibbs appearance model," in *Proceedings of International Conference Medical Image Computing and Computer-Assisted Intervention, (MICCAI'12)*, Nice, France, October 1–5, 2012, pp. 114–121.

97. A. El-Baz, M. Nitzken, A. Elnakib, F. Khalifa, G. Gimel'farb, R. Falk, and M. A. El-Ghar, "3D shape analysis for early diagnosis of malignant lung nodules," in *Proceedings of International Conference Medical Image Computing and Computer-Assisted Intervention, (MICCAI'11)*, Toronto, Canada, September 18–22, 2011, pp. 175–182.

98. A. El-Baz, M. Nitzken, G. Gimelfarb, E. Van Bogaert, R. Falk, M. A. El-Ghar, and J. Suri, "Three-dimensional shape analysis using spherical harmonics for early assessment of detected lung nodules," in A. El-Baz and J. Suri, Eds., *Lung Imaging and Computer Aided Diagnosis*, CRC Press, 2011, ch. 19, pp. 421–438.

99. A. El-Baz, M. Nitzken, F. Khalifa, A. Elnakib, G. Gimel'farb, R. Falk, and M. A. El-Ghar, "3D shape analysis for early diagnosis of malignant lung nodules," in *Proceedings of International Conference on Information Processing in Medical Imaging, (IPMI'11)*, Monastery Irsee, Germany (Bavaria), July 3–8, 2011, pp. 772–783.

100. A. El-Baz, M. Nitzken, E. Vanbogaert, G. Gimel'Farb, R. Falk, and M. Abo El-Ghar, "A novel shape-based diagnostic approach for early diagnosis of lung nodules," in *2011 IEEE International Symposium on Biomedical Imaging: From Nano to Macro*, IEEE, 2011, pp. 137–140.

101. A. El-Baz, P. Sethu, G. Gimel'farb, F. Khalifa, A. Elnakib, R. Falk, and M. A. El-Ghar, "Elastic phantoms generated by microfluidics technology: Validation of an imaged-based approach for accurate measurement of the growth rate of lung nodules," *Biotechnology Journal*, vol. 6, no. 2, pp. 195–203, 2011.

102. A. El-Baz, P. Sethu, G. Gimel'farb, F. Khalifa, A. Elnakib, R. Falk, and M. A. El-Ghar, "A new validation approach for the growth rate measurement using elastic phantoms generated by state-of-the-art microfluidics technology," in *Proceedings of IEEE International Conference on Image Processing, (ICIP'10)*, Hong Kong, September 26–29, 2010, pp. 4381–4383.

103. A. El-Baz, P. Sethu, G. Gimel'farb, F. Khalifa, A. Elnakib, R. Falk, and M. A. E.-G. J. Suri, "Validation of a new imaged-based approach for the accurate estimating of the growth rate of detected lung nodules using real CT images and elastic phantoms generated by

state-of-the-art microfluidics technology," in *Handbook of Lung Imaging and Computer Aided Diagnosis*, A. El-Baz and J. S. Suri, Eds. Taylor & Francis, New York, 2011, vol. 1, ch. 18, pp. 405–420.

104. A. El-Baz, A. Soliman, P. McClure, G. Gimel'farb, M. A. El-Ghar, and R. Falk, "Early assessment of malignant lung nodules based on the spatial analysis of detected lung nodules," in *Proceedings of IEEE International Symposium on Biomedical Imaging: From Nano to Macro, (ISBI'12)*, IEEE, 2012, pp. 1463–1466.

105. A. El-Baz, S. E. Yuksel, S. Elshazly, and A. A. Farag, "Non-rigid registration techniques for automatic follow-up of lung nodules," in *Proceedings of Computer Assisted Radiology and Surgery, (CARS'05)*, vol. 1281, Elsevier, 2005, pp. 1115–1120.

106. A. S. El-Baz and J. S. Suri, *Lung Imaging and Computer Aided Diagnosis*. CRC Press, 2011.

107. A. Soliman, F. Khalifa, N. Dunlap, B. Wang, M. El-Ghar, and A. El-Baz, "An iso-surfaces based local deformation handling framework of lung tissues," in *2016 IEEE 13th International Symposium on Biomedical Imaging (ISBI)*, IEEE, 2016, pp. 1253–1259.

108. A. Soliman, F. Khalifa, A. Shaffie, N. Dunlap, B. Wang, A. Elmaghraby, and A. El-Baz, "Detection of lung injury using 4D-CT chest images," in *2016 IEEE 13th International Symposium on Biomedical Imaging (ISBI)*, IEEE, 2016, pp. 1274–1277.

109. A. Soliman, F. Khalifa, A. Shaffie, N. Dunlap, B. Wang, A. Elmaghraby, G. Gimel'farb, M. Ghazal, and A. El-Baz, "A comprehensive framework for early assessment of lung injury," in *2017 IEEE International Conference on Image Processing (ICIP)*, IEEE, 2017, pp. 3275–3279.

110. A. Shaffie, A. Soliman, M. Ghazal, F. Taher, N. Dunlap, B. Wang, A. Elmaghraby, G. Gimel'farb, and A. El-Baz, "A new framework for incorporating appearance and shape features of lung nodules for precise diagnosis of lung cancer," in *2017 IEEE International Conference on Image Processing (ICIP)*, IEEE, 2017, pp. 1372–1376.

111. A. Soliman, F. Khalifa, A. Shaffie, N. Liu, N. Dunlap, B. Wang, A. Elmaghraby, G. Gimel'farb, and A. El-Baz, "Image-based CAD system for accurate identification of lung injury," in *2016 IEEE International Conference on Image Processing (ICIP)*, IEEE, 2016, pp. 121–125.

112. B. Dombroski, M. Nitzken, A. Elnakib, F. Khalifa, A. El-Baz, and M. F. Casanova, "Cortical surface complexity in a population-based normative sample," *Translational Neuroscience*, vol. 5, no. 1, pp. 17–24, 2014.

113. A. El-Baz, M. Casanova, G. Gimel'farb, M. Mott, and A. Switala, "An MRI-based diagnostic framework for early diagnosis of dyslexia," *International Journal of Computer Assisted Radiology and Surgery*, vol. 3, no. 3–4, pp. 181–189, 2008.

114. A. El-Baz, M. Casanova, G. Gimel'farb, M. Mott, A. Switala, E. Vanbogaert, and R. McCracken, "A new CAD system for early diagnosis of dyslexic brains," in *Proceedings of International Conference on Image Processing (ICIP'2008)*, IEEE, 2008, pp. 1820–1823.

115. A. El-Baz, M. F. Casanova, G. Gimel'farb, M. Mott, and A. E. Switwala, "A new image analysis approach for automatic classification of autistic brains," in *Proceedings of IEEE International Symposium on Biomedical Imaging: From Nano to Macro (ISBI'2007)*, IEEE, 2007, pp. 352–355.

116. A. El-Baz, A. Elnakib, F. Khalifa, M. A. El-Ghar, P. McClure, A. Soliman, and G. Gimel'farb, "Precise segmentation of 3-D magnetic resonance angiography," *IEEE Transactions on Biomedical Engineering*, vol. 59, no. 7, pp. 2019–2029, 2012.

117. A. El-Baz, A. Farag, G. Gimel'farb, M. A. El-Ghar, and T. Eldiasty, "Probabilistic modeling of blood vessels for segmenting MRA images," in *18th International Conference on Pattern Recognition (ICPR'06)*, vol. 3, IEEE, 2006, pp. 917–920.

118. A. El-Baz, A. A. Farag, G. Gimelfarb, M. A. El-Ghar, and T. Eldiasty, "A new adaptive probabilistic model of blood vessels for segmenting mra images," in *Medical Image Computing and Computer-Assisted Intervention–MICCAI 2006*, vol. 4191, Springer, 2006, pp. 799–806.

119. A. El-Baz, A. A. Farag, G. Gimelfarb, and S. G. Hushek, "Automatic cerebrovascular segmentation by accurate probabilistic modeling of TOF-MRA images," in *Medical Image Computing and Computer-Assisted Intervention–MICCAI 2005*, Springer, 2005, pp. 34–42.

120. A. El-Baz, A. Farag, A. Elnakib, M. F. Casanova, G. Gimel'farb, A. E. Switala, D. Jordan, and S. Rainey, "Accurate automated detection of autism related corpus callosum abnormalities," *Journal of Medical Systems*, vol. 35, no. 5, pp. 929–939, 2011.

121. A. El-Baz, A. Farag, and G. Gimelfarb, "Cerebrovascular segmentation by accurate probabilistic modeling of TOF-MRA images," in *Image Analysis*, vol. 3540, Springer, 2005, pp. 1128–1137.

122. A. El-Baz, G. Gimelfarb, R. Falk, M. A. El-Ghar, V. Kumar, and D. Heredia, "A novel 3D joint Markov-gibbs model for extracting blood vessels from PC–MRA images," in *Medical Image Computing and Computer-Assisted Intervention–MICCAI 2009*, vol. 5762, Springer, 2009, pp. 943–950.

123. A. Elnakib, A. El-Baz, M. F. Casanova, G. Gimel'farb, and A. E. Switala, "Image-based detection of corpus callosum variability for more accurate discrimination between dyslexic and normal brains," in *Proceedings of IEEE International Symposium on Biomedical Imaging: From Nano to Macro (ISBI'2010)*, IEEE, 2010, pp. 109–112.

124. A. Elnakib, M. F. Casanova, G. Gimel'farb, A. E. Switala, and A. El-Baz, "Autism diagnostics by centerline-based shape analysis of the corpus callosum," in *Proceedings of IEEE International Symposium on Biomedical Imaging: From Nano to Macro (ISBI'2011)*, IEEE, 2011, pp. 1843–1846.

125. A. Elnakib, M. Nitzken, M. Casanova, H. Park, G. Gimel'farb, and A. El-Baz, "Quantification of age-related brain cortex change using 3D shape analysis," in *2012 21st International Conference on Pattern Recognition (ICPR)*, IEEE, 2012, pp. 41–44.

126. M. Mostapha, A. Soliman, F. Khalifa, A. Elnakib, A. Alansary, M. Nitzken, M. F. Casanova, and A. El-Baz, "A statistical framework for the classification of infant DT images," in *2014 IEEE International Conference on Image Processing (ICIP)*, IEEE, 2014, pp. 2222–2226.

127. M. Nitzken, M. Casanova, G. Gimel'farb, A. Elnakib, F. Khalifa, A. Switala, and A. El-Baz, "3D shape analysis of the brain cortex with application to dyslexia," in *2011 18th IEEE International Conference on Image Processing (ICIP)*, Brussels, Belgium, IEEE, September 2011, pp. 2657–2660, (Selected for oral presentation. Oral acceptance rate is 10 percent and the overall acceptance rate is 35 percent).

128. F. E.-Z. A. El-Gamal, M. M. Elmogy, M. Ghazal, A. Atwan, G. N. Barnes, M. F. Casanova, R. Keynton, and A. S. El-Baz, "A novel CAD system for local and global early diagnosis of Alzheimer's disease based on PIB-PET scans," in *2017 IEEE International Conference on Image Processing (ICIP)*, IEEE, 2017, pp. 3270–3274.

129. M. Ismail, A. Soliman, M. Ghazal, A. E. Switala, G. Gimelfarb, G. N. Barnes, A. Khalil, and A. El-Baz, "A fast stochastic framework for automatic MR brain images segmentation," *PloS one*, vol. 12, no. 11, e0187391, 2017.

130. M. M. Ismail, R. S. Keynton, M. M. Mostapha, A. H. ElTanboly, M. F. Casanova, G. L. Gimel'farb, and A. El-Baz, "Studying autism spectrum disorder with structural and diffusion magnetic resonance imaging: A survey," *Frontiers in Human Neuroscience*, vol. 10, p. 211, 2016.

131. A. Alansary, M. Ismail, A. Soliman, F. Khalifa, M. Nitzken, A. Elnakib, M. Mostapha, et al., "Infant brain extraction in T1-weighted MR images using BET and refinement using LCDG and MGRF models," *IEEE Journal of Biomedical and Health Informatics*, vol. 20, no. 3, pp. 925–935, 2016.

132. M. Ismail, A. Soliman, A. ElTanboly, A. Switala, M. Mahmoud, F. Khalifa, G. Gimel'farb, M. F. Casanova, R. Keynton, and A. El-Baz, "Detection of white matter abnormalities in MR brain images for diagnosis of autism in children," in *2016 IEEE 13th International Symposium on Biomedical Imaging (ISBI)*, Prague, Czech Republic, IEEE, pp. 6–9, 2016.

133. M. Ismail, M. Mostapha, A. Soliman, M. Nitzken, F. Khalifa, A. Elnakib, G. Gimel'farb, M. Casanova, and A. El-Baz, "Segmentation of infant brain MR images based on adaptive shape prior and higher-order MGRF," in *2015 IEEE International Conference on Image Processing (ICIP)*, Québec City, QC, pp. 4327–4331, 2015.

134. E. H. Asl, M. Ghazal, A. Mahmoud, A. Aslantas, A. Shalaby, M. Casanova, G. Barnes, G. Gimelfarb, R. Keynton, and A. El-Baz, "Alzheimers disease diagnostics by a 3D deeply supervised adaptable convolutional network," *Frontiers in Bioscience (Landmark Edition)*, vol. 23, pp. 584–596, 2018.

135. A. Mahmoud, A. El-Barkouky, H. Farag, J. Graham, and A. Farag, "A non-invasive method for measuring blood flow rate in superficial veins from a single thermal image," in *Proceedings of the IEEE Conference on Computer Vision and Pattern Recognition Workshops*, Portland, OR, 2013, pp. 354–359.

136. A. El-baz, A. Shalaby, F. Taher, M. El-Baz, M. Ghazal, M. A. El-Ghar, A. Takieldeen, and J. Suri, "Probabilistic modeling of blood vessels for segmenting magnetic resonance angiography images," *Medical Research Archives*, vol. 5, no. 3, 2017.

137. A. S. Chowdhury, A. K. Rudra, M. Sen, A. Elnakib, and A. El-Baz, "Cerebral white matter segmentation from MRI using probabilistic graph cuts and geometric shape priors," in *2010 IEEE International Conference on Image Processing (ICIP)*, Hong Kong, 2010, pp. 3649–3652.

138. Y. Gebru, G. Giridharan, M. Ghazal, A. Mahmoud, A. Shalaby, and A. El-Baz, "Detection of cerebrovascular changes using magnetic resonance angiography," in A. El-Baz and J. Suri, Eds., *Cardiovascular Imaging and Image Analysis*. CRC Press, 2018, pp. 1–22.

139. A. Mahmoud, A. Shalaby, F. Taher, M. El-Baz, J. S. Suri, and A. El-Baz, "Vascular tree segmentation from different image modalities," in A. El-Baz and J. Suri, Eds., *Cardiovascular Imaging and Image Analysis*. CRC Press, 2018, pp. 43–70.

140. F. Taher, A. Mahmoud, A. Shalaby, and A. El-Baz, "A review on the cerebrovascular segmentation methods," in *2018 IEEE International Symposium on Signal Processing and Information Technology (ISSPIT)*, IEEE, 2018, pp. 359–364.

141. H. Kandil, A. Soliman, L. Fraiwan, A. Shalaby, A. Mahmoud, A. ElTanboly, A. Elmaghraby, G. Giridharan, and A. El-Baz, "A novel MRA framework based on integrated global and local analysis for accurate segmentation of the cerebral vascular system," in *2018 IEEE 15th International Symposium on Biomedical Imaging (ISBI 2018)*, IEEE, 2018, pp. 1365–1368.

Chapter 10

Overview of Deep Learning Algorithms Applied to Medical Images

Behnaz Abdollahi, Ayman El-Baz, and Hermann B. Frieboes

10.1 Deep Learning Algorithms

10.1.1 Machine Learning as the Foundation for Deep Learning

Machine learning algorithms require input data to learn and approximate mathematical functions, as these algorithms are trained to learn the correlation of the input data to the corresponding output. The desired function is thus trained and evaluated on the provided dataset to make it generalizable on unknown data. For optimal performance, the design, development, and evaluation of these algorithms require a large amount of training data. According to specific objectives, machine learning methods include supervised learning, unsupervised learning, semi-supervised learning, reinforcement learning, transfer learning, and deep learning, which is based on artificial neural networks (1, 2). Generally machine learning models are designed and developed on 70–80% of the collected data, which is defined as the training dataset. The developed model is evaluated on a smaller subset of the dataset, which was not part of the training set and is called the validation set. The model's performance is measured on the validation set, and the hyperparameters are fine tuned. Lastly, the tuned model is tested on another subset of the dataset called the test set. The purpose of the test set is to evaluate the generalizability of the tuned model on unknown data.

Assuming that the objective is to estimate a function that classifies the input data with high accuracy on a training set with the goal to classify unknown data, three properties typically evaluated include robustness, stability, and generalizability (3–5). A general mathematical representation of a machine learning model can be written as:

$$f : X \rightarrow Y,$$

where f is the function from X to Y that is learned based on the relation between the input and output data.

If the target value of every input data is provided, then supervised learning or classification methods can be applied to the dataset. For example, if a dataset of brain cancer MR images is provided and each image is a picture of a different type of brain tumor, and the label of each image is given, then the trained classification model can learn to label new images as particular brain tumors. Considering medical images, the lack of labeled data represents a serious obstacle. If the target label of the input data is not provided, the developed model is called unsupervised learning or clustering, in which the input data are clustered based on common characteristics. Thus, unsupervised learning algorithms could be applied to cluster images of brain tumors into different categories. If the target values of some of the input data are given, then semi-supervised algorithms could be applicable, representing a combination of both supervised and unsupervised approaches.

Regarding the probabilistic modeling of objective functions, machine learning algorithms are also categorized as discriminative and generative models (1, 6). Discriminative models learn the boundaries between classes, modeling joint distribution of the input data. The probabilistic representation of discriminative learning models maximizes conditional data likelihood and minimizes the error between the actual function and the approximated function:

$$W \leftarrow \arg\max \ln \left(\prod_{i=1}^{n} P(y_i \mid x_i, w) \right)$$

$$W \leftarrow \arg\min \left(\sum_{i=1}^{n} \left(y_i - \hat{f}(x_i) \right) \right)^2,$$

where $P(Y_i \mid x_i, w)$ is the conditional probability of each output given the input and the parameters, and $\hat{f}(x_i)$ is the estimated mapping function from input to output using the designed model. Generative models such as naïve Bayes classifier assume some prior distribution for $P(y)$ and $P(X \mid Y)$ and use Bayes' rule to calculate $P(Y \mid X)$. In contrast, generative models learn the distribution of each class and require the modeling of the conditional probabilities. It is possible to use Bayes' rule to change the definition of a discriminative model into a generative model (1, 6). Generative models are also used for generating synthetic data to compensate for a low number of samples, with generative adversarial networks as an example of this approach.

Conventional machine learning algorithms are successfully developed on datasets large enough to include different varieties of input data. For instance, a dataset of images might include noisy images, blurry or even distorted images. Preprocessing and data cleaning are primarily applied before the feature extraction and feature selection steps. Choosing whether to apply the preprocessing step and deciding the required pre-processing modules depends on the given data. For example, if the input dataset is structured data and saved as records in a relational database, then there might not be a need for the preprocessing step; however if the data are extracted from unstructured datasets such as clinical reports or medical images, then data cleaning can be a substantial step. The output of the feature extraction and feature selection steps is basically different representations of the input data. Feature extraction methods vary based on the dataset domain and require expert information in the domain. The next step is feature selection which is applied to extracted features to collect the best optimized features from all the extracted features. After applying feature extraction and selection, the input data is mapped to a new feature space with lower or higher dimensions. Conventional machine learning methods have focused on extracting and selecting optimized features (7–11).

Finally, the optimized feature vector, which represents the original input data in the new feature space, is used to train the appropriate machine learning model. In practice, appropriate models are applied on the training set and validated on the test set, and then the model with the minimum error rate on the training and fine tuned on validation set is selected. Based on several parameters, such as the number of samples, statistical information of the features, the objective of the defined problem, etc., some of the models are eliminated before even applying them. For example, if the data are linearly separable then support vector machines are able to classify the data, and thus there is no need to choose a non-linear model. The accuracy of the developed model is evaluated based on the training and cross validation. The test set is chosen from unseen data and helps to evaluate the accuracy of the developed models.

Feature extraction is a substantial step in developing conventional machine learning algorithms. As manually extracting the best features from the input dataset is challenging and sometimes impossible, recently deep artificial neural networks, also called deep learning, have been used to extract non-linear features automatically. Deep learning methods eliminate manual feature extraction and feature selection, which enables the training with large and complicated datasets with high accuracy. In fact, deep learning methods can automatically extract non-linear features by adding more hidden layers. Biomedical data characteristics are complicated, and, hence, deep learning algorithms are an appropriate choice for biomedical data and, particularly, for medical images.

Deep learning algorithms in the form of shallow neural networks have been previously utilized (11). The number of layers indicates whether a neural network is shallow or deep. This number depends on the complexity of the approximate function. Shallow neural networks cannot solve complicated functions, and thus deeper neural networks are required to solve non-linear or complicated functions. Training deep neural networks may incur a high computational cost, as very large number of training data with different varieties may be required (12). Only recently has computer technology been able to provide the computational power needed to make the implementation of deep neural networks practically feasible.

10.1.2 Deep Learning as an Extension of Artificial Neural Networks

Artificial neural networks were first developed to simulate the functionality of the human brain, as a combination of neurons arranged in sequential layers. Each neuron is connected with several other neurons, and any neural network architecture can be represented as a non-linear mathematical function (also called activation function), which maps its inputs to its output. Layers of neurons between the input and output neuron layers are called hidden layers. The number of neurons in each layer and the structure of the networks vary based on the input data and the desired function.

A typical neuron activation function is defined as:

$$a = \text{activation function } (w.T + b),$$

where a is the output of the applied activation function on its input, $w.T$ is the transpose of the weights in the layer, and b is a scalar bias parameter. Activation functions are non-linear functions that decide whether a neuron is active. The functions are generally chosen as one of sigmoid function, ReLU function, *Tanh*, Leaky ReLU function, and softmax (13).

Choosing an appropriate activation function depends on the structure of the proposed neural network and the goals of the implementation. Different activation functions might be chosen in different layers of the neural network. Let's assume that the cost function is chosen as mean square error which is widely used and the mean square difference between the estimated function and the real function is calculated. The two parameters w and b are the inputs to the neurons. The cost function J is defined as:

$$J(w,b) = \frac{1}{m} \sum_{i=1}^{m} (\hat{y} - y)^2,$$

where $\hat{y}$ is the function estimated by applying the machine learning model, and y is the actual function on the real data.

Cost functions are categorized into two types based on their mathematical shape: convex and non-convex. In order to minimize a mathematical function, the first derivative of the function is calculated to determine its minimum values. Non-convex cost functions do not have just one global minimum but several local minima. Convex cost functions have only one global minimum, and the minimization algorithm converges to find the global minimum (14). In order to estimate the neural network parameters, the partial derivative of the parameters is calculated in each iteration, and the parameters are updated until they converge to the optimized solution, based on the model accuracy with the training and validation sets.

Gradient descent is an iterative optimization technique widely used to optimize machine learning cost functions. It is an iterative method that updates the parameters until they converge to the optimized solution values. Assuming a model with cost function J and two parameters w and b, the first two equations below calculate the partial derivatives of w and b, while the second two equations update the parameters.

$$dw = dJ \,/\, dw$$

$$db = dJ \,/\, db$$

$$w = w - \text{learning rate} \cdot dw$$

$$b = b - \text{learning rate} \cdot db.$$

Learning rate is a hyperparameter that defines the speed of convergence of the optimization algorithm. If the rate is small, the optimization algorithm such as gradient descent will converge slowly, meaning that several iterations are needed for training the model; however, if the rate is large, the algorithm may be unable to converge to the optimized point. One way to address this issue is to dynamically change the learning rate through the optimization iteration using momentum. In the first few iterations, the rate is chosen large so that the algorithm advances faster towards the optimal point, and then the rate is reduced so as not to miss the convergence point. Hence, the rate assumes different values depending on the distance of the current solution to the convergence point(14). The above equations represent the basic mathematical representation of each step of gradient descent; more in-depth information is provided in (1). Calculating the partial derivative of the parameters is called backpropagation. The gradient descent algorithm converges when the parameters do not change significantly compared to the previous iteration. Gradient descent optimization can be categorized in several types, such as stochastic gradient descent (SGD), mini batch gradient descent, and batch gradient descent (15).

Neural networks are able to solve specific problems such as detecting handwritten characters (16), face recognition (17), and speech recognition (18). The number of neural networks layers is related to the non-linearity of the approximated function. A neural network structure with several (i.e., hundreds or even thousands) hidden layers is called a deep neural network, or deep learning for short. The "learning" occurs by breaking down the input into more basic components and then reconstructing the output "bottom-up" from them. Recently, deep learning algorithms have become a leading-edge technology in machine learning, excelling in some applications such as speech recognition and object recognition in images (19). Deep learning is commonly used in analyzing large datasets of natural images (20–23). Deep learning algorithms are categorized into representative models, including autoencoders, deep belief network, feedforward/backpropagate neural network, convolutional neural network, and recurrent neural network (RNN). Applications include image segmentation, object detection, disease diagnosis, and image de-noising. A multiscale deep neural network was developed to detect early Alzheimer's disease in Lu et al. (2018) and applied to structural MRI and FDG-PET. The combination of these two datasets helped to develop optimized features using multiscale feature engineering (24).

10.1.3 Advantages and Disadvantages of Deep Learning Methods

A significant advantage of deep learning models over conventional machine learning algorithms is their independence from manual feature extraction and selection. Deep learning algorithms automatically extract non-linear and complicated features based on the architecture of the neural network. The output of each layer is the non-linear and compressed representation of the input data. Each layer represents an abstract feature of the input data. The backpropagation step tunes the parameters and updates them by calculating the derivative of the parameters calculated in the forward propagation. Further, in image analysis applications, deep learning techniques are more robust to noise and invariant to color and rotation compared to conventional machine learning algorithms.

However, these methods may employ thousands or even millions of parameters, so the computational cost is high to train a very large and deep neural network. Layers may incur the vanishing and exploding of parameters problems because each layer multiplies the parameters for the next layer. If the numbers are small, their multiplications in each layer result in smaller numbers approaching zero, and if the numbers are large, their multiplication in several layers results in huge numbers. A new deep learning structure called ResNet has addressed this issue. (25)

Another disadvantage is that the structure of deep learning models is based on the assumption that the human brain learns similarly by breaking information down into basic components and then reconstructing it. However, this assumption is not universally accepted and alternative (non-deep learning) models are under development (26).

10.2   Challenges in Developing Machine Learning Algorithms

10.2.1   Parameters and Hyperparameters

Machine learning models have several variables that require tuning and adjustment during the learning process. These variables are categorized as parameters and hyperparameters. Parameters are the variables that are defined directly based on the input data, while hyperparameters are variables that depend on the design of the model and cannot be extracted from the input data. The values assigned to each connection between nodes in a neural network are called neural network parameters. In this case, hyperparameters are variables that determine the structure of the neural network, such as the number of the hidden layers, the number of the gradient descent epochs, the choosing of the best activation function for each layer, and the learning rate parameters. Hyperparameters and parameters are tuned and optimized during the training phase. The model is trained with the best and optimized values for each hyperparameter and parameter. The accuracy level of the model reflects the optimized combination of the hyperparameters and parameters (27).

10.2.2   Bias–Variance Trade-Off

One of the criteria to find the best optimized model is acquiring low bias and variance based on the fine tuned parameters and hyperparameters. There is a trade-off between bias and variance. High-bias models present high error on training and test sets; technically, these models are under-fitted. High-variance models are overfitted; the error rate is low on the training set while being high on the test set. An optimized machine learning model is equivalent to a low bias and low variance model. In practice, it is challenging to find an optimized model using conventional machine learning algorithms. Since deep learning models are implemented on large datasets, they have been shown to successfully find a balanced model in terms of low bias and low variance (28, 29).

10.2.3   Regularization Methods

In this section, we summarize some of the most common regularization techniques. More complicated algorithms include those described in (30–33). In order to deal with high-bias models, adding more samples and applying more complicated models are two basic solutions that can yield a low-bias model (19). As high-variance models are the result of an overfitted model, a common solution to avoid overfitting is regularization. The original cost function is changed by adding a regularization term to the cost function J in order to eliminate the overfitting. There are two types of regularization terms: L1 norm and L2 norm. L1 norm mostly zeroes out the terms, so the network and the parameters will become relatively sparse and a more linear model is attained. L2 norm involves multiplying the parameters with power two, to prevent the sparsity and decrease the effect of some of the parameters without zeroing them out. In practice, L2 norm is mostly used (34, 35).

If the cost function is not regularized, then in each iteration the cost function should decrease monotonically. However, by adding the L2 regularization term, the decreasing function becomes less noisy while approaching the solution. L2 regularization terms are defined as:

$$L2 = \frac{\lambda}{2m} W^2,$$

where λ is the scalar, m is the number the samples and W is the parameter vector of the neural network. The regularization term will force the values to become small, so the network will change to a simpler network, which prevents overfitted models.

Another category of regularization methods is called dropout where some of the neurons are dropped out. The dropout rate of neurons is delineated by a probability, which decides the number of neurons that are kept or zeroed out in each layer, and thus in each training iteration only a specific number of neurons are activated, leading to a simpler and linear network (30). A modified version of the dropout regularization is to use an adaptive probability dependent on the size of each layer. This approach avoids eliminating some of the layers and oversimplifying the model to avoid high bias. The higher probability is equivalent to keeping more layers. The dropout should not be used on the test set, since a random network during the test phase is undesirable.

Data augmentation is used as one of the regularization method. Images are linearly or non-linearly transformed, so more images are generated without collecting new data. For example, if one wishes to classify tumors in a dataset of images that includes different types of tumors, one solution to increase the model accuracy is to collect more images from every tumor type, which might be time consuming or impossible to obtain in a clinical domain. A more practical solution that is employed is to generate more images from every tumor type by rotating, scaling, or non-linearly transforming them (31, 32).

10.3 Deep Learning Methods in Medical Image Analysis

10.3.1 Autoencoders

Autoencoders are categorized as an unsupervised learning method, also commonly used as a non-linear feature extraction technique. The basic autoencoder structure called vanilla has two principal components: encoder and decoder. The goal of autoencoder is to generate an output as a good approximation of the input (19) while learning the non-linear representation of the original input. Thus, autoencoders are categorized as one of the unsupervised methods in which the output layer values result in the same value as the input:

$$\hat{y}(W,b) = x, \text{ so the output } \hat{x} \text{ is equal to the input } x.$$

The encoder learns the compressed version of the input data. Input data are mapped to the hidden layer which has lower dimensional data than the input data, so the input data are compressed or actually encoded, while the output of the hidden layer is mapped to the output, set the same as the input, thus decoding the encoded data. Autoencoders get the transition of the input and generate non-linear features automatically. More specialized autoencoders structures have been developed, such as stacked autoencoders, convolutional autoencoders, denoising autoencoders, sparse autoencoders, and variational autoencoders (36, 37).

In the case of convolutional autoencoders (CAE), the encoding and decoding layers are convolutional, while in the original autoencoders the data are not convolved and the original data are given as input. CAE employs unsupervised learning, which learns the best optimized filter parameters by reconstructing original images with the minimum error. Chen et al. (2017) developed a customized version of CAE on computed tomography (CT) images in order to classify and detect lung nodules. The proposed method applies convolutional autoencoders on a dataset of unlabeled CT images, thus learning the features of lung nodules. The optimized features extracted from the images based on CAE were compared with supervised learning algorithms that extract features manually, and the features based on CAE indicated accurate results (38).

10.3.2   Deep Feedforward/Backpropagation Neural Networks

Shallow neural networks are typically comprised of an input layer that considers the dataset, a few hidden layers, and one output layer. An example of their application is to binary classify MRI images into images that have tumors and images without tumors. A dataset of images with and without tumors is needed. The image pixels are used as input, and the output layer is chosen with one neuron to learn and decide whether the input image is of a tumor or otherwise. The number of the hidden layers between the input and the output layer is tuned based on the given dataset. The hidden layers determine the performance and impact model accuracy, since the addition of hidden layers helps to solve non-linear functions with higher accuracy. Each layer develops non-linear representations of its input. Input data are typically noisy and complicated, and thus a non-linear model is usually indicated. Adding more hidden layers may improve the performance while increasing the computational cost, and hence the model may require more powerful hardware to train several layers of neural networks.

Feed forward neural networks are considered a category of deep neural network in which the parameters are calculated only in the forward steps and no backpropagation is calculated. The inputs are given, and the hidden layers weights are calculated, and the results show up in the output layer. If the derivative of the output layer is calculated relative to its previous layer and iteratively relative to the input data, the whole process is called backpropagation. Feed forward neural networks include an input layer and an output layer based on the objective function, with a number of hidden layers depending on the complexity of the approximated mathematical function. The partial derivative of the parameters is calculated in the backpropagation step. This step is responsible for fine tuning the parameters using the derivative of the parameters. Depending on the dataset, a specific cost function such as mean square error is chosen, and an optimization technique such as gradient descent is applied to iteratively move toward the global minimum. The speed of the learning depends on the learning rate which is one of the hyperparameters defined on the given dataset. Mohsen et al. (2017) developed a deep feedforward neural network that classifies brain tumor MR images into normal, glioblastoma, sarcoma, and metastatic bronchogenic carcinoma (39).

10.3.3   Convolutional Neural Networks

Convolutional neural networks (CNN) are one of the most popular types of deep learning models employed for medical image analysis. The convolution operation on each image is applied in several layers. The layers are categorized into three types: convolution, pooling, and fully connected. Convolutional layers extract features from the input image by convolving different types of filters with the provided images. CNN, in general, have two advantages over feedforward/backpropagation neural networks. One advantage is the use of parameter sharing; for example, an edge detection filter can detect all the edges in the images by convolving the filter with the images, and a new filtering is not needed to detect edges in the images. The second advantage is sparsity, which means that each layer only depends on a smaller number of inputs. The first layer of the deep neural network is mostly chosen as a convolutional layer in which filters are convolved with the input image, and activation functions such as Sigmoid, ReLU, or other suitable functions are applied after the convolution. The second layer is mostly defined as a pooling layer which compresses the output of the convolutional layer. The final layers are fully connected.

Havaei et al. (2017) proposed a multi-scale CNN, in which the input data are provided at multiple scales instead of individual images. A new CNN was developed to segment brain MRI images with lesions such as traumatic brain injuries, brain tumors, and ischemic stroke (40). Histology images mostly called whole slide images capture for cell's structure to detect the abnormal section of the images and classify the cancer type, since they are high resolution images it is challenging to apply an end to end deep learning model, so some of the research papers use spatially constrained CNN was developed to detect nuclei in histology images (41), some of them studies break each whole slide image into several smaller images and use a rule based method to find the correlation between each subsection of the image (42). A novel CNN structure was proposed on CT images to classify interstitial lung diseases on CT images (43) and nodule detection (44) using CNN as a better mathematical machine learning model. To capture anatomical shape variations in structural brain MRI scans to help diagnose Alzheimer's disease, Hosseini-Asl et al. (2016) proposed a deep 3D convolutional neural network (3D-CNN), built upon a 3D CAE. The results showed that the 3D-CNN outperformed several conventional classifiers (45).

10.3.4   Transfer Learning

Deep neural networks incur high computational cost to implement and may require calculating millions of parameters. To mitigate this challenge, a variety of pre-trained classic neural networks have been developed with predefined network structures, such as LeNet 5, AlexNet, and VGG 16. As training a new neural network structure from the ground up is sometimes computationally challenging, pre-trained neural networks may present a viable, albeit less specific, alternative to solve a desired problem. These networks have been developed by companies such as Google, obviating the need to train all the layers. Using pre-trained neural networks is also called transfer learning (46). Shin et al. (2016) used pre-trained ImageNet and evaluated the influences of the data size on the accuracy of lung disease classification (46). Although transfer learning is widely employed, the design and development of models with customized training is still required in order to achieve high accuracy (47).

10.3.5 Generative Adversarial Networks

There are several limitations that make medical image analysis challenging. Deep learning models need lots of data for training, which is sometimes impossible to acquire. Healthcare data might take years to collect for one type of disease. Besides the data collection, labeling the data can be costly and time-consuming. Machine learning classifiers that are designed to classify diseases based on images need to have approximately the same number of data examples in each class, which is difficult since it requires collecting images from many patients with the same disease. Recently, generative adversarial networks (GAN) have been proposed (19) as a data augmentation technique in the medical domain. The idea is to generate data that are fake but similar to the original data, to help balance the number of samples in each class. The model has two neural networks that compete with each other. One of them is generative, generating fake data, and the second one is discriminative, discriminating between the fake and the real data. The model is responsible for finding the equilibrium between these two networks. The discriminator learns based on the original data and improves its accuracy while the generator generates fake data by inserting noise in the first step. While the discriminator detects and classifies the fake data, the generator objective function is updated to generate more accurate data so that the discriminator is able to detect whether the data is fake or real.

10.3.6 Summary of Models

Models that have been typically applied to medical image analysis are summarized in Table 10.1.

TABLE 10.1: Summary of Machine Learning Models Employed with Different Modalities and Their Applications in Medical Image Analysis

Machine Learning Model	Modality	Application
Deep Belief Network	MRI	• Image classification • Schizophrenia and Huntington's disease classification • Schizophrenia/NH classification • AD/MCI/HC classification
Autoencoder	X-ray DCE-MRI MRI and FDG-PET	• Image content retrieval needs • Alzheimer's disease
Convolutional Neural Network	Brain MRI CT images Histology images Prostate MRI images	• Skin lesion classification • Brain tissue classification • Brain lesion such as traumatic brain injuries, brain tumors, and ischemic stroke • Nucleus detection • Nodule detection • Interstitial lung disease classification • Melanoma lesion detection • 3D classification
Transfer Learning	CT images	• Thoracic-abdominal lymph nodes • Interstitial lung disease classification

10.4 Conclusion

Deep learning algorithms are a new approach building on artificial neural networks, offering high accuracy to analyze challenging medical images. The approach helps to avoid the time-consuming process of manually labeling images. Deep autoencoders have been developed for the pre-trained step of the network. Convolutional neural networks are used to classify images in order to localize tumors or to classify normal from abnormal images. Pre-trained neural networks may provide a less specific alternative to fully training a customized network. The lack of adequately large datasets is a major limitation to the application of these methods in medical imaging. Generative adversarial networks and generative models may represent a future direction tailored to addressing this issue. To address these issues with the goal to offer clinical utility, the design of deep learning algorithms could benefit from further research in the domain of medical image analysis.

References

1. Alpaydin E, *Introduction to Machine Learning*. 2014: MIT press.
2. Cohen PR, *Empirical Methods for Artificial Intelligence*. Vol. 139. 2017: MIT press, Cambridge, MA.
3. Xu H, Mannor S, Robustness and generalization. *Machine Learning* 2012;86(3):391–423.
4. Saeys Y, Abeel T, Van de Peer Y, Robust feature selection using ensemble feature selection techniques, in *Joint European Conference on Machine Learning and Knowledge Discovery in Databases*. 2008. Springer.
5. Neyshabur B, Bhojanapalli S, McAllester D, Srebro N, Exploring generalization in deep learning, in *Advances in Neural Information Processing Systems*, pp. 5947–5956. 2017.
6. Jebara T, *Machine Learning: Discriminative and Generative*. Vol. 755. 2012: Springer Science & Business Media.
7. Zheng A, Casari A, *Feature Engineering for Machine Learning: Principles and Techniques for Data Scientists*. 2018: O'Reilly Media, Inc.
8. Bradley AP, Machine learning for medical diagnostics: Techniques for feature extraction, classification, and evaluation. PhD thesis, School of Computer Science and Electrical Engineering, 1998, University of Queensland: Queensland, Asutralia.
9. Guyon I, Gunn S, Nikravesh M, Zadeh LA, *Feature Extraction: Foundations and Applications*. Vol. 207. 2008: Springer.
10. Liu H, Motoda H, *Computational Methods of Feature Selection*. 2007: CRC Press.
11. Liu H, Yu L, Toward integrating feature selection algorithms for classification and clustering. *IEEE Transactions on Knowledge and Data Engineering* 2005;17(4):491–502.
12. Mhaskar HN, Poggio T, Deep vs. Shallow networks: An approximation theory perspective. *Analysis and Applications* 2016;14(06):829–848.
13. Ramachandran P, Zoph B, Le QV, Searching for activation functions, in ICLR2018: Vancouver, British Columbia.
14. Jain P, Kar P, Non-convex optimization for machine learning. *Foundations and Trends® in Machine Learning* 2017;10(3–4):142–336.
15. Ruder S, An overview of gradient descent optimization algorithms, 2016: arXiv.org.
16. LeCun Y, Boser B, Denker JS, Henderson D, Howard RE, Hubbard W, Jackel LD,Backpropagation applied to handwritten zip code recognition. *Neural Computation* 1989;1(4):541–551.
17. Cottrell GW, Extracting features from faces using compression networks: Face, identity, emotion, and gender recognition using holons, in *Connectionist Models—Proceedings of the 1990 Summer School*. 1991. Elsevier. pp. 328–337.

18. Dahl GE, Yu D, Deng L, Acero A, Context-dependent pre-trained deep neural networks for large-vocabulary speech recognition. *IEEE Transactions on Audio, Speech, Language Processing* 2012;20(1):30–42.

19. Goodfellow I, Bengio Y, Courville A, *Deep Learning*. 2016: MIT Press.

20. Farabet C, Couprie C, Najman L, LeCun Y, Learning hierarchical features for scene labeling. *IEEE Transactions on Pattern Analysis and Machine Intelligence* 2013;35(8):1915–1929.

21. Szegedy C, Liu W, Jia Y, Sermanet P, Reed S, Anguelov D, Erhan D et al., Going deeper with convolutions, in *Proceedings of the IEEE Conference on Computer Vision and Pattern Recognition*. 2015. Boston. MA.

22. Simonyan K, Zisserman A, Very deep convolutional networks for large-scale image recognition, *arXiv* 2014;1409:1556 .

23. Krizhevsky A, Sutskever I, Hinton GE, Imagenet classification with deep convolutional neural networks, in *Advances in Neural Information Processing Systems* 2012;25(2):1097–1105.

24. Lu D, Popuri K, Ding GW, Balachandar R, Beg MF, Alzheimer's Disease Neuroimaging I, Multiscale deep neural network based analysis of fdg-pet images for the early diagnosis of alzheimer's disease. *Medical Image Analysis* 2018;46:26–34.

25. He K, Zhang X, Ren S, Sun J, Deep residual learning for image recognition, in *Proceedings of the IEEE Conference on Computer Vision and Pattern Recognition*. 2016. Las Vegas, NV.

26. Hawkins J, Lewis M, Klukas M, Purdy S, Ahmad S, A framework for intelligence and cortical function based on grid cells in the neocortex. *Front Neural Circuits* 2018;12:121.

27. Probst P, Bischl B, Boulesteix A-L, Tunability: Importance of hyperparameters of machine learning algorithms, 2018: arXiv.org.

28. Domingos P, A unified bias-variance decomposition, in *Proceedings of 17th International Conference on Machine Learning*. 2000. Stanford, CA.

29. Ng A, *Machine Learning Yearning*, 96. 2017. http://www.mlyearning.org/.

30. Neyshabur B, Tomioka R, Salakhutdinov R, Srebro N, Geometry of optimization and implicit regularization in deep learning, in *31st Conference on Neural Information Processing Systems*. 2017. Long Beach, CA.

31. Inoue H, Data augmentation by pairing samples for images classification, 2018: arXiv.org.

32. Taylor L, Nitschke G, Improving deep learning using generic data augmentation, 2017: arXiv.org.

33. Kukačka J, Golkov V, Cremers D, Regularization for deep learning: A taxonomy, 2017: arXiv.org.

34. Srivastava N, Hinton G, Krizhevsky A, Sutskever I, Salakhutdinov R, Dropout: A simple way to prevent neural networks from overfitting. *The Journal of Machine Learning Research* 2014;15(1):1929–1958.

35. Goodfellow IJ, Warde-Farley D, Mirza M, Courville A, Bengio Y, Maxout networks. *Proceedings of Machine Learning Research* 2013;28(3): 1319–1327.

36. Rezende DJ, Mohamed S, Wierstra D, Stochastic backpropagation and approximate inference in deep generative models. *Proceedings of Machine Learning Research* 2014;32(2): 1278–1286.

37. Kingma DP, Welling M, Auto-encoding variational Bayes, 2013: arXiv.org.

38. Chen M, Shi X, Zhang Y, Wu D, Guizani M, Deep features learning for medical image analysis with convolutional autoencoder neural network. *IEEE Transactions on Big Data*. 2017.

39. Mohsen H, El-Dahshan E-SA, El-Horbaty E-SM, Salem A-BM, Classification using deep learning neural networks for brain tumors. *Future Computing Informatics Journal* 2018;3(1):68–71.

40. Havaei M, Guizard N, Larochelle H, Jodoin P-M, Deep learning trends for focal brain pathology segmentation in MRI, in Holzinger A, ed., *Machine Learning for Health Informatics* 2016. Springer, p. 125–148.

41. Sirinukunwattana K, Raza SeA, Tsang Y-W, Snead DR, Cree IA, Rajpoot NM, Locality sensitive deep learning for detection and classification of nuclei in routine colon cancer histology images. *IEEE Transactions on Medical Imaging* 2016;35(5):1196–1206.

42. Komura Daisuke, Shumpei Ishikawa, Machine learning methods for histopathological image analysis. *Computational and Structural Biotechnology Journal* 2018;16:34–42.

43. Anthimopoulos M, Christodoulidis S, Ebner L, Christe A, Mougiakakou S, Lung pattern classification for interstitial lung diseases using a deep convolutional neural network. *IEEE Transactions on Medical Imaging* 2016;35(5):1207–1216.

44. Setio AAA, Ciompi F, Litjens G, Gerke P, Jacobs C, Van Riel SJ, Wille MMW et al., Pulmonary nodule detection in ct images: False positive reduction using multi-view convolutional networks. *IEEE Transactions on Medical Imaging* 2016;35(5):1160–1169.

45. Hosseini-Asl E, Ghazal M, Mahmoud A, Aslantas A, Shalaby AM, Casanova MF, Barnes GN et al., Alzheimer's disease diagnostics by a 3D deeply supervised adaptable convolutional network. *Front Biosci (Landmark Ed)* 2018;23:584–596.

46. Shin H-C, Roth HR, Gao M, Lu L, Xu Z, Nogues I, Yao J et al., Deep convolutional neural networks for computer-aided detection: CNN architectures, dataset characteristics and transfer learning. *IEEE Transactions on Medical Imaging* 2016;35(5):1285–1298.

47. Tajbakhsh N, Shin JY, Gurudu SR, Hurst RT, Kendall CB, Gotway MB, Liang J, Convolutional neural networks for medical image analysis: Full training or fine tuning? *IEEE Transactions on Medical Imaging* 2016;35(5):1299–1312.

Chapter 11

Big Data in Prostate Cancer

Islam Reda, Ashraf Khalil, Mohammed Ghazal,
Ahmed Shalaby, Mohammed Elmogy, Ahmed Aboelfetouh,
Ali Mahmoud, Mohamed Abou El-Ghar, and Ayman El-Baz

11.1   Introduction

Prostate cancer is the second most common cancer and is the fifth most common cause of cancer-related deaths in men worldwide [1]. Prostate cancer is also the second most common cancer in American men after skin cancer and is the second cause of cancer-related deaths in American men after lung cancer. More than 174,000 new cases will be diagnosed with prostate cancer, and more than 31,000 deaths because of prostate cancer among Americans are expected in 2019 [2]. By 2030, it is expected that the number of cases that will be diagnosed with prostate cancer will increase globally to 1.7 million, and prostate cancer will cause up to 0.5 million male deaths per year [3]. Fortunately, the introduction and spread of different screening tests and improvements in treatment procedures have resulted in decreasing the mortality rates, especially when prostate cancer is detected in its early stages.

Although many researchers have investigated the different causes of prostate cancer, only an indistinct list of risk factors has been recognized. Those risk factors include, for example, the family history, genetic factors, race, and body mass index (BMI) [4]. The likelihood to develop prostate cancer for a man with a first degree relative who suffers from prostate cancer was found to be twice as high as the likelihood to develop prostate cancer for a man with no relatives affected [5, 6]. According to the research conducted by Agalliu et al. [7] on 979 cases of prostate cancer, men who suffer from protein-truncating mutations BRCA2 genes have been associated with high Gleason score prostate cancer. It was shown that the race of a person has an effect on the probability of developing prostate cancer. African Americans have a 1.6 times higher chance of developing prostate cancer than European Americans [8]. Rodriguez et al. [9] investigated the association

239

between BMI and weight change and the incidence of prostate cancer. They found that there was a positive correlation between BMI and the risk of aggressive prostate cancer.

11.1.1 Current Diagnostic Techniques

Currently, the well-established techniques used for diagnosing prostate cancer are digital rectal exam (DRE) [10], prostate specific antigen (PSA) [11], and transrectal ultrasound (TRUS)-guided needle biopsy [12]. In the DRE screening, a physician manually examines the prostate through the rectum to find any anomalies in its size or hardness. Through this screening, some peripheral zone tumors can be detected. However, tumors that are not large enough to be palpated, in addition to most of the central zone and the transitional zone tumors, cannot be detected through the DRE. Therefore, the experience and the skills of the physician are of significant effect on the accuracy of the DRE.

PSA screening is a blood-based screening that measures the PSA level in the blood. An increased value of PSA indicates a higher probability for prostate cancer. However, elevated levels of PSA may also signify other conditions, such as prostatitis or benign prostatic hyperplasia. If the blood PSA levels exceed four nanograms per millimeter (4 ng/mL), patients undergo further screening, such as biopsy, to confirm the presence or absence of the prostate cancer. Generally, the sensitivity and specificity of PSA screening are higher than the DRE screening [10].

In a TRUS-guided biopsy, small tissue specimens are acquired from the prostate gland to be examined by a pathologist. To analyze the sample, a number of scoring systems have been developed, including the Gleason [13], the modified Gleason and Mellinger [14], and the International Society of Urological Pathology (ISUP) modified Gleason system [15, 16]. The ISUP modified Gleason system involves the pathologist assigning a score between 1 and 5 based on the degree of each of the two tumor patterns, i.e., architectural and neoplasm patterns. A score of 5 for the architectural pattern indicates that the tissue is the least differentiated typical of cancerous tissue, while a neoplasm score of 5 signifies that the tumor resembles the most prevalent neoplasm pattern. Summation of these two scores indicates the severity of the neoplasm where a score between 6–10 means the tumor is cancerous [17]. TRUS-guided biopsy is an accurate technique for detecting cancer and determining its aggressiveness. However, it is an expensive, highly invasive, and painful procedure. Moreover, as a result of the random strategy for acquiring the tissue samples, there is a possibility of missing some aggressive tumors. Thus, it is imperative that an accurate non-invasive method with high selectivity and specificity be developed.

Some of the major drawbacks with current prostate cancer screening are over-diagnosis and over-treatment [18]. For instance, in the investigation conducted by Schröder et al. [19], which covered 11 years of follow-up to determine the role of PSA and TRUS-guided biopsy in decreasing the mortality rate caused by prostate cancer, they found that to obviate one death from prostate cancer, more than 1000 men would need to be examined and 37 would need to be treated. These problems of over-diagnosis and over-treatment motivated research towards developing techniques to visualize such structures and their associated abnormalities.

11.1.2 Different Modalities for Imaging Prostate

The field of medical imaging technology has provided excellent tools for visualizing different body structures and their correlated anomalies. Particularly, ultrasound and magnetic resonance imaging (MRI) have been used extensively to visualize the prostate to determine the severity of the disorders. As a consequence to the drawbacks associated with the current screening techniques previously described, in-vivo image-based computer aided diagnosis (CAD) systems have been utilized to identify and localize the size and extent of prostate cancer. In the following paragraphs, the uses and the advantages of both ultrasound and MRI are briefly described.

TRUS is the most commonly used prostate imaging technique since it is used primarily in guiding needle biopsies and identifying the prostate volume [20, 21]. The fundamental advantages of TRUS are: its portability, low cost compared to other imaging modalities, its ability to generate real-time imaging data, and absence of any type of radiation. On the other hand, it has some drawbacks: it produces low contrast images that contain speckles, has a low signal-to-noise (SNR) ratio, and generates shadow artifacts [22]. Consequently, it is hard to detect tumors with a high level of accuracy and/or identify the stage of cancer using TRUS imaging techniques.

MRI offers the best soft tissue contrast compared to other image modalities, such as CT and TRUS. MRI does not involve radiation and is useful for determining the stage of cancer. However, MRI is not portable, is sensitive to noise and image artifacts, has difficulties implementing real-time imaging due to its relatively long and complex acquisition, and has a relatively high cost [20, 23].

Several different MRI techniques have been extensively used in the prostate cancer CAD systems, such as T2-MRI, dynamic contrast-enhanced (DCE)-MRI, and diffusion-weighted imaging (DWI). Even though T2-MRI provides good contrast between soft tissues, it lacks functional information. DCE-MRI is a technique that provides detailed information on the anatomy and function of different tissues. It has gained wide attention due to the increased spatial resolution, the ability to yield information about the hemodynamics (i.e., perfusion), micro-vascular permeability, and extracellular leakage space [24]. DCE-MRI has been extensively used in many clinical applications [25] in addition to the detection of prostate cancer [26–29]. In DCE-MRI, a series of MR images is taken prior to and after administering a contrast agent into the bloodstream. Contrast agents significantly enhance the contrast between the different tissue types and ease visualization of the anatomical structures which have alternating magnetic properties in their vicinity. The acquired signal intensity is proportional to the concentration of contrast agent in each voxel. Several types of MRI contrast agents can be used depending on the application, such as paramagnetic agents, superparamagnetic agents, extracellular fluid space agents, and tissue (organ)-specific agents. DWI is another MRI modality that can be acquired in a short time, and unlike DCE-MRI, it does not involve the use of contrast agents and has been attracting researchers recently. DWI is based on the measurement of micro movements of water molecules inside the body [30].

11.1.3 Related Work

Developing CAD systems for detecting prostate cancer is an ongoing research area [31, 32]. Those CAD systems vary in their accuracy, speed, and level of automation. The first multiparametric CAD system was proposed by Chan et al. [33] using T2-MRI, T2-mapping, and line scan diffusion imaging (LSDI). Intensity and textural features were extracted from manually localized prostate region and fed into a support vector machine (SVM) classifier or Fisher linear discriminant (FLD) classifier to detect prostate cancer in the peripheral zone (PZ) of the prostate. The area under the curve (AUC) was 0.76 ± 0.04 for the SVM and 0.84 ± 0.06 for the FLD. Another multiparametric CAD system that employed T2-MRI, DCE-MRI, and DW-MRI was proposed by Litjens et al. [34]. In this system, an SVM classifier used apparent diffusion coefficients (ADCs) and pharmacokinetic features extracted from the segmented prostate to determine malignant and benign regions. Vos et al. [35] proposed another multiparametric CAD system that utilized the same MRI modalities used in [34]. In their system, a linear discriminant analysis (LDA) classifier employed a set of features (e.g., texture-based, ADC maps) to differentiate between malignant and benign prostates. Firjani et al. [36] developed a DWI-based CAD system in which the prostate segmentation was based on a maximum a posteriori estimation that utilizes shape, spatial, and appearance information. A k-nearest-neighbor (KNN) classifier used three intensity features to classify the prostate into benign or malignant. Madabhushi et al. [37] proposed a fully automated CAD system for prostate cancer detection. MR images were first preprocessed to correct background inhomogeneity and nonstandardness. Then 3D texture features were extracted and used as input to a Bayesian classifier to assign a likelihood of malignancy for each image voxel of each feature individually. Those generated likelihood images were then integrated by a combination technique of weighted feature. The assessment of their CAD system was done using 33 2D MR slices from five different 3D MR prostate studies. Artan et al. [38] developed an automated technique for localizing prostate cancer from multi-spectrum MRI data using cost-sensitive SVM and conditional-random-fields (CRF). The evaluation of their technique using multi-spectrum MRI datasets acquired from 21 subjects showed that cost-sensitive SVM and cost-sensitive CRF resulted in better cancer localization accuracy than conventional SVM.

The main focus of this chapter is to build an accurate CAD system for diagnosing prostate cancer using a deep learning technique that fuses the blood test results with DWI-based results. The chapter is organized as follows: Section 11.2 describes the proposed framework focusing on the stage of extracting discriminatory features (Subsection 11.2.1) and the classification stage (Subsection 11.2.2). Then, Section 11.3 reports the experimental results. Finally, Section 11.4 provides conclusions and future trends.

11.2 Methods

The proposed CAD system summarized in Figure 11.1 performs sequentially three steps. First, the prostate is segmented using our previously developed geometric deformable model (level-set) as described in [39]. This model is guided by a stochastic speed function that is derived using non-negative matrix factorization (NMF). The NMF attributes are calculated using information from the MRI intensity, a probabilistic shape model, and the spatial interactions between prostate voxels. The proposed approach reaches 86.89% overall Dice similarity coefficient and an average Hausdorff distance of 5.72 mm, indicating high segmentation accuracy. Details of this approach and comparisons with other segmentation approaches can be found in [39]. Afterwards, global features describing the water diffusion inside the prostate tissue are extracted based on the cumulative distribution functions of the ADC maps. Finally, a two-stage structure of stacked non-negativity constraint auto-encoder (SNCAE) is trained to classify the prostate tumor as benign or malignant based on the CDFs constructed in the previous step and the blood test-based probabilities. The latter two steps of the proposed CAD system are discussed in the following sections.

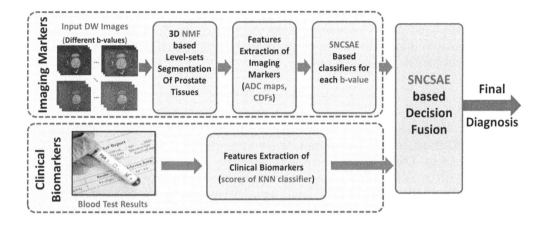

FIGURE 11.1: Steps of the proposed CAD system.

11.2.1 Feature Extraction

The most recent MRI modality used for the detection and the diagnosis of prostate cancer is DWI. DWI is a functional MRI that measures the freedom with which water molecules spread within tissues. This supplies information about the cellular nature of that tissue in addition to the features of the space within and between tissue cells. This information is beneficial in distinguishing cancerous from healthy tissues and recognizing the cancer degree. DWI is different from DCE-MRI as it can be collected rapidly and it does not encompass the use of any contrast materials. Contrast materials have harmful effects on patients with kidney problems. Most of DWI concentrates on fluid between cells to test whether or not a certain tissue region has abnormally limited spread. In other words, water molecules have a bigger impediment to spread for long distances in comparison to what that tissue should exhibit in case of normality. This limited spread of fluid is caused by the reduction in the size of space between cells. DWI relies on the variations in the motion of water molecules within tissues to produce diffusion images. This motion is random, and the amount of randomness is positively correlated with the signal loss of DWI. The signal loss is defined by [40]:

$$S_d \sim e^{-b \times ADC} \tag{11.1}$$

where ADC is the apparent diffusion coefficient and b is a variable depending on the magnitude and timing of gradient pulses. The use of gradient pulses resulted in improved sensitivity to spread in comparison to the steady state gradients [41]. The b-value is given by:

$$b = \gamma^2 G^2 \delta^2 \left(\Delta - \frac{\delta}{3} \right)$$

where γ is the gyromagnetic ratio which is a constant value related to the magnet strength. G^2 and δ^2 are the magnitude and width of the two gradient pulses, which are part of the sequence coefficients, respectively. Δ is the interval between the two gradient pulses which is the time between the beginning of the acquisition and the time of registering the echo.

The voxels of a DW image collected at a given b-value (S_b) have signal intensities with values that equal those intensities of the conforming voxels of the image collected without diffusion weighting ($b=0$ s/mm^2) decreased by the signal loss of Equation 11.1). Those signal intensities are defined by:

$$S_b = S_0 \times e^{-b \times ADC} \tag{11.2}$$

Because of the low specificity and the low quality of DWI [42], many researchers decide to employ ADC maps, which are quantitative maps calculated from DWI, for diagnosing prostate cancer. The basic idea behind the discriminative abilities of ADC maps is that tissues with prostate cancer have lower ADC values than noncancerous/healthy tissues. In order to produce the ADC map, two DW images are needed: the first one is acquired at the baseline, b_0 ($b=0$ s/mm^2), whereas the other is acquired at a higher b-value ($b_1 > 0$ s/mm^2). The ADC maps are computed using the following equation:

$$ADC = -\frac{\ln S_{b_1} - \ln S_0}{b_1} \tag{11.3}$$

In Equation 11.3), the signal of DW image $\left(S_{b_1}\right)$ is divided by the signal of b_0 image (S_0) pixelwise, and then the natural logarithm is taken to remove the T2-effect in order to produce an unadulterated map.

Then, all ADC maps at a certain b-value for all subjects are normalized with respect to the maximum value of all of these maps to make all calculated ADC maps in the same range (between 0 and 1) in order to use a unique color coding for all of them. The calculated ADC values are refined using a generalized Gauss–Markov random field (GGMRF) image model with a 26-voxel neighborhood to remove any data inconsistency and preserve continuity. Continuity of the constructed 3D volume is amplified by using their maximum a posteriori (MAP) estimates. The CDFs of the normalized ADCs of each subject are constructed. These CDFs are considered as global features distinguishing between benign and malignant cases. Instead of using the whole ADC volume, the resultant CDFs are used to train an SNCAE classifier using the deep learning approach. A KNN model is used to convert the results of the blood test into probabilities which are fed to the second stage of SNCAE. Figure 11.2 summarizes the feature extraction from DWI and the conversion of the blood test results into probabilities.

It is worth noting that conventional classification methods, employing directly the voxel-wise ADCs of the entire prostate volume as discriminative features, encounter at least two serious difficulties. Various input data sizes require unification by either data truncation for large prostate volumes or zero padding for small ones. Both ways may decrease the accuracy of the classification. In addition, large ADC data volumes lead to considerable time expenditures for training and classification. Contrastingly, our SNCAE classifier exploits only the 100-component CDFs to describe the entire 3D ADC maps estimated at each b-value. This fixed data size helps overcome the above challenges and notably expedites the classification.

11.2.2 A Two-Stage Classification

To classify the prostate tumor, our CAD system employs a deep neural network with two-stage structure of SNCAE. In the first stage, seven SNCAE classifiers, one

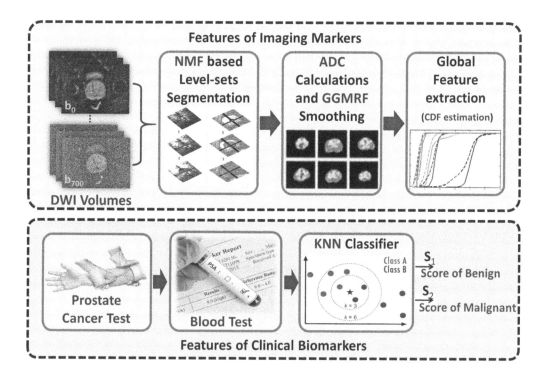

FIGURE 11.2: Feature extraction from DWI and conversion of blood test results into probabilities.

classifier for each of seven distinct-values (100 to 700/mm²), are utilized to estimate initial probabilities that are concatenated in addition to the blood test-based probabilities and fed in the second stage into another SNCAE to estimate the final decision.

Each auto-encoder (AE) compresses its input data (100-component CDFs at some b-value) to capture the most prominent variations and is built separately by greedy unsupervised pre-training [43]. A softmax output layer, stacked after AE layers, facilitates the subsequent supervised back-propagation-based fine tuning of the entire classifier by minimizing the total loss (negative log-likelihood) for given training labeled data. Using the AEs with a non-negativity constraint (NCAE) [44] yields both more reasonable data codes (features) during its unsupervised pre-training and better classification performance after the supervised refinement.

For each SNCAE, let $\mathbf{W} = \left\{ \mathbf{W}_j^{e}, \mathbf{W}_i^{d} : j = 1, \ldots, s; i = 1, \ldots, n \right\}$ denote a set of column vectors of weights for encoding (e) and decoding (d) layers of a single AE. Let $\top$ denote vector transposition. The AE converts an n-dimensional column vector $\mathbf{u} = [u_1, \ldots, u_n]^{\top}$ of input signals into an s-dimensional column vector $\mathbf{h} = [h_1, \ldots, h_s]^{\top}$ of hidden codes (features, or activations), such that $s \ll n$, by uniform nonlinear transformation of s weighted linear combinations of signals:

$$h_j = \sigma\left(\left(\mathbf{W}_j^{e}\right)^{\top} \mathbf{u} \right) \equiv \sigma\left(\sum_{i=1}^{n} w_{j:i}^{e} u_i \right) \tag{11.4}$$

where $\sigma(\ldots)$ is a certain sigmoid, i.e., a differentiable monotone scalar function with values in the range [0,1].

The classifier is built by stacking the NCAE layers with an output softmax layer, that computes a softmax regression, generalizing the common logistic regression to more than two classes as shown in Figure 11.3(a). Each NCAE is pre-trained separately in the unsupervised mode, by using the activation vector of a lower layer as the input to the upper layer. In our case, the initial input data $\mathbf{u}_f$; $f = 1, \ldots, 7$ consisted of the 100-component CDFs, each of size 100. The bottom NCAE compresses the input vector to $s_1 = 50$ first-level activators, compressed by the next NCAE to $s_2 = 5$ second-level activators, which are reduced in turn by the output softmax layer to $s^o = 2$ values.

The activations of the second NCAE layer, $\mathbf{h}^{[2]} = \sigma\left(\mathbf{W}_{[2]}^{e}{}^{\top}\mathbf{h}^{[1]}\right)$, are inputs of the softmax classification layer, as sketched in Figure 11.3(a), to compute the plausibility of a decision in favor of each particular output class, $c = 1, 2$:

$$p(c; \mathbf{W}_{o:c}) = \frac{\exp\left(\mathbf{W}_{o:c}^{\top}\mathbf{h}^{[2]}\right)}{\exp\left(\mathbf{W}_{o:1}^{\top}\mathbf{h}^{[2]}\right) + \exp\left(\mathbf{W}_{o:2}^{\top}\mathbf{h}^{[2]}\right)}; c = 1, 2;$$

(11.5)

$$\sum_{c=1}^{2} p\left(c; \mathbf{W}_{o:c}; \mathbf{h}^{[2]}\right) = 1.$$

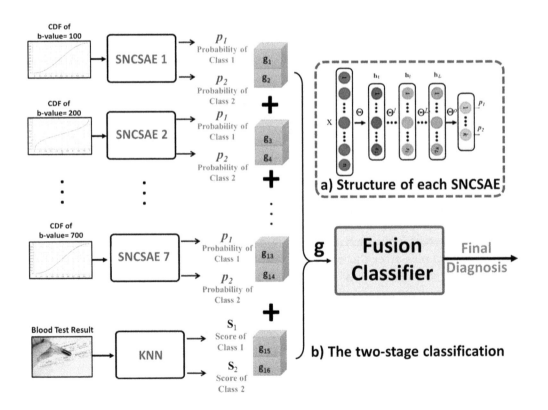

FIGURE 11.3: Structure of (a) SNCAE (b) two stages of classification.

Finally, the entire stacked NCAE classifier (SNCAE) is fine-tuned on the labeled training data by the conventional error back-propagation through the network and penalizing only the negative weights of the softmax layer. The parameters that specify relative contributions of the non-negativity and sparsity constraints to the overall loss were chosen based on comparative experiments.

In the second stage, each SNCAE's output probabilities in addition to the blood test-based probabilities are fused by concatenation, resulting in a vector of fused probabilities $\mathbf{u}_t = [\mathbf{g}_1, ..., \mathbf{g}_{14}]$ as shown in Figure 11.3(b). To enhance the classification accuracy, this vector ($\mathbf{u}_t$) is fed into a new SNCAE to estimate the final classification as a class probability using the following equation,

$$p_t\left(c; \mathbf{W}_{o:c}^t\right) = \frac{\exp\left(\mathbf{W}_{o:c}^{t\ \top} \mathbf{g}_t\right)}{\displaystyle\sum_{c=1}^{C} \exp\left(\mathbf{W}_{o:c}^{t\ \top} \mathbf{g}_t\right)}; c = 1, 2 \tag{11.6}$$

11.3 Experimental Results

Experiments were conducted on 18 DWI datasets obtained using a body coil Signa Horizon GE scanner in axial plane with the following parameters: magnetic field strength: 1.5 T; TE: 84.6 ms; TR: 8000 ms; bandwidth: 142.86 kHz; FOV: 34 cm; slice thickness: 3 mm; inter-slice gap: 0 mm; acquisition sequence: conventional EPI; diffusion weighting directions: mono direction; the used range of b-values is from 0 to 700 s/mm². On average, 26 slices were obtained in 120 s to cover the prostate in each patient with voxel size of $1.25 \times 1.25 \times 3.00$ mm³. Half of the subjects are benign, and the other half are malignant. All the subjects were diagnosed using a biopsy. The cases were evaluated as a whole and not per tumor. Samples of the acquired DWI from two patients and the calculated ADC values for these DWI volumes are shown in Figures 11.4 and 11.5, respectively.

To learn the statistical characteristics of both benign and malignant subjects, we trained in the first stage seven different SNCAEs, one for each b-value, using the DWI datasets. All training was done inside a leave-one-subject-out cross-validation framework. The features involved for classification are the CDFs of the normalized ADC maps at the seven distinct-values of the segmented prostates. The overall diagnostic accuracy at the distinct-values are summarized in Table 11.1.

In a similar experiment, the blood test results are converted to probabilities by a KNN model using a leave-one-subject-out scenario. This experiment resulted in accuracy, sensitivity, and specificity of 78%, 56%, and 100%, respectively.

An SNCAE is used in the second stage to fuse the DWI probabilities together with the blood test-based probabilities to calculate the final classification. The accuracy, sensitivity, and specificity after this fusion are shown in Table 11.2.

To highlight the merit of using SNCAE, a comparison between SNCAE and two classifiers (Random Tree and Random Forest) [45] is summarized in Table 11.3. As demonstrated in Table 11.3, SNCAE outperforms the other alternatives. The corresponding AUC of the receiver operating characteristics of those classifiers are shown in Figure 11.6. The AUC of the proposed classifier is 0.98.

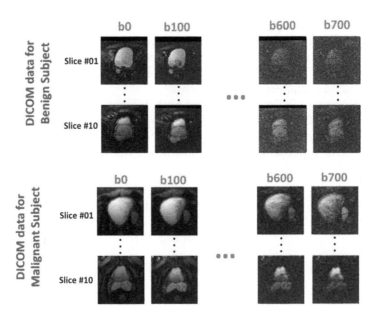

FIGURE 11.4: Samples of the acquired DWI from two patients.

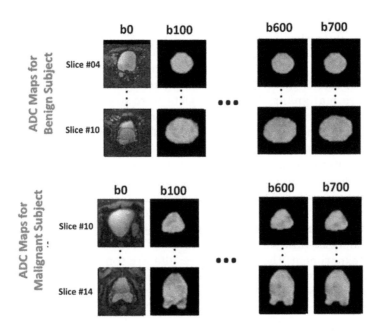

FIGURE 11.5: Samples of the ADC maps from two patients.

TABLE 11.1: Classification Accuracy, Sensitivity, and Specificity of Seven SNCAE Classifiers at Distinct-Values

b-value	Accuracy	Sensitivity	Specificity
100 smm^{-2}	78%	78%	78%
200 smm^{-2}	67%	78%	56%
300 smm^{-2}	72%	78%	67%
400 smm^{-2}	72%	78%	67%
500 smm^{-2}	72%	78%	67%
600 smm^{-2}	83%	89%	78%
700 smm^{-2}	83%	89%	78%

TABLE 11.2: Classification Accuracy, Sensitivity, and Specificity of the Second-Stage SNCAE

Model	Accuracy	Sensitivity	Specificity
SNCAE	94%	89%	100%

TABLE 11.3: Classification Accuracy of SNCAE, Random Tree, and Random Forest

Model	Accuracy	Sensitivity	Specificity
SNCAE	94%	89%	100%
Random Tree	89%	100%	78%
Random Forest	89%	89%	89%

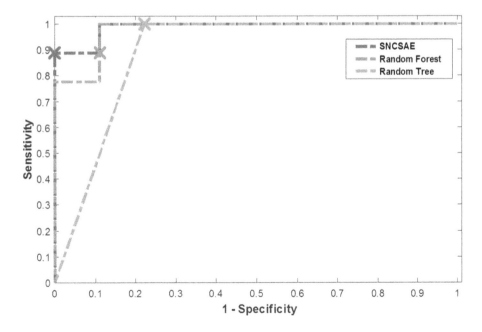

FIGURE 11.6: ROC curve for the proposed CNN-based classifier, SVM, and RT at b-value = 400 smm^{-2}.

11.4   Conclusion

In this chapter, a CAD system for diagnosing prostate cancer from DWI is proposed. The system starts with segmenting the prostate using a level set model that utilizes NMF to integrate three different types of information for accurate segmentation. Then, the ADC maps are calculated for the segmented prostate at seven b-values. These ADC values are normalized and refined using GGMRF. The CDFs of these ADC values are constructed as global features that can distinguish between benign and malignant tumors. These CDFs at the distinct b-values are fed to SNCAE to get preliminary probabilities. A KNN is used to produce blood test-based probabilities that are fused with the DWI preliminary probabilities and fed to the prediction stage SNCAE to calculate the final classification. The experiments on 18 DWI datasets show that the system accuracy improved by fusing the blood test results with DWI results.

In addition to the prostate [46–51, 26–30], this work could also be applied to various other applications in medical imaging, such as the kidney, the heart, the lung, and the retina, as well as several non-medical applications [52–55].

One application is renal transplant functional assessment, especially with developing noninvasive CAD systems for renal transplant function assessment, utilizing different image modalities (e.g., ultrasound, computed tomography (CT), MRI, etc.). Accurate assessment of renal transplant function is critically important for graft survival. Although transplantation can improve a patient's wellbeing, there is a potential post-transplantation risk of kidney dysfunction that, if not treated in a timely manner, can lead to the loss of the entire graft and even patient death. In particular, dynamic and diffusion MRI-based systems have been clinically used to assess transplanted kidneys with the advantage of providing information on each kidney separately. For more details about renal transplant functional assessment, please read [56–84].

The heart is also an important application to this work. The clinical assessment of myocardial perfusion plays a major role in the diagnosis, management, and prognosis of ischemic heart disease patients. Thus, there have been ongoing efforts to develop automated systems for accurate analysis of myocardial perfusion using first-pass images [85–102].

Another application for this work could be the detection of retinal abnormalities. A majority of ophthalmologists depend on visual interpretation for the identification of disease types. However, inaccurate diagnosis will affect the treatment procedure which may lead to fatal results. Hence, there is a crucial need for computer automated diagnosis systems that yield highly accurate results. Optical coherence tomography (OCT) has become a powerful modality for the non-invasive diagnosis of various retinal abnormalities such as glaucoma, diabetic macular edema, and macular degeneration. The problem with diabetic retinopathy (DR) is that the patient is not aware of the disease until the changes in the retina have progressed to a level at which treatment tends to be less effective. Therefore, automated early detection could limit the severity of the disease and assist ophthalmologists in investigating and treating it more efficiently [103–105].

Abnormalities of the lung could also be another promising area of research and a related application of this work. Radiation-induced lung injury is the main side effect of radiation therapy for lung cancer patients. Although higher radiation doses increase the radiation therapy effectiveness for tumor control, this can lead to lung injury as a greater quantity of normal lung tissues is included in the treated area. Almost one-third of patients who undergo radiation therapy develop lung injury following radiation treatment. The severity of radiation-induced lung injury ranges from ground-glass opacities and consolidation at the early phase to fibrosis and traction bronchiectasis in the late phase. Early detection of lung injury will thus help to improve management of the treatment [105–150].

This work can also be applied to other brain abnormalities, such as dyslexia, in addition to autism. Dyslexia is one of the most complicated developmental brain disorders that affect children's learning abilities. Dyslexia leads to the failure to develop age-appropriate reading skills in spite of normal intelligence level and adequate reading instructions. Neuropathological studies have revealed an abnormal anatomy of some structures, such as the corpus callosum, in dyslexic brains. There has been a lot of work in the literature that aims at developing CAD systems for diagnosing this disorder, along with other brain disorders [151–173].

For the vascular system [174], this work could also be applied for the extraction of blood vessels, e.g., from phase contrast (PC) magnetic resonance angiography (MRA). Accurate cerebrovascular segmentation using non-invasive MRA is crucial for the early diagnosis and timely treatment of intracranial vascular diseases [156, 157, 175–180].

References

1. Global Cancer Observatory: Cancer fact sheets by cancer site. https://gco.iarc.fr/today/data/factsheets/cancers/27-Prostate-fact-sheet.pdf (2018) [Online; accessed 04-April-2019].
2. American Cancer Society: Key statistics for prostate cancer. https://www.cancer.org/cancer/prostate-cancer/about/key-statistics.html (2019) [Online; accessed 04-April-2019].
3. Du, W., Liu, Y.P., Wang, S., Peng, Y., Oto, A.: Features extraction of prostate with graph spectral method for prostate cancer detection. In: *17th IEEE/ACIS International Conference on Software Engineering, Artificial Intelligence, Networking and Parallel/Distributed Computing (SNPD)*, IEEE (2016) 663–668.
4. Cokkinides, V., Albano, J., Samuels, A., Ward, M., Thum, J.: *American Cancer Society: Cancer Facts and Figures 2016*. Atlanta: American Cancer Society (2016).
5. Kalish, L.A., McDougal, W.S., McKinlay, J.B.: Family history and the risk of prostate cancer. *Urology* **56**(5) (2000) 803–806.
6. Steinberg, G.D., Carter, B.S., Beaty, T.H., Childs, B., Walsh, P.C.: Family history and the risk of prostate cancer. *The Prostate* **17**(4) (1990) 337–347.
7. Agalliu, I., Gern, R., Leanza, S., Burk, R.D.: Associations of high-grade prostate cancer with BRCA1 and BRCA2 founder mutations. *Clinical Cancer Research* **15**(3) (2009) 1112–1120.
8. Hoffman, R.M., Gilliland, F.D., Eley, J.W., Harlan, L.C., Stephenson, R.A., Stanford, J.L., Albertson, P.C., Hamilton, A.S., Hunt, W.C., Potosky, A.L.: Racial and ethnic differences in advanced-stage prostate cancer: The prostate cancer outcomes study. *Journal of the National Cancer Institute* **93**(5) (2001) 388–395.
9. Rodriguez, C., Freedland, S.J., Deka, A., Jacobs, E.J., McCullough, M.L., Patel, A.V., Thun, M.J., Calle, E.E.: Body mass index, weight change, and risk of prostate cancer in the cancer prevention study ii nutrition cohort. *Cancer Epidemiology and Prevention Biomarkers* **16**(1) (2007) 63–69.
10. Mistry, K., Cable, G.: Meta-analysis of prostate-specific antigen and digital rectal examination as screening tests for prostate carcinoma. *The Journal of the American Board of Family Practice* **16**(2) (2003) 95–101.
11. Dijkstra, S., Mulders, P., Schalken, J.: Clinical use of novel urine and blood based prostate cancer biomarkers: A review. *Clinical Biochemistry* **47**(10) (2014) 889–896.
12. Davis, M., Sofer, M., Kim, S.S., Soloway, M.S.: The procedure of transrectal ultrasound guided biopsy of the prostate: A survey of patient preparation and biopsy technique. *The Journal of Urology* **167**(2) (2002) 566–570.

13. Gleason, D.F.: Classification of prostatic carcinomas. *Cancer Chemotherapy Reports*. Part 1 **50**(3) (1966) 125–128.
14. Gleason, D.F., Mellinger, G.T.: Prediction of prognosis for prostatic adenocarcinoma by combined histological grading and clinical staging. *The Journal of Urology* **111**(1) (1974) 58–64.
15. Montironi, R., Cheng, L., Lopez-Beltran, A., Scarpelli, M., Mazzucchelli, R., Mikuz, G., Kirkali, Z., Montorsi, F.: Original Gleason system versus 2005 ISUP modified Gleason system: The importance of indicating which system is used in the patient's pathology and clinical reports. *European Urology* **58**(3) (2010) 369–373.
16. Epstein, J.I., Allsbrook, W.C., Jr, Amin, M.B., Egevad, L.L., ISUP Grading Committee: The 2005 International Society of Urological Pathology (ISUP) consensus conference on Gleason grading of prostatic carcinoma. *The American Journal of Surgical Pathology* **29**(9) (2005) 1228–1242.
17. DeMarzo, A.M., Nelson, W.G., Isaacs, W.B., Epstein, J.I.: Pathological and molecular aspects of prostate cancer. *The Lancet* **361**(9361) (2003) 955–964.
18. Schroder, F.H., Hugosson, J., Roobol, M.J., Tammela, T.L., Ciatto, S., Nelen, V., Kwiatkowski, M., Lujan, M., Lilja, H., Zappa, M., et al.: Screening and prostate-cancer mortality in a randomized European study. *New England Journal of Medicine* **360**(13) (2009) 1320–1328.
19. Schrooder, F.H., Hugosson, J., Roobol, M.J., Tammela, T.L., Ciatto, S., Nelen, V., Kwiatkowski, M., Lujan, M., Lilja, H., Zappa, M., et al.: Prostate-cancer mortality at 11 years of follow-up. *New England Journal of Medicine* **366**(11) (2012) 981–990.
20. Hricak, H., Choyke, P.L., Eberhardt, S.C., Leibel, S.A., Scardino, P.T.: Imaging prostate cancer: A multidisciplinary perspective 1. *Radiology* **243**(1) (2007) 28–53.
21. Fichtinger, G., Krieger, A., Susil, R.C., Tanacs, A., Whitcomb, L.L., Atalar, E.: Transrectal prostate biopsy inside closed MRI scanner with remote actuation, under real-time image guidance. In: *Medical Image Computing and Computer-Assisted Intervention MICCAI 2002*, Springer (2002) 91–98.
22. Applewhite, J.C., Matlaga, B., McCullough, D., Hall, M.: Transrectal ultrasound and biopsy in the early diagnosis of prostate cancer. *Cancer Control: Journal of the Moffitt Cancer Center* **8**(2) (2000) 141–150.
23. Fuchsjager, M., Shukla-Dave, A., Akin, O., Barentsz, J., Hricak, H.: Prostate cancer imaging. *Acta Radiologica* **49**(1) (2008) 107–120.
24. Collins, D.J., Padhani, A.R.: Dynamic magnetic resonance imaging of tumor perfusion. *IEEE Engineering in Medicine and Biology Magazine* **23**(5) (2004) 65–83.
25. Khalifa, F., Beache, G.M., Gimel'farb, G., El-Baz, A.: A novel CAD system for analyzing cardiac first-pass MR images. In: *21st International Conference on Pattern Recognition (ICPR)*, IEEE (2012) 77–80.
26. Reda, I., Shalaby, A., Elmogy, M., Elfotouh, A.A., Khalifa, F., El-Ghar, M.A., Hosseini-Asl, E., Gimel'farb, G., Werghi, N., El-Baz, A.: A comprehensive non-invasive framework for diagnosing prostate cancer. *Computers in Biology and Medicine* **81** (2017) 148–158.
27. Reda, I., Shalaby, A., Elmogy, M., Aboulfotouh, A., Khalifa, F., El-Ghar, M.A., Gimelfarb, G., El-Baz, A.: Image-based computer-aided diagnostic system for early diagnosis of prostate cancer. In: *International Conference on Medical Image Computing and Computer-Assisted Intervention*, Springer (2016) 610–618.
28. Reda, I., Shalaby, A., El-Ghar, M.A., Khalifa, F., Elmogy, M., Aboulfotouh, A., Hosseini-Asl, E., El-Baz, A., Keynton, R.: A new NMF-autoencoder based CAD system for early diagnosis of prostate cancer. In: *2016 IEEE 13th International Symposium on Biomedical Imaging (ISBI)*, IEEE (2016) 1237–1240.
29. Reda, I., Shalaby, A., Khalifa, F., Elmogy, M., Aboulfotouh, A., El-Ghar, M.A., Hosseini-Asl, E., Werghi, N., Keynton, R., El-Baz, A.: Computer-aided diagnostic tool for early detection of prostate cancer. In: *2016 IEEE International Conference on Image Processing (ICIP)*, IEEE (2016) 2668–2672.
30. Reda, I., Khalil, A., Elmogy, M., Abou El-Fetouh, A., Shalaby, A., Abou El-Ghar, M., Elmaghraby, A., Ghazal, M., El-Baz, A.: Deep learning role in early diagnosis of prostate cancer. *Technology in Cancer Research and Treatment* **17** (2018) 1533034618775530.

31. Liu, L., Tian, Z., Zhang, Z., Fei, B.: Computer-aided detection of prostate cancer with MRI: Technology and applications. *Academic Radiology* **23**(8) (2016) 1024–1046.

32. McClure, P., Elnakib, A., El-Ghar, M.A., Khalifa, F., Soliman, A., El-Diasty, T., Suri, J.S., Elmaghraby, A., El-Baz, A.: In-vitro and in-vivo diagnostic techniques for prostate cancer: A review. *Journal of Biomedical Nanotechnology* **10**(10) (2014) 2747–2777.

33. Chan, I., Wells, W., III, Mulkern, R.V., Haker, S., Zhang, J., Zou, K.H., Maier, S.E., Tempany, C.M.: Detection of prostate cancer by integration of line-scan diffusion, T2-mapping and T2-weighted magnetic resonance imaging; a multichannel statistical classifier. *Medical Physics* **30**(9) (2003) 2390–2398.

34. Litjens, G., Vos, P., Barentsz, J., Karssemeijer, N., Huisman, H.: Automatic computer aided detection of abnormalities in multi-parametric prostate MRI. In: *Proceedings of SPIE Medical Imaging 2011: Computer-Aided Diagnosis*. Volume **7963**, International Society for Optics and Photonics (2011) 79630T.

35. Vos, P., Barentsz, J., Karssemeijer, N., Huisman, H.: Automatic computer-aided detection of prostate cancer based on multiparametric magnetic resonance image analysis. *Physics in Medicine and Biology* **57**(6) (2012) 1527.

36. Firjani, A., Elnakib, A., Khalifa, F., Gimel'farb, G., El-Ghar, M.A., Elmaghraby, A., El-Baz, A.: A diffusion-weighted imaging based diagnostic system for early detection of prostate cancer. *Journal of Biomedical Science and Engineering* **6**(3) (2013) 346.

37. Madabhushi, A., Feldman, M.D., Metaxas, D.N., Tomaszeweski, J., Chute, D.: Automated detection of prostatic adenocarcinoma from high-resolution ex vivo MRI. *IEEE Transactions on Medical Imaging* **24**(12) (2005) 1611–1625.

38. Artan, Y., Haider, M.A., Langer, D.L., Van der Kwast, T.H., Evans, A.J., Yang, Y., Wernick, M.N., Trachtenberg, J., Yetik, I.S.: Prostate cancer localization with multispectral MRI using cost-sensitive support vector machines and conditional random fields. *IEEE Transactions on Image Processing* **19**(9) (2010) 2444–2455.

39. McClure, P., Khalifa, F., Soliman, A., El-Ghar, M.A., Gimelfarb, G., Elmagraby, A., El-Baz, A.: A novel NMF guided level-set for DWI prostate segmentation. *Journal of Computer Science and Systems Biology* **7**(6) (2014) 209–216.

40. Huisman, T.A.: Diffusion-weighted imaging: Basic concepts and application in cerebral stroke and head trauma. *European Radiology* **13**(10) (2003) 2283–2297.

41. Hrabe, J., Kaur, G., Guilfoyle, D.N.: Principles and limitations of NMR diffusion measurements. *Journal of Medical Physics/Association of Medical Physicists of India* **32**(1) (2007) 34.

42. Choi, Y.J., Kim, J.K., Kim, N., Kim, K.W., Choi, E.K., Cho, K.S.: Functional MR imaging of prostate cancer. *RadioGraphics* **27**(1) (2007) 63–75.

43. Bengio, Y., Lamblin, P., Popovici, D., Larochelle, H.: Greedy layer-wise training of deep networks. *Advances in Neural Information Processing Systems* **19** (2007) 153.

44. Hosseini-Asl, E., Zurada, J., Nasraoui, O.: Deep learning of part-based representation of data using sparse autoencoders with nonnegativity constraints. *IEEE Transactions on Neural Networks and Learning Systems* **99** (2015) 1–13.

45. Hall, M., Frank, E., Holmes, G., Pfahringer, B., Reutemann, P., Witten, I.H.: The WEKA data mining software: An update. *ACM SIGKDD Explorations* **11**(1) (2009) 10–18.

46. Reda, I., Ghazal, M., Shalaby, A., Elmogy, M., Aboulfotouh, A., El-Ghar, M.A., Elmaghraby, A., Keynton, R., El-Baz, A.: A computer-aided system for prostate cancer diagnosis. In: *2018 IEEE International Symposium on Signal Processing and Information Technology (ISSPIT)*, IEEE (2018) 465–469.

47. Reda, I., Shalaby, A., Elmogy, M., Ghazal, M., Aboulfotouh, A., El-Ghar, M.A., Elmaghraby, A., Keynton, R., El-Baz, A.: A new fast framework for early detection of prostate cancer without prostate segmentation. In: *2018 IEEE International Conference on Imaging Systems and Techniques (IST)*, IEEE (2018) 1–5.

48. Reda, I., Ghazal, M., Shalaby, A., Elmogy, M., AbouEl-Fetouh, A., Ayinde, B.O., AbouEl-Ghar, M., Elmaghraby, A., Keynton, R., El-Baz, A.: A novel adcs-based cnn classification system for precise diagnosis of prostate cancer. In: *2018 24th International Conference on Pattern Recognition (ICPR)*, IEEE (2018) 3923–3928.

49. Shalaby, A., Hajjdiab, H., Ghazal, M., Reda, I., Elmogy, M., Aboulfotouh, A., Mahmoud, A., El-giziri, A., Elmaghraby, A., El-Baz, A.: Computer-aided diagnosis of prostate cancer on diffusion weighted imaging: A technical review. In: *2018 IEEE International Conference on Imaging Systems and Techniques (IST)*, IEEE (2018) 1–6.

50. Reda, I., Ayinde, B.O., Elmogy, M., Shalaby, A., El-Melegy, M., El-Ghar, M.A., El-fetouh, A.A., Ghazal, M., El-Baz, A.: A new CNN-based system for early diagnosis of prostate cancer. In: *2018 IEEE 15th International Symposium on Biomedical Imaging (ISBI 2018)*, IEEE (2018) 207–210.

51. Reda, I., Elmogy, M., Aboulfotouh, A., Ismail, M., El-Baz, A., Keynton, R.: Prostate segmentation using deformable model-based methods. In: El-Baz, A., Jiang, X., Suri, J.S., eds.: *Biomedical Image Segmentation: Advances and Trends*, CRC Press (2016) 293–308.

52. Mahmoud, A.H. *Utilizing Radiation for Smart Robotic Applications Using Visible, Thermal, and Polarization Images*. PhD thesis, University of Louisville (2014).

53. Mahmoud, A., El-Barkouky, A., Graham, J., Farag, A.: Pedestrian detection using mixed partial derivative based his togram of oriented gradients. In: *2014 IEEE International Conference on Image Processing (ICIP)*, IEEE (2014) 2334–2337.

54. El-Barkouky, A., Mahmoud, A., Graham, J., Farag, A.: An interactive educational drawing system using a humanoid robot and light polarization. In: *2013 IEEE International Conference on Image Processing*, IEEE (2013) 3407–3411.

55. Mahmoud, A.H., El-Melegy, M.T., Farag, A.A.: Direct method for shape recovery from polarization and shading. In: *2012 19th IEEE International Conference on Image Processing*, IEEE (2012) 1769–1772.

56. Shehata, M., Ghazal, M., Khalifa, F., El-Ghar, M.A., Khalil, A., Dwyer, A.C., El-giziri, A., El-Melegy, M., El-Baz, A.: A novel CAD system for detecting acute rejection of renal allografts based on integrating imaging-markers and laboratory biomarkers. In: *2018 IEEE International Conference on Imaging Systems and Techniques (IST)*, IEEE (2018) 1–6.

57. Shehata, M., Mahmoud, A., Soliman, A., Khalifa, F., Ghazal, M., El-Ghar, M.A., El-Melegy, M., El-Baz, A.: 3D kidney segmentation from abdominal diffusion MRI using an appearance-guided deformable boundary. *PLoS One* 13(7) (2018) e0200082.

58. Ali, A.M., Farag, A.A., El-Baz, A.: Graph cuts framework for kidney segmentation with prior shape constraints. In: *Proceedings of International Conference on Medical Image Computing and Computer-Assisted Intervention (MICCAI'07)*. Volume 1, Brisbane, Australia, October 29–November 2 (2007) 384–392.

59. Chowdhury, A.S., Roy, R., Bose, S., Elnakib, F.K.A., El-Baz, A.: Non-rigid biomedical image registration using graph cuts with a novel data term. In: *Proceedings of IEEE International Symposium on Biomedical Imaging: From NANO to Macro (ISBI'12)*, Barcelona, Spain, May 2–5 (2012) 446–449.

60. El-Baz, A., Farag, A., Fahmi, R., Yuksel, S., El-Ghar, M.A., Eldiasty, T.: Image analysis of renal DCE MRI for the detection of acute renal rejection. In: *Proceedings of IAPR International Conference on Pattern Recognition (ICPR'06)*, Hong Kong, August 20–24 (2006) 822–825.

61. El-Baz, A., Farag, A.A., Yuksel, S.E., El-Ghar, M.E., Eldiasty, T.A., Ghoneim, M.A.: Application of deformable models for the detection of acute renal rejection. In: *Deformable Models*. New York, NY: Springer (2007) 293–333.

62. El-Baz, A., Farag, A., Fahmi, R., Yuksel, S., Miller, W., El-Ghar, M.A., El-Diasty, T., Ghoneim, M.: A new CAD system for the evaluation of kidney diseases using DCE-MRI. In: *Proceedings of International Conference on Medical Image Computing and Computer-Assisted Intervention (MICCAI'08)*, Copenhagen, Denmark, October 1–6 (2006) 446–453.

63. El-Baz, A., Gimel'farb, G., El-Ghar, M.A.: A novel image analysis approach for accurate identification of acute renal rejection. In: *Proceedings of IEEE International Conference on Image Processing (ICIP'08)*, San Diego, California, USA, October 12–15 (2008) 1812–1815.

64. El-Baz, A., Gimel'farb, G., El-Ghar, M.A.: Image analysis approach for identification of renal transplant rejection. In: *Proceedings of IAPR International Conference on Pattern Recognition (ICPR'08)*, Tampa, Florida, USA, December 8–11 (2008) 1–4.

65. El-Baz, A., Gimel'farb, G., El-Ghar, M.A.: New motion correction models for automatic identification of renal transplant rejection. In: *Proceedings of International Conference on Medical Image Computing and Computer-Assisted Intervention (MICCAI'07)*, Brisbane, Australia, October 29–November 2 (2007) 235–243.

66. Farag, A., El-Baz, A., Yuksel, S., El-Ghar, M.A., Eldiasty, T.: A framework for the detection of acute rejection with Dynamic Contrast Enhanced magnetic resonance imaging. In: *Proceedings of IEEE International Symposium on Biomedical Imaging: From NANO to Macro (ISBI'06)*, Arlington, Virginia, USA, April 6–9 (2006) 418–421.

67. Khalifa, F., Beache, G.M., El-Ghar, M.A., El-Diasty, T., Gimel'farb, G., Kong, M., El-Baz, A.: Dynamic contrast-enhanced MRI-based early detection of acute renal transplant rejection. *IEEE Transactions on Medical Imaging* **32**(10) (2013) 1910–1927.

68. Khalifa, F., El-Baz, A., Gimel'farb, G., El-Ghar, M.A.: Non-invasive image-based approach for early detection of acute renal rejection. In: *Proceedings of International Conference Medical Image Computing and Computer-Assisted Intervention (MICCAI'10)*, Beijing, China, September 20–24 (2010) 10–18.

69. Khalifa, F., El-Baz, A., Gimel'farb, G., Ouseph, R., El-Ghar, M.A.: Shape-appearance guided level-set deformable model for image segmentation. In: *Proceedings of IAPR International Conference on Pattern Recognition (ICPR'10)*, Istanbul, Turkey, August 23–26 (2010) 4581–4584.

70. Khalifa, F., El-Ghar, M.A., Abdollahi, B., Frieboes, H., El-Diasty, T., El-Baz, A.: A comprehensive non-invasive framework for automated evaluation of acute renal transplant rejection using DCE-MRI. *NMR in Biomedicine* **26**(11) (2013) 1460–1470.

71. Khalifa, F., El-Ghar, M.A., Abdollahi, B., Frieboes, H.B., El-Diasty, T., El-Baz, A.: Dynamic contrast-enhanced MRI-based early detection of acute renal transplant rejection. In: *2014 Annual Scientific Meeting and Educational Course Brochure of the Society of Abdominal Radiology (SAR'14)*, Boca Raton, Florida, March 23–28 (2014) CID: 1855912.

72. Khalifa, F., Elnakib, A., Beache, G.M., Gimel'farb, G., El-Ghar, M.A., Sokhadze, G., Manning, S., McClure, P., El-Baz, A.: 3D kidney segmentation from CT images using a level set approach guided by a novel stochastic speed function. In: *Proceedings of International Conference Medical Image Computing and Computer-Assisted Intervention (MICCAI'11)*, Toronto, Canada, September 18–22 (2011) 587–594.

73. Khalifa, F., Gimel'farb, G., El-Ghar, M.A., Sokhadze, G., Manning, S., McClure, P., Ouseph, R., El-Baz, A.: A new deformable model-based segmentation approach for accurate extraction of the kidney from abdominal CT images. In: *Proceedings of IEEE International Conference on Image Processing (ICIP'11)*, Brussels, Belgium, September 11–14 (2011) 3393–3396.

74. Mostapha, M., Khalifa, F., Alansary, A., Soliman, A., Suri, J., El-Baz, A.: Computer-aided diagnosis systems for acute renal transplant rejection: Challenges and methodologies. In: El-Baz, A., Saba, L., Suri, J., eds.: *Abdomen and Thoracic Imaging*, Springer (2014) 1–35.

75. Shehata, M., Khalifa, F., Hollis, E., Soliman, A., Hosseini-Asl, E., El-Ghar, M.A., El-Baz, M., Dwyer, A.C., El-Baz, A., Keynton, R.: A new non-invasive approach for early classification of renal rejection types using diffusion-weighted MRI. In: *2016 IEEE International Conference on Image Processing (ICIP)*, IEEE (2016) 136–140.

76. Khalifa, F., Soliman, A., Takieldeen, A., Shehata, M., Mostapha, M., Shaffie, A., Ouseph, R., Elmaghraby, A., El-Baz, A.: Kidney segmentation from CT images using a 3D NMF-guided active contour model. In: *2016 IEEE 13th International Symposium on Biomedical Imaging (ISBI)*, IEEE (2016) 432–435.

77. Shehata, M., Khalifa, F., Soliman, A., Takieldeen, A., El-Ghar, M.A., Shaffie, A., Dwyer, A.C., Ouseph, R., El-Baz, A., Keynton, R.: 3D diffusion MRI-based CAD system for early diagnosis of acute renal rejection. In: *2016 IEEE 13th International Symposium on Biomedical Imaging (ISBI)*, IEEE (2016) 1177–1180.

78. Shehata, M., Khalifa, F., Soliman, A., Alrefai, R., El-Ghar, M.A., Dwyer, A.C., Ouseph, R., El-Baz, A.: A level set-based framework for 3D kidney segmentation from diffusion MR images. In: *2015 IEEE International Conference on Image Processing (ICIP)*, IEEE (2015) 4441–4445.

79. Shehata, M., Khalifa, F., Soliman, A., El-Ghar, M.A., Dwyer, A.C., Gimelfarb, G., Keynton, R., El-Baz, A.: A promising non-invasive CAD system for kidney function assessment. In: *International Conference on Medical Image Computing and Computer-Assisted Intervention*, Springer (2016) 613–621.

80. Khalifa, F., Soliman, A., Elmaghraby, A., Gimelfarb, G., El-Baz, A.: 3D kidney segmentation from abdominal images using spatial-appearance models. *Computational and Mathematical Methods in Medicine* **2017** (2017) 1–10.

81. Hollis, E., Shehata, M., Khalifa, F., El-Ghar, M.A., El-Diasty, T., El-Baz, A.: Towards non-invasive diagnostic techniques for early detection of acute renal transplant rejection: A review. *The Egyptian Journal of Radiology and Nuclear Medicine* **48**(1) (2016) 257–269.

82. Shehata, M., Khalifa, F., Soliman, A., El-Ghar, M.A., Dwyer, A.C., El-Baz, A.: Assessment of renal transplant using image and clinical-based biomarkers. In: *Proceedings of 13th Annual Scientific Meeting of American Society for Diagnostics and Interventional Nephrology (ASDIN'17)*, New Orleans, LA, USA, February 10–12 (2017).

83. Shehata, M., Khalifa, F., Soliman, A., El-Ghar, M.A., Dwyer, A.C., El-Baz, A.: Early assessment of acute renal rejection. In: *Proceedings of 12th Annual Scientific Meeting of American Society for Diagnostics and Interventional Nephrology (ASDIN'16)*, Pheonix, AZ, USA, February 19–21, 2016 (2017).

84. Eltanboly, A., Ghazal, M., Hajjdiab, H., Shalaby, A., Switala, A., Mahmoud, A., Sahoo, P., El-Azab, M., El-Baz, A.: Level sets-based image segmentation approach using statistical shape priors. *Applied Mathematics and Computation* **340** (2019) 164–179.

85. Sliman, H., Khalifa, F., Elnakib, A., Soliman, A., Beache, G., Gimel'farb, G., Emam, A., Elmaghraby, A., El-Baz, A.: Accurate segmentation framework for the left ventricle wall from cardiac cine MRI. In: *AIP Conference Proceedings*. Volume **1559**, AIP (2013) 287–296.

86. Sliman, H., Khalifa, F., Elnakib, A., Soliman, A., El-Baz, A., Beache, G.M., Elmaghraby, A., Gimel'farb, G.: Myocardial borders segmentation from cine MR images using bidirectional coupled parametric deformable models. *Medical Physics* **40**(9) (2013).

87. Khalifa, F., Beache, G., El-Baz, A., Gimel'farb, G.: Deformable model guided by stochastic speed with application in cine images segmentation. In: *Proceedings of IEEE International Conference on Image Processing (ICIP'10)*, Hong Kong, September 26–29 (2010) 1725–1728.

88. Khalifa, F., Beache, G.M., Elnakib, A., Sliman, H., Gimel'farb, G., Welch, K.C., El-Baz, A.: A new shape-based framework for the left ventricle wall segmentation from cardiac first-pass perfusion MRI. In: *Proceedings of IEEE International Symposium on Biomedical Imaging: From NANO to Macro (ISBI'13)*, San Francisco, CA, April 7–11 (2013) 41–44.

89. Khalifa, F., Beache, G.M., Elnakib, A., Sliman, H., Gimel'farb, G., Welch, K.C., El-Baz, A.: A new nonrigid registration framework for improved visualization of transmural perfusion gradients on cardiac first-pass perfusion MRI. In: *Proceedings of IEEE International Symposium on Biomedical Imaging: From NANO to Macro (ISBI'12)*, Barcelona, Spain, May 2–5 (2012) 828–831.

90. Khalifa, F., Beache, G.M., Firjani, A., Welch, K.C., Gimel'farb, G., El-Baz, A.: A new non-rigid registration approach for motion correction of cardiac first-pass perfusion MRI. In: *Proceedings of IEEE International Conference on Image Processing (ICIP'12)*, Lake Buena Vista, Florida, September 30–October 3 (2012) 1665–1668.

91. Khalifa, F., Beache, G.M., Gimel'farb, G., El-Baz, A.: A novel CAD system for analyzing cardiac first-pass MR images. In: *Proceedings of IAPR International Conference on Pattern Recognition (ICPR'12)*, Tsukuba Science City, Japan, November 11–15 (2012) 77–80.

92. Khalifa, F., Beache, G.M., Gimel'farb, G., El-Baz, A.: A novel approach for accurate estimation of left ventricle global indexes from short-axis cine MRI. In: *Proceedings of IEEE International Conference on Image Processing (ICIP'11)*, Brussels, Belgium, September 11–14 (2011) 2645–2649.

93. Khalifa, F., Beache, G.M., Gimel'farb, G., Giridharan, G.A., El-Baz, A.: A new image-based framework for analyzing cine images. In: El-Baz, A., Acharya, U.R., Mirmedhdi, M., Suri, J.S., eds.: *Handbook of Multi Modality State-Of-the-Art Medical Image Segmentation and Registration Methodologies*. Volume **2**, New York: Springer (2011) 69–98.

94. Khalifa, F., Beache, G.M., Gimel'farb, G., Giridharan, G.A., El-Baz, A.: Accurate automatic analysis of cardiac cine images. *IEEE Transactions on Bio-Medical Engineering* **59**(2) (2012) 445–455.
95. Khalifa, F., Beache, G.M., Nitzken, M., Gimel'farb, G., Giridharan, G.A., El-Baz, A.: Automatic analysis of left ventricle wall thickness using short-axis cine CMR images. In: *Proceedings of IEEE International Symposium on Biomedical Imaging: From NANO to Macro (ISBI'11)*, Chicago, Illinois, March 30–April 2 (2011) 1306–1309.
96. Nitzken, M., Beache, G., Elnakib, A., Khalifa, F., Gimel'farb, G., El-Baz, A.: Accurate modeling of tagged CMR 3D image appearance characteristics to improve cardiac cycle strain estimation. In: *2012 19th IEEE International Conference on Image Processing (ICIP)*, IEEE, Orlando, Florida, USA, September 2012 (2012) 521–524.
97. Nitzken, M., Beache, G., Elnakib, A., Khalifa, F., Gimel'farb, G., El-Baz, A.: Improving full-cardiac cycle strain estimation from tagged CMR by accurate modeling of 3D image appearance characteristics. In: *2012 9th IEEE International Symposium on Biomedical Imaging (ISBI)*, IEEE, Barcelona, Spain, May 2012 (2012) 462–465 (Selected for oral presentation).
98. Nitzken, M.J., El-Baz, A.S., Beache, G.M.: Markov-Gibbs random field model for improved full-cardiac cycle strain estimation from tagged CMR. *Journal of Cardiovascular Magnetic Resonance* **14**(1) (2012) 1–2.
99. Sliman, H., Elnakib, A., Beache, G., Elmaghraby, A., El-Baz, A.: Assessment of myocardial function from cine cardiac MRI using a novel 4D tracking approach. *Journal of Computer Science and Systems Biology* **7**(5) (2014) 169–173.
100. Sliman, H., Elnakib, A., Beache, G.M., Soliman, A., Khalifa, F., Gimel'farb, G., Elmaghraby, A., El-Baz, A.: A novel 4D PDE-based approach for accurate assessment of myocardium function using cine cardiac magnetic resonance images. In: *Proceedings of IEEE International Conference on Image Processing (ICIP'14)*, Paris, France, October 27–30 (2014) 3537–3541.
101. Sliman, H., Khalifa, F., Elnakib, A., Beache, G.M., Elmaghraby, A., El-Baz, A.: A new segmentation-based tracking framework for extracting the left ventricle cavity from cine cardiac MRI. In: *Proceedings of IEEE International Conference on Image Processing (ICIP'13)*, Melbourne, Australia, September 15–18 (2013) 685–689.
102. Sliman, H., Khalifa, F., Elnakib, A., Soliman, A., Beache, G.M., Elmaghraby, A., Gimel'farb, G., El-Baz, A.: Myocardial borders segmentation from cine MR images using bi-directional coupled parametric deformable models. *Medical Physics* **40**(9) (2013) 1–13.
103. Eladawi, N., Elmogy, M., Ghazal, M., Helmy, O., Aboelfetouh, A., Riad, A., Schaal, S., El-Baz, A.: Classification of retinal diseases based on oct images. *Frontiers in Bioscience (Landmark Edition)* **23** (2018) 247–264.
104. ElTanboly, A., Ismail, M., Shalaby, A., Switala, A., El-Baz, A., Schaal, S., Gimel-farb, G., El-Azab, M.: A computer-aided diagnostic system for detecting diabetic retinopathy in optical coherence tomography images. *Medical Physics* **44**(3) (2017) 914–923.
105. Sandhu, H.S., El-Baz, A., Seddon, J.M.: Progress in automated deep learning for macular degeneration. *JAMA Ophthalmology* **136**(12) (2018) 1366–1367.
106. Abdollahi, B., Civelek, A.C., Li, X.F., Suri, J., El-Baz, A.: PET/CT nodule segmentation and diagnosis: A survey. In: Saba, L., Suri, J.S., eds.: *Multi Detector CT Imaging*, Taylor & Francis (2014) 639–651.
107. Abdollahi, B., El-Baz, A., Amini, A.A.: A multi-scale non-linear vessel enhancement technique. In: *2011 Annual International Conference of the IEEE Engineering in Medicine and Biology Society (EMBC)*, IEEE (2011) 3925–3929.
108. Abdollahi, B., Soliman, A., Civelek, A., Li, X.F., Gimel'farb, G., El-Baz, A.: A novel gaussian scale space-based joint MGRF framework for precise lung segmentation. In: *Proceedings of IEEE International Conference on Image Processing (ICIP'12)*, IEEE (2012) 2029–2032.
109. Abdollahi, B., Soliman, A., Civelek, A., Li, X.F., Gimelfarb, G., El-Baz, A.: A novel 3D joint MGRF framework for precise lung segmentation. In: *Machine Learning in Medical Imaging*, Springer (2012) 86–93.
110. Ali, A.M., El-Baz, A.S., Farag, A.A.: A novel framework for accurate lung segmentation using graph cuts. In: *Proceedings of IEEE International Symposium on Biomedical Imaging: From NANO to Macro (ISBI'07)*, IEEE (2007) 908–911.

111. El-Baz, A., Beache, G.M., Gimel'farb, G., Suzuki, K., Okada, K.: Lung imaging data analysis. *International Journal of Biomedical Imaging* **2013** (2013) 1–2.

112. El-Baz, A., Beache, G.M., Gimel'farb, G., Suzuki, K., Okada, K., Elnakib, A., Soliman, A., Abdollahi, B.: Computer-aided diagnosis systems for lung cancer: Challenges and methodologies. *International Journal of Biomedical Imaging* **2013** (2013) 1–46.

113. El-Baz, A., Elnakib, A., Abou El-Ghar, M., Gimel'farb, G., Falk, R., Farag, A.: Automatic detection of 2D and 3D lung nodules in chest spiral CT scans. *International Journal of Biomedical Imaging* **2013** (2013) 1–11.

114. El-Baz, A., Farag, A.A., Falk, R., La Rocca, R.: A unified approach for detection, visualization, and identification of lung abnormalities in chest spiral CT scans. In: *International Congress Series*. Volume **1256**, Elsevier (2003) 998–1004.

115. El-Baz, A., Farag, A.A., Falk, R., La Rocca, R.: Detection, visualization and identification of lung abnormalities in chest spiral CT scan: Phase-I. In: *Proceedings of International Conference on Biomedical Engineering*. Volume **12**, Cairo, Egypt (2002).

116. El-Baz, A., Farag, A., Gimel'farb, G., Falk, R., El-Ghar, M.A., Eldiasty, T.: A framework for automatic segmentation of lung nodules from low dose chest CT scans. In: *Proceedings of International Conference on Pattern Recognition (ICPR'06)*. Volume **3**, IEEE (2006) 611–614.

117. El-Baz, A., Farag, A., Gimelfarb, G., Falk, R., El-Ghar, M.A.: A novel level set-based computer-aided detection system for automatic detection of lung nodules in low dose chest computed tomography scans. *Lung Imaging and Computer Aided Diagnosis* **10** (2011) 221–238.

118. El-Baz, A., Gimel'farb, G., Abou El-Ghar, M., Falk, R.: Appearance-based diagnostic system for early assessment of malignant lung nodules. In: *Proceedings of IEEE International Conference on Image Processing (ICIP'12)*, IEEE (2012) 533–536.

119. El-Baz, A., Gimel'farb, G., Falk, R.: A novel 3D framework for automatic lung segmentation from low dose CT images. In: El-Baz, A., Suri, J.S., eds.: *Lung Imaging and Computer Aided Diagnosis*, Taylor & Francis (2011) 1–16.

120. El-Baz, A., Gimel'farb, G., Falk, R., El-Ghar, M.: Appearance analysis for diagnosing malignant lung nodules. In: *Proceedings of IEEE International Symposium on Biomedical Imaging: From NANO to Macro (ISBI'10)*, IEEE (2010) 193–196.

121. El-Baz, A., Gimel'farb, G., Falk, R., El-Ghar, M.A.: A novel level set-based CAD system for automatic detection of lung nodules in low dose chest CT scans. In: El-Baz, A., Suri, J.S., eds.: *Lung Imaging and Computer Aided Diagnosis*. Volume **1**, Taylor & Francis (2011) 221–238.

122. El-Baz, A., Gimel'farb, G., Falk, R., El-Ghar, M.A.: A new approach for automatic analysis of 3D low dose CT images for accurate monitoring the detected lung nodules. In: *Proceedings of International Conference on Pattern Recognition (ICPR'08)*, IEEE (2008) 1–4.

123. El-Baz, A., Gimel'farb, G., Falk, R., El-Ghar, M.A.: A novel approach for automatic follow-up of detected lung nodules. In: *Proceedings of IEEE International Conference on Image Processing (ICIP'07)*. Volume **5**, IEEE (2007) V-501.

124. El-Baz, A., Gimel'farb, G., Falk, R., El-Ghar, M.A.: A new CAD system for early diagnosis of detected lung nodules. In: *2007 IEEE International Conference on Image Processing (ICIP)*. Volume **2**, IEEE (2007) II-461.

125. El-Baz, A., Gimel'farb, G., Falk, R., El-Ghar, M.A., Refaie, H.: Promising results for early diagnosis of lung cancer. In: *Proceedings of IEEE International Symposium on Biomedical Imaging: From NANO to Macro (ISBI'08)*, IEEE (2008) 1151–1154.

126. El-Baz, A., Gimel'farb, G.L., Falk, R., Abou El-Ghar, M., Holland, T., Shaffer, T.: A new stochastic framework for accurate lung segmentation. In: *Proceedings of Medical Image Computing and Computer-Assisted Intervention (MICCAI'08)* (2008) 322–330.

127. El-Baz, A., Gimel'farb, G.L., Falk, R., Heredis, D., Abou El-Ghar, M.: A novel approach for accurate estimation of the growth rate of the detected lung nodules. In: *Proceedings of International Workshop on Pulmonary Image Analysis* (2008) 33–42.

128. El-Baz, A., Gimel'farb, G.L., Falk, R., Holland, T., Shaffer, T.: A framework for unsupervised segmentation of lung tissues from low dose computed tomography images. In: *Proceedings of British Machine Vision (BMVC'08)* (2008) 1–10.

129. El-Baz, A., Gimelfarb, G., Falk, R., El-Ghar, M.A.: 3D MGRF-based appearance modeling for robust segmentation of pulmonary nodules in 3D LDCT chest images. In: El-Baz, A., Suri, J.S., eds.: *Lung Imaging and Computer Aided Diagnosis*, CRC Press (2011) 51–63.

130. El-Baz, A., Gimelfarb, G., Falk, R., El-Ghar, M.A.: Automatic analysis of 3D low dose CT images for early diagnosis of lung cancer. *Pattern Recognition* **42**(6) (2009) 1041–1051.

131. El-Baz, A., Gimelfarb, G., Falk, R., El-Ghar, M.A., Rainey, S., Heredia, D., Shaffer, T.: Toward early diagnosis of lung cancer. In: *Proceedings of Medical Image Computing and Computer-Assisted Intervention (MICCAI'09)*, Springer (2009) 682–689.

132. El-Baz, A., Gimelfarb, G., Falk, R., El-Ghar, M.A., Suri, J.: Appearance analysis for the early assessment of detected lung nodules. In: El-Baz, A., Suri, J.S., eds.: *Lung Imaging and Computer Aided Diagnosis*, CRC Press (2011) 395–404.

133. El-Baz, A., Khalifa, F., Elnakib, A., Nitkzen, M., Soliman, A., McClure, P., Gimel'farb, G., El-Ghar, M.A.: A novel approach for global lung registration using 3D Markov Gibbs appearance model. In: *Proceedings of International Conference Medical Image Computing and Computer-Assisted Intervention (MIC-CAI'12)*, Nice, France, October 1–5 (2012) 114–121.

134. El-Baz, A., Nitzken, M., Elnakib, A., Khalifa, F., Gimel'farb, G., Falk, R., El-Ghar, M.A.: 3D shape analysis for early diagnosis of malignant lung nodules. In: *Proceedings of International Conference Medical Image Computing and Computer-Assisted Intervention (MICCAI'11)*, Toronto, Canada, September 18–22 (2011) 175–182.

135. El-Baz, A., Nitzken, M., Gimelfarb, G., Van Bogaert, E., Falk, R., El-Ghar, M.A., Suri, J.: Three-dimensional shape analysis using spherical harmonics for early assessment of detected lung nodules. In: El-Baz, A., Suri, J.S., eds.: *Lung Imaging and Computer Aided Diagnosis*, CRC Press (2011) 421–438.

136. El-Baz, A., Nitzken, M., Khalifa, F., Elnakib, A., Gimel'farb, G., Falk, R., El-Ghar, M.A.: 3D shape analysis for early diagnosis of malignant lung nodules. In: *Proceedings of International Conference on Information Processing in Medical Imaging (IPMI'11)*, Monastery Irsee, Germany (Bavaria), July 3–8 (2011) 772783.

137. El-Baz, A., Nitzken, M., Vanbogaert, E., Gimel'Farb, G., Falk, R., Abo El-Ghar, M.: A novel shape-based diagnostic approach for early diagnosis of lung nodules. In: *2011 IEEE International Symposium on Biomedical Imaging: From Nano to Macro*, IEEE (2011) 137–140.

138. El-Baz, A., Sethu, P., Gimel'farb, G., Khalifa, F., Elnakib, A., Falk, R., El-Ghar, M.A.: Elastic phantoms generated by microfluidics technology: Validation of an imaged-based approach for accurate measurement of the growth rate of lung nodules. *Biotechnology Journal* **6**(2) (2011) 195–203.

139. El-Baz, A., Sethu, P., Gimel'farb, G., Khalifa, F., Elnakib, A., Falk, R., El-Ghar, M.A.: A new validation approach for the growth rate measurement using elastic phantoms generated by state-of-the-art microfluidics technology. In: *Proceedings of IEEE International Conference on Image Processing (ICIP'10)*, Hong Kong, September 26–29 (2010) 4381–4383.

140. El-Baz, A., Sethu, P., Gimel'farb, G., Khalifa, F., Elnakib, A., Falk, R., Suri, M.A.E.G.J.: Validation of a new imaged-based approach for the accurate estimating of the growth rate of detected lung nodules using real CT images and elastic phantoms generated by state-of-the-art microfluidics technology. In: El-Baz, A., Suri, J.S., eds.: *Handbook of Lung Imaging and Computer Aided Diagnosis*. Volume 1, New York: Taylor & Francis (2011) 405–420.

141. El-Baz, A., Soliman, A., McClure, P., Gimel'farb, G., El-Ghar, M.A., Falk, R.: Early assessment of malignant lung nodules based on the spatial analysis of detected lung nodules. In: *Proceedings of IEEE International Symposium on Biomedical Imaging: From NANO to Macro (ISBI'12)*, IEEE (2012) 1463–1466.

142. El-Baz, A., Yuksel, S.E., Elshazly, S., Farag, A.A.: Non-rigid registration techniques for automatic follow-up of lung nodules. In: *Proceedings of Computer Assisted Radiology and Surgery (CARS'05)*. Volume **1281**, Elsevier (2005) 1115–1120.

143. El-Baz, A.S., Suri, J.S. *Lung Imaging and Computer Aided Diagnosis*. CRC Press (2011).

144. Soliman, A., Khalifa, F., Dunlap, N., Wang, B., El-Ghar, M., El-Baz, A.: An iso-surfaces based local deformation handling framework of lung tissues. In: *2016 IEEE 13th International Symposium on Biomedical Imaging (ISBI)*, IEEE (2016) 1253–1259.

145. Soliman, A., Khalifa, F., Shaffie, A., Dunlap, N., Wang, B., Elmaghraby, A., El-Baz, A.: Detection of lung injury using 4D-CT chest images. In: *2016 IEEE 13th International Symposium on Biomedical Imaging (ISBI)*, IEEE (2016) 1274–1277.

146. Soliman, A., Khalifa, F., Shaffie, A., Dunlap, N., Wang, B., Elmaghraby, A., Gimel'farb, G., Ghazal, M., El-Baz, A.: A comprehensive framework for early assessment of lung injury. In: *2017 IEEE International Conference on Image Processing (ICIP)*, IEEE (2017) 3275–3279.

147. Shaffie, A., Soliman, A., Ghazal, M., Taher, F., Dunlap, N., Wang, B., El-maghraby, A., Gimel'farb, G., El-Baz, A.: A new framework for incorporating appearance and shape features of lung nodules for precise diagnosis of lung cancer. In: *2017 IEEE International Conference on Image Processing (ICIP)*, IEEE (2017) 1372–1376.

148. Soliman, A., Khalifa, F., Shaffie, A., Liu, N., Dunlap, N., Wang, B., Elmaghraby, A., Gimel'farb, G., El-Baz, A.: Image-based CAD system for accurate identification of lung injury. In: *2016 IEEE International Conference on Image Processing (ICIP)*, IEEE (2016) 121–125.

149. Shaffie, A., Soliman, A., Ghazal, M., Taher, F., Dunlap, N., Wang, B., Van Berkel, V., Gimelfarb, G., Elmaghraby, A., El-Baz, A.: A novel autoencoder-based diagnostic system for early assessment of lung cancer. In: *2018 25th IEEE International Conference on Image Processing (ICIP)*, IEEE (2018) 1393–1397.

150. Shaffie, A., Soliman, A., Fraiwan, L., Ghazal, M., Taher, F., Dunlap, N., Wang, B., van Berkel, V., Keynton, R., Elmaghraby, A., et al.: A generalized deep learning-based diagnostic system for early diagnosis of various types of pulmonary nodules. *Technology in Cancer Research and Treatment* **17** (2018) 1533033818798800.

151. Dombroski, B., Nitzken, M., Elnakib, A., Khalifa, F., El-Baz, A., Casanova, M.F.: Cortical surface complexity in a population-based normative sample. *Translational Neuroscience* **5**(1) (2014) 17–24.

152. El-Baz, A., Casanova, M., Gimel'farb, G., Mott, M., Switala, A.: An MRI-based diagnostic framework for early diagnosis of dyslexia. *International Journal of Computer Assisted Radiology and Surgery* **3**(3–4) (2008) 181–189.

153. El-Baz, A., Casanova, M., Gimel'farb, G., Mott, M., Switala, A., Vanbogaert, E., McCracken, R.: A new CAD system for early diagnosis of dyslexic brains. In: *Proceedings of the International Conference on Image Processing (ICIP'2008)*, IEEE (2008) 1820–1823.

154. El-Baz, A., Casanova, M.F., Gimel'farb, G., Mott, M., Switwala, A.E.: A new image analysis approach for automatic classification of autistic brains. In: *Proceedings of the IEEE International Symposium on Biomedical Imaging: From NANO to Macro (ISBI'2007)*, IEEE (2007) 352–355.

155. El-Baz, A., Elnakib, A., Khalifa, F., El-Ghar, M.A., McClure, P., Soliman, A., Gimel'farb, G.: Precise segmentation of 3-D magnetic resonance angiography. *IEEE Transactions on Biomedical Engineering* **59**(7) (2012) 2019–2029.

156. El-Baz, A., Farag, A., Gimel'farb, G., El-Ghar, M.A., Eldiasty, T.: Probabilistic modeling of blood vessels for segmenting MRA images. In: *18th International Conference on Pattern Recognition (ICPR'06)*. Volume 3, IEEE (2006) 917–920.

157. El-Baz, A., Farag, A.A., Gimelfarb, G., El-Ghar, M.A., Eldiasty, T.: A new adaptive probabilistic model of blood vessels for segmenting MRA images. In: *Medical Image Computing and Computer-Assisted Intervention (MICCAI)*. Volume **4191**, Springer (2006) 799–806.

158. El-Baz, A., Farag, A.A., Gimelfarb, G., Hushek, S.G.: Automatic cerebrovascular segmentation by accurate probabilistic modeling of TOF-MRA images. In: *Medical Image Computing and Computer-Assisted Intervention (MICCAI 2005)*, Springer (2005) 34–42.

159. El-Baz, A., Farag, A., Elnakib, A., Casanova, M.F., Gimel'farb, G., Switala, A.E., Jordan, D., Rainey, S.: Accurate automated detection of autism related corpus callosum abnormalities. *Journal of Medical Systems* **35**(5) (2011) 929–939.

160. El-Baz, A., Farag, A., Gimelfarb, G.: Cerebrovascular segmentation by accurate probabilistic modeling of TOF-MRA images. In: *Scandinavian Conference on Image Analysis*. Volume **3540**, Springer (2005) 1128–1137.

161. El-Baz, A., Gimelfarb, G., Falk, R., El-Ghar, M.A., Kumar, V., Heredia, D.: A novel 3D joint Markov-Gibbs model for extracting blood vessels from PC-MRA images. In: *Medical Image Computing and Computer-Assisted Intervention (MICCAI 2009)*. Volume **5762**, Springer (2009) 943–950.

162. Elnakib, A., El-Baz, A., Casanova, M.F., Gimel'farb, G., Switala, A.E.: Image-based detection of corpus callosum variability for more accurate discrimination between dyslexic and normal brains. In: *Proceedings of the IEEE International Symposium on Biomedical Imaging: From NANO to Macro (ISBI'2010)*, IEEE (2010) 109–112.

163. Elnakib, A., Casanova, M.F., Gimel'farb, G., Switala, A.E., El-Baz, A.: Autism diagnostics by centerline-based shape analysis of the corpus callosum. In: *Proceedings of the IEEE International Symposium on Biomedical Imaging: From NANO to Macro (ISBI'2011)*, IEEE (2011) 1843–1846.

164. Elnakib, A., Nitzken, M., Casanova, M., Park, H., Gimel'farb, G., El-Baz, A.: Quantification of age-related brain cortex change using 3D shape analysis. In: *2012 21st International Conference on Pattern Recognition (ICPR)*, IEEE (2012) 41–44.

165. Mostapha, M., Soliman, A., Khalifa, F., Elnakib, A., Alansary, A., Nitzken, M., Casanova, M.F., El-Baz, A.: A statistical framework for the classification of infant DT images. In: *2014 IEEE International Conference on Image Processing (ICIP)*, IEEE (2014) 2222–2226.

166. Nitzken, M., Casanova, M., Gimel'farb, G., Elnakib, A., Khalifa, F., Switala, A., El-Baz, A.: 3D shape analysis of the brain cortex with application to dyslexia. In: *2011 18th IEEE International Conference on Image Processing (ICIP)*, Brussels, Belgium: IEEE, September 2011 (2011) 2657–2660 (Selected for oral presentation. Oral acceptance rate is 10 percent and the overall acceptance rate is 35 percent).

167. El-Gamal, F.E.Z.A., Elmogy, M.M., Ghazal, M., Atwan, A., Barnes, G.N., Casanova, M.F., Keynton, R., El-Baz, A.S.: A novel CAD system for local and global early diagnosis of Alzheimer's disease based on PIB-PET scans. In: *2017 IEEE International Conference on Image Processing (ICIP)*, IEEE (2017) 3270–3274.

168. Ismail, M., Soliman, A., Ghazal, M., Switala, A.E., Gimelfarb, G., Barnes, G.N., Khalil, A., El-Baz, A.: A fast stochastic framework for automatic MR brain images segmentation. *PloS one* **12**(11) (2017) e0187391.

169. Ismail, M.M., Keynton, R.S., Mostapha, M.M., ElTanboly, A.H., Casanova, M.F., Gimel'farb, G.L., El-Baz, A.: Studying autism spectrum disorder with structural and diffusion magnetic resonance imaging: A survey. *Frontiers in Human Neuro-Science* **10** (2016) 211.

170. Alansary, A., Ismail, M., Soliman, A., Khalifa, F., Nitzken, M., Elnakib, A., Mostapha, M., Black, A., Stinebruner, K., Casanova, M.F., et al.: Infant brain extraction inT1-weighted MR images using BET and refinement using LCDG and MGRF models. *IEEE Journal of Biomedical and Health Informatics* **20**(3) (2016) 925–935.

171. Ismail, M., Soliman, A., ElTanboly, A., Switala, A., Mahmoud, M., Khalifa, F., Gimel'farb, G., Casanova, M.F., Keynton, R., El-Baz, A.: Detection of white matter abnormalities in MR brain images for diagnosis of autism in children. In: *2016 IEEE 13th International Symposium on Biomedical Imaging (ISBI)*, IEEE (2016) 6–9.

172. Ismail, M., Mostapha, M., Soliman, A., Nitzken, M., Khalifa, F., Elnakib, A., Gimel'farb, G., Casanova, M., El-Baz, A.: Segmentation of infant brain MR images based on adaptive shape prior and higher-order MGRF. In: *2015 IEEE 13th International Symposium on Biomedical Imaging (ISBI)*, IEEE (2015) 4327–4331.

173. Asl, E.H., Ghazal, M., Mahmoud, A., Aslantas, A., Shalaby, A., Casanova, M., Barnes, G., Gimelfarb, G., Keynton, R., El-Baz, A.: Alzheimers disease diagnostics by a 3D deeply supervised adaptable convolutional network. *Frontiers in Bioscience (Landmark Edition)* **23** (2018) 584–596.

174. Mahmoud, A., El-Barkouky, A., Farag, H., Graham, J., Farag, A.: A non-invasive method for measuring blood flow rate in superficial veins from a single thermal image. In: *Proceedings of the IEEE Conference on Computer Vision and Pattern Recognition Workshops* (2013) 354–359.

175. El-baz, A., Shalaby, A., Taher, F., El-Baz, M., Ghazal, M., El-Ghar, M.A., Takieldeen, A., Suri, J.: Probabilistic modeling of blood vessels for segmenting magnetic resonance angiography images. *Medical Research Archives* **5** (2017).

176. Chowdhury, A.S., Rudra, A.K., Sen, M., Elnakib, A., El-Baz, A.: Cerebral white matter segmentation from MRI using probabilistic graph cuts and geometric shape priors. In: *ICIP* (2010) 3649–3652.
177. Gebru, Y., Giridharan, G., Ghazal, M., Mahmoud, A., Shalaby, A., El-Baz, A.: Detection of cerebrovascular changes using magnetic resonance angiography. In: El-Baz, A., Suri J.S., eds.: *Cardiovascular Imaging and Image Analysis*, CRC Press (2018) 1–22.
178. Mahmoud, A., Shalaby, A., Taher, F., El-Baz, M., Suri, J.S., El-Baz, A.: Vascular tree segmentation from different image modalities. In: El-Baz, A., Suri J.S., eds.: *Cardiovascular Imaging and Image Analysis*, CRC Press (2018) 43–70.
179. Taher, F., Mahmoud, A., Shalaby, A., El-Baz, A.: A review on the cerebrovascular segmentation methods. In: *2018 IEEE International Symposium on Signal Processing and Information Technology (ISSPIT)*, IEEE (2018) 359–364.
180. Kandil, H., Soliman, A., Fraiwan, L., Shalaby, A., Mahmoud, A., ElTanboly, A., Elmaghraby, A., Giridharan, G., El-Baz, A.: A novel MRA framework based on integrated global and local analysis for accurate segmentation of the cerebral vascular system. In: *2018 IEEE 15th International Symposium on Biomedical Imaging (ISBI 2018)*, IEEE (2018) 1365–1368.

Chapter 12

Automatic Detection of Early Signs of Diabetic Retinopathy Based on Feature Fusion from OCT and OCTA Scans

Nabila Eladawi, Ahmed ElTanboly, Mohammed Elmogy,
Mohammed Ghazal, Ali Mahmoud, Ahmed Aboelfetouh,
Alaa Riad, Magdi El-Azab, Jasjit S. Suri,
Guruprasad Giridharan, and Ayman El-Baz

12.1 Introduction

Diabetic retinopathy (DR) affects the blood circular system in the eye. It is the most common cause of vision loss among people with diabetes. In addition, it is the leading cause of vision impairment and blindness among working-age people. The earlier stage of DR is called non-proliferative diabetic retinopathy (NPDR). In the NPDR stage, abnormalities, such as microaneurysms, vessel dilation and tortuosity, foveal avascular zone (FAZ) enlargement, and capillary dropout, start to appear [1–8]. The more advanced stage of the disease is proliferative diabetic retinopathy (PDR). PDR is characterized by retinal and/or optic nerve neovascularization. Ophthalmologists aim to prevent and treat DR to avoid vision loss, not to restore it. To be able to do so, early detection of DR is needed. Since microvascular pathology causes DR, we need imaging techniques that can visualize retinal vasculature. The standard clinical techniques to visualize the ocular vasculature are fluorescein angiography (FA) and indocyanine green angiography (ICGA). Due to their invasiveness and cost, they cannot be used routinely to examine DR. Optical coherence tomography angiography (OCTA) depends on repeated B-scans that are taken in rapid succession. Due to the noninvasiveness character of OCTA, it is ideal for monitoring and detecting DR in diabetic patients. OCTA can easily detect capillary dropout among other abnormalities that appear in the early stage of DR [4, 9–15].

OCTA provides us with a comprehensive vasculature network in superficial, deep, and capillary layers in the retina giving us the ability to understand the ischemic processes that affect different layers of the retina, for example, the deep plexus sometimes affected by paracentral acute middle maculopathy, and the cotton-wool spots that affect the superficial plexus. OCTA provides us, in addition to this anatomic detail, data on vascular flow. So, in the absence of apparent morphological change, this data will affect our understanding of tissue perfusion. For example, a flow index of the optic nerve head can be used to ascertain disc perfusion. OCTA is sensitive enough to detect a small increase in the parafoveal capillary flow. With the anatomic details and the detailed perfusion data provided, OCTA can help in the prediction of a variety of ophthalmic diseases.

Many research groups are working in detecting the early signs of DR. For example, Mastropasqua et al. [16] investigated the difference in FAZ and the vessel densities in diabetic patients with or without DR using OCTA images. They found a significant enlargement in the FAZ area in a subject with PDR and moderate and severe NPDR. They concluded that as the severity of DR increases, the FAZ area increases while the vessel density decreases. Hwang et al. [11] investigated the features of DR using OCTA images. They were able to detect an enlargement in the FAZ, retinal capillary dropout, and pruning of arteriolar branches.

Matsunaga et al. [17] used a set of OCTA images to investigate the ability of OCTA in detecting the clinical vascular changes in DR patients. They were able to detect microaneurysms, impaired vascular perfusion, some forms of intraretinal fluid, vascular loops, intraretinal microvascular abnormalities, neo-vascularization, and cotton-wool spots. Samara et al. [18] used 55 DR patients and 34 control eyes to quantify FAZ and vessel density in DR patients using OCTA images. They found that the area of FAZ is larger in diabetic eyes than the control eyes while the vessel density is less. They concluded that OCTA could be used to monitor DR disease and helped in identifying the parameters that affect visual acuity. Durbin et al. [19] investigated the ability of OCTA to distinguish eyes with DR disease. They were able to measure the size of FAZ and vessel density. Schottenhamml et al. [9] developed a fully automated algorithm using OCTA images to detect capillary dropout in diabetic patients. Using 26 OCTA images for subjects ranging from healthy controls to PDR eyes, they have noticed an enlargement in the intercapillary area that increases as DR progresses. Hwang et al. [20] investigated the quantification of capillary nonperfusion as a sign for DR using OCTA images. In the DR patients, they noticed that the vessel density had been reduced, and they also noticed an enlargement in the FAZ.

From the literature above and some surveys [21–27] on DR, we found that no study integrates the analysis of retinal layers using OCT images with the analysis of blood vessels using OCTA images. In this chapter, we propose a computer-aided diagnosis (CAD) system that was able to segment the blood vessels from OCTA images and 12 different retinal layers from OCT scans. Then, five features were extracted from both segmented OCTA maps and OCT layers. Finally, these features are fused to generate a non-invasive diagnostic tool for DR.

12.2 Methods

We propose a comprehensive CAD system that can detect the early signs of DR. First, it segments the blood vessels from retinal superficial and deep plexuses of the OCTA images. Also, it segments 12 different retinal layers from OCT scans. Then, two different local features are extracted from the segmented OCTA images. Then, three various features are extracted from the segmented OCT layers. Finally, these features are fused to generate a comprehensive non-invasive diagnostic tool for detecting the

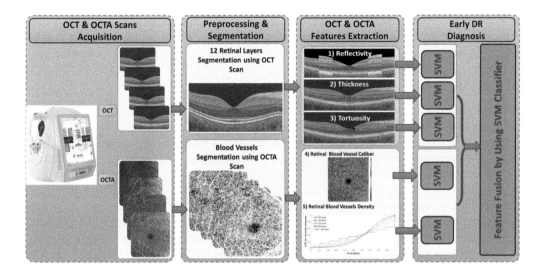

FIGURE 12.1: The proposed CAD system to diagnose early signs of DR based on fusing OCT and OCTA scans.

early signs of DR. Figure 12.1 shows the structure of the proposed CAD system. We will discuss the stages of the CAD system in the following subsections.

12.2.1 Retinal Blood Vessels and Layers Segmentation

In the beginning, we need to segment the blood vessels from OCTA images and the retinal layers from OCT scans. The first step for OCTA images is to preprocess the images of the retina for both superficial and deep maps to decrease the noise and to improve the homogeneity. To make sure that the gray level in the image is uniformly distributed, we used the regional dynamic histogram equalization (RDHE). Then, we used a combination of generalized Gauss–Markov random field (GGMRF) model and an adaptive gray level threshold estimation technique to improve the homogeneity in the images. The second step for OCTA images is to segment the blood vessels. Three models were used to segment these blood vessels, which are the prior intensity model, the current intensity model, and the higher-order spatial model. The prior intensity model was generated from a set of manually segmented images by three experts. An enhanced version of the Expectation Maximization (EM) algorithm was used to generate the current intensity model. A higher-order Markov–Gibbs random field (HO-MGRF) was used to calculate the higher-order spatial model. Finally, a Naïve Bayes (NB) is applied to label and analyze the connected components to obtain the final segmentation. For more detail about the segmentation technique, please see [28].

For OCT scans, the OCT layer segmentation is based on a joint model that combines intensity, appearance, and spatial information. The shape prior (Atlas) is constructed using 12 co-aligned OCT segmented images (six males and six females), which were selected to capture the biological variability among the whole dataset. The intensity model is built using a linear combination of discrete Gaussians (LCDG) model [29]. The spatial information is constructed using a second-order Markov–Gibbs random field (MGRF), which also accounts for noise and inhomogeneities [30]. Twelve different retina layers were extracted from these OCT images. The approach accuracy has been compared to the ground truth (GT) and validated by three different experts. More details about each component of this model and comparison with other algorithms are completely covered in [31].

12.2.2 Features Extraction

In this stage, we need some features from segmented OCT and OCTA images to help our system in the diagnosis of DR. We were able to extract five features from both OCT and OCTA images. Two features were extracted from OCTA images, which are the blood vessels density and blood vessel caliber. Figure 12.2 shows some example of the these extracted features. For OCT segmented images, we were able to extract three features, which are intensity, curvature, and thickness. Figure 12.3 shows some example of these extracted features.

We have chosen the blood vessel density because according to our ophthalmologists, the blood vessels density will change in the case of DR. Blood vessels density feature shows the changes that will happen to the blood vessels after being affected by DR. To extract the blood vessels density, a Parzen window technique was used. The Parzen window technique calculates the density ($P_{PW}(\mathbf{B_r})$) at specific location $\mathbf{r}$ depending on the neighboring pixels in the segmented OCTA images ($\mathbf{B_r}$) using a given window size.

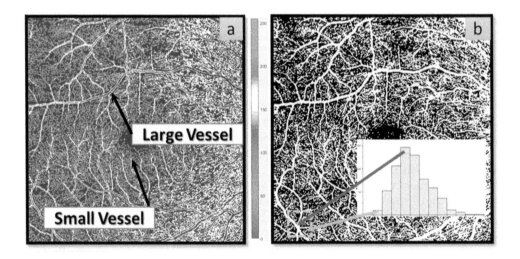

FIGURE 12.2: The OCTA features: (a) blood vessels density and (b) retinal blood vessel caliber.

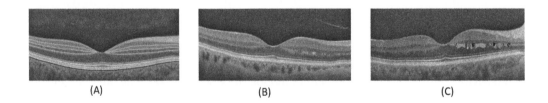

FIGURE 12.3: The OCT features: (a) intensity, (b) thickness, and (c) curvature.

We tested our system using five different window sizes (3×3, 5×5, 7×7, 9×9, and 11×11) to make sure that the results were not sensitive to the choice of the window size. To be able to use this as a feature and feed it to the classifier, we used the cumulative distribution function (CDF) to represent it for each tested window size.

The second feature, retinal blood vessel caliber, is extracted to analyze the appearance and intensity of the blood vessels. As we mentioned previously, the features are extracted from both superficial and deep maps. Blood vessels caliber is used to differentiate between small and large vessels. To extract the intensity values, in the beginning, we multiply the segmented image by the original image. Then, CDFs are generated from these intensities to allow to distinguish large vessels from small ones. An incremental value of 0.02 is chosen to generate these CDFs, and each one will be represented as a vector of 128 elements.

For the OCT images, three different global features, which are the curvature, intensity, and thickness, are extracted from each segmented layer. Using unpaired t-test results, we found that the curvature of the Inner Nuclear layer (INL), the intensity of the Myoid Zone (MZ), and the thickness of the Nerve Fiber layer (NFL) have a significant effect on the results in distinguishing between normal and diseased cases. So, these results motivated us to focus only on these three features and their effect on the classification approach [24].

Layer intensity demonstrates significant changes throughout life and between genders. For each scan, the intensity is measured from the thickest portions of the retina on the nasal and temporal sides of the foveola. A normalized reflectivity scale (NRS) is defined to take a value of 0 arbitrary units (AU) inside the vitreous and 1000 AU in the RPE layer. The average intensity within a segment is calculated according to Huber's M-estimates, which are robust to possible outlying values, such as very bright pixels in the innermost segment that accidentally belong to the internal limiting membrane, rather than the NFL.

The second feature is the curvature of the retinal layers, which combines all Menger curvature values calculated for each point across the layer. First, a locally weighted polynomial is applied for smoothing the surface, then Menger curvature is calculated for each point.

Finally, one of the quantitative evaluations in the retinal diseases is measuring retinal thickness. The thickness measurements and topographic thickness maps provide valuable information on retinal cell layers. So, measuring each layer thickness and validating it within normal ranges could indicate the existence of such abnormalities or retinal disease. The thickness of a retinal layer is calculated from the streamlines of Laplace's equation between the two surfaces of the layer.

12.2.3 Diagnosis

The diagnosis stage is the last stage in our system. This step is responsible for the classifying of our subjects according to the extracted features. A two-stage support vector machines (SVM) with a linear kernel is used to classify these subjects and to distinguish between normal and early DR cases. The first stage classifies the extracted feature from both modalities separately. Then, the second stage fuses the resulting probabilities from the first stage.

12.3 Results

Both OCT and OCTA images were extracted by a ZEISS AngioPlex OCT angiography machine [32]. This machine can generate five different OCTA maps of blood vessels in addition to a complete OCT B-scan. The extracted OCTA and OCT images had a size of 1024×1024 pixels. OCTA images were taken on 6×6 mm² sections centered on the fovea. The OCT data were exported for analysis as grayscale raw files with five slices, where the field of view was 6 mm nasal-temporal (N-T) and 2 mm posterior–anterior (P–A), and the slice spacing was 0.25 mm.

The system was trained and tested on 46 subjects that have both OCT and OCTA scans (23 normal cases and 23 DR cases). To evaluate our proposed system, we calculated three various performance evaluation metrics, which are the accuracy (ACC), the sensitivity (Sens.), and the specificity (Spec.). To overcome the overfitting, we used four-fold and two-fold cross-validation methods. Table 12.1 lists the results of our proposed system as tested on different feature combinations. First, we tested the proposed system on each extracted feature from both OCT and OCTA scans. We found that the average ACCs are 67% and 70% for OCT and OCTA features, respectively. Then, we combined the extracted features from each modality and evaluated the diagnosis of DR for each combined feature. We found that the average ACC for the combined OCT features is 92% and the average ACC for the combined OCTA features is 96%. Finally, we fused all extracted features from both modalities by using a two-stage SVM classifier with a linear kernel. We got the ACC of 97.2% for all combined features. Therefore, when combining all extracted features from both OCT and OCTA scans, it gives promising results in the diagnosis of early signs of DR.

Also, the proposed system was evaluated against two different state-of-the-art classifiers, which are the SVM with a polynomial kernel, and K-nearest neighbor (KNN) techniques. For four-fold cross-validation, Table 12.2 shows the results after combining the extracted features from both imaging modalities using different classifiers. It shows

TABLE 12.1: The Performance Evaluation of the Proposed CAD System Using Different Feature Combinations With Four-Fold Cross-Validation

Modality	Feature	ACC (%)	Sens. (%)	Spec. (%)
	NFL thickness	76.3	88.7	47.8
OCT	MZ intensity	59.2	75.5	21.7
	INL curvature	65.8	88.7	13
	All OCT features	92	84	100
	Superficial density	77.6	81.1	69.6
OCTA	Deep density	73.7	83	52.2
	Superficial caliber	69.7	84.9	34.8
	Deep caliber	73.7	90.6	34.8
	All OCTA features	96.1	96.9	94.7
Both	All five features	97.2	100	94.7

TABLE 12.2: The Results of the Proposed CAD System Using Different State-Of-The-Art Classifiers Based on Fusing Both OCT and OCTA Features with Four-Fold Cross-Validation

Classifier	ACC (%)	Sens. (%)	Spec. (%)
SVM (linear)	97.2	100	94.7
SVM (polynomial)	95.4	96.1	93.7
KNN	95.7	98.1	93.7

that the SVM with a linear kernel outperforms the other classifiers and provides promising results. Similarly, the system achieved ACC of 97.2%. These results show the capability of the proposed CAD system to differentiate early DR cases from normal cases.

12.4 Conclusion

In this chapter, we were able to develop a CAD system that uses both OCT and OCTA images to detect the early signs of DR. This system was able to segment 12 layers from OCT scans and the blood vessels from OCTA ones. Five features were extracted from both scans to help in the early detection of DR. These features are used by an SVM to distinguish between normal and diseased subjects. Using four-fold cross-validation, our proposed system achieved an average ACC of 97.2%. In the future, we will try to increase the number of cases in our dataset. Also, we will try to improve our system by integrating the demographical and clinical data to be able to diagnose different stages of DR.

This work could also be applied to various other applications in medical imaging, such as the kidney, the heart, the prostate, and the lung, as well as several non-medical applications [33–36].

One application is renal transplant functional assessment, especially with developing noninvasive CAD systems for renal transplant function assessment, utilizing different image modalities (e.g., ultrasound, computed tomography [CT], MRI, etc.). Accurate assessment of renal transplant function is critically important for graft survival. Although transplantation can improve a patient's wellbeing, there is a potential post-transplantation risk of kidney dysfunction that, if not treated in a timely manner, can lead to the loss of the entire graft and even patient death. In particular, dynamic and diffusion MRI-based systems have been clinically used to assess transplanted kidneys with the advantage of providing information on each kidney separately. For more details about renal transplant functional assessment, please read [37–54], [54–64].

The heart is also an important application for this work. The clinical assessment of myocardial perfusion plays a major role in the diagnosis, management, and prognosis of ischemic heart disease patients. Thus, there have been ongoing efforts to develop automated systems for accurate analysis of myocardial perfusion using first-pass images [65–81].

Moreover, the work could be applied for prostate cancer which is the most common cancer in American men, and its related mortality rate is the second after lung cancer. Fortunately, the mortality rate can be reduced if prostate cancer is detected in its early stages. Early detection enables physicians to treat prostate cancer before it develops to clinically significant disease [82, 83].

Abnormalities of the lung could also be another promising area of research and a related application of this work. Radiation-induced lung injury is the main side effect of radiation therapy for lung cancer patients. Although higher radiation doses increase the radiation therapy effectiveness for tumor control, this can lead to lung injury as a greater quantity of normal lung tissues is included in the treated area. Almost one-third of patients who undergo radiation therapy develop lung injury following radiation treatment. The severity of radiation-induced lung injury ranges from ground-glass opacities and consolidation at the early phase to fibrosis and traction bronchiectasis in the late phase. Early detection of lung injury will thus help to improve management of the treatment [84–126].

This work can also be applied to other brain abnormalities, such as dyslexia in addition to autism. Dyslexia is one of the most complicated developmental brain disorders that affect children's learning abilities. Dyslexia leads to the failure to develop age-appropriate reading skills in spite of normal intelligence level and adequate reading instructions. Neuropathological studies have revealed an abnormal anatomy of some structures, such as the corpus callosum, in dyslexic brains. There has been a lot of work in the literature that aims at developing CAD systems for diagnosing this disorder, along with other brain disorders [127–149].

For the vascular system [150], this work could also be applied for the extraction of blood vessels, e.g., from phase contrast (PC) magnetic resonance angiography (MRA). Accurate cerebrovascular segmentation using non-invasive MRA is crucial for the early diagnosis and timely treatment of intracranial vascular diseases [132, 133, 151–156].

References

1. N. Unoki, K. Nishijima, A. Sakamoto, M. Kita, D. Watanabe, M. Hangai, T. Kimura, N. Kawagoe, M. Ohta, and N. Yoshimura, "Retinal sensitivity loss and structural disturbance in areas of capillary nonperfusion of eyes with diabetic retinopathy," *American Journal of Ophthalmology*, vol. 144, no. 5, pp. 755–760.e1, 2007.

2. M. Tyrberg, V. Ponjavic, and M. Lövestam-Adrian, "Multifocal electroretinogram (MFERG) in patients with diabetes mellitus and an enlarged foveal avascular zone (FAZ)," *Documenta Ophthalmologica*, vol. 117, no. 3, pp. 185–189, 2008.

3. M. Adhi, E. Brewer, N. K. Waheed, and J. S. Duker, "Analysis of morphological features and vascular layers of choroid in diabetic retinopathy using spectral-domain optical coherence tomography," *JAMA Ophthalmology*, vol. 131, no. 10, pp. 1267–1274, 2013.

4. A. Ishibazawa, T. Nagaoka, A. Takahashi, T. Omae, T. Tani, K. Sogawa, H. Yokota, and A. Yoshida, "Optical coherence tomography angiography in diabetic retinopathy: A prospective pilot study," *American Journal of Ophthalmology*, vol. 160, no. 1, pp. 35–44.e1, 2015.

5. R. Varma, N. M. Bressler, Q. V. Doan, M. Gleeson, M. Danese, J. K. Bower, E. Selvin, C. Dolan, J. Fine, S. Colman, and A. Turpcu, "Prevalence of and risk factors for diabetic macular edema in the united states," *JAMA Ophthalmology*, vol. 132, no. 11, pp. 1334–1340, 2014.

6. M. J. Elman, L. P. Aiello, R. W. Beck, N. M. Bressler, S. B. Bressler, A. R. Edwards, F. L. Ferris, S. M. Friedman, A. R. Glassman, K. M. Miller, I. U. Scott, C. R. Stockdale, and J. K. Sun, "Randomized trial evaluating ranibizumab plus prompt or deferred laser or triamcinolone plus prompt laser for diabetic macular edema," *Ophthalmology*, vol. 117, no. 6, pp. 1064–1077.e35, 2010.

7. "Photocoagulation for diabetic macular edema: Early treatment diabetic retinopathy study report number 1 early treatment diabetic retinopathy study research group," *Archives of Ophthalmology*, vol. 103, no. 12, pp. 1796–1806, 1985.

8. Y. Miwa, T. Murakami, K. Suzuma, A. Uji, S. Yoshitake, M. Fujimoto, T. Yoshitake, Y. Tamura, and N. Yoshimura, "Relationship between functional and structural changes in diabetic vessels in optical coherence tomography angiography," *Scientific Reports*, vol. 6, p. 29064, 2016.

9. J. Schottenhamml, E. M. Moult, S. Ploner, B. Lee, E. A. Novais, E. Cole, S. Dang, C. D. Lu, L. Husvogt, N. K. Waheed, J. S. Duker, J. Hornegger, and J. G. Fujimoto, "An automatic, intercapillary area-based algorithm for quantifying diabetes-related capillary dropout using optical coherence tomography angiography," *RETINA*, vol. 36, p. 593, 2016.

10. D. A. Salz, E. Talisa, M. Adhi, E. Moult, W. Choi, C. R. Baumal, A. J. Witkin, J. S. Duker, J. G. Fujimoto, and N. K. Waheed, "Select features of diabetic retinopathy on swept-source optical coherence tomographic angiography compared with fluorescein angiography and normal eyes," *JAMA Ophthalmology*, vol. 134, no. 6, pp. 644–650, 2016.

11. T. S. Hwang, Y. Jia, S. S. Gao, S. T. Bailey, A. K. Lauer, C. J. Flaxel, D. J. Wilson, and D. Huang, "Optical coherence tomography angiography features of diabetic retinopathy," *Retina (Philadelphia, Pa.)*, vol. 35, no. 11, p. 2371, 2015.

12. T. E. de Carlo, M. A. B. Filho, A. T. Chin, M. Adhi, D. Ferrara, C. R. Baumal, A. J. Witkin, E. Reichel, J. S. Duker, and N. K. Waheed, "Spectral-domain optical coherence tomography angiography of choroidal neovascularization," *Ophthalmology*, vol. 122, no. 6, pp. 1228–1238, 2015.

13. J. M. B. de Barros Garcia, D. L. C. Isaac, and M. Avila, "Diabetic retinopathy and oct angiography: clinical findings and future perspectives," *International Journal of Retina and Vitreous*, vol. 3, no. 1, p. 14, 2017.

14. K. Tarassoly, A. Miraftabi, M. Soltan Sanjari, and M. M. Parvaresh, "The relationship between foveal avascular zone area, vessel density, and cystoid changes in diabetic retinopathy: An optical coherence tomography angiography study," *RETINA*, vol. 38, no. 8, pp. 1613–1619, 2018.

15. M. Soares, C. Neves, I. P. Marques, I. Pires, C. Schwartz, M. Â. Costa, T. Santos, M. Durbin, and J. Cunha-Vaz, "Comparison of diabetic retinopathy classification using fluorescein angiography and optical coherence tomography angiography," *British Journal of Ophthalmology*, vol. 101, no. 1, pp. 62–68, 2016.

16. R. Mastropasqua, L. Toto, A. Mastropasqua, R. Aloia, C. De Nicola, P. A. Mattei, G. Di Marzio, M. Di Nicola, and L. Di Antonio, "Foveal avascular zone area and parafoveal vessel density measurements in different stages of diabetic retinopathy by optical coherence tomography angiography," *International Journal of Ophthalmology*, vol. 10, no. 10, p. 1545, 2017.

17. D. R. Matsunaga, J. Y. Jack, L. O. De Koo, H. Ameri, C. A. Puliafito, and A. H. Kashani, "Optical coherence tomography angiography of diabetic retinopathy in human subjects," *Ophthalmic Surgery, Lasers and Imaging Retina*, vol. 46, no. 8, pp. 796–805, 2015.

18. W. A. Samara, A. Shahlaee, M. K. Adam, M. A. Khan, A. Chiang, J. I. Maguire, J. Hsu, and A. C. Ho, "Quantification of diabetic macular ischemia using optical coherence tomography angiography and its relationship with visualacuity," *Ophthalmology*, vol. 124, no. 2, pp. 235–244, 2017.

19. M. K. Durbin, L. An, N. D. Shemonski, M. Soares, T. Santos, M. Lopes, C. Neves, and J. Cunha-Vaz, "Quantification of retinal microvascular density in optical coherence tomographic angiography images in diabetic retinopathy," *JAMA Ophthalmology*, vol. 135, no. 4, pp. 370–376, 2017.

20. T. S. Hwang, S. S. Gao, L. Liu, A. K. Lauer, S. T. Bailey, C. J. Flaxel, D. J. Wilson, D. Huang, and Y. Jia, "Automated quantification of capillary nonperfusion using optical coherence tomography angiography in diabetic retinopathy," *JAMA Ophthalmology*, vol. 134, no. 4, pp. 367–373, 2016.

21. M. D. Abramoff, M. K. Garvin, and M. Sonka, "Retinal imaging and image analysis," *IEEE Reviews in Biomedical Engineering*, vol. 3, pp. 169–208, 2010.

22. N. Eladawi, M. Elmogy, L. Fraiwan, F. Pichi, M. Ghazal, A. Aboelfetouh, A. Riad, R. Keynton, S. Schaal, and A. El-Baz, "Early diagnosis of diabetic retinopathy in octa images based on local analysis of retinal blood vessels and foveal avascular zone," in *International Conference in Pattern Recognition (ICPR 2018)*, Beijing, China, 2018.

23. H. S. Sandhu, N. Eladawi, M. Elmogy, R. Keynton, O. Helmy, S. Schaal, and A. El-Baz, "Automated diabetic retinopathy detection using optical coherence tomography angiography: a pilot study," *British Journal of Ophthalmology*, 2018.

24. A. ElTanboly, M. Ismail, A. Shalaby, A. Switala, A. El-Baz, S. Schaal, G. Gime´lfarb, and M. El-Azab, "A computer-aided diagnostic system for detecting diabetic retinopathy in optical coherence tomography images," *Medical Physics*, vol. 44, no. 3, pp. 914–923, 2017.

25. A. ElTanboly, A. Placio, A. Shalaby, A. Switala, O. Helmy, S. Schaal, and A. El-Baz, "An automated approach for early detection of diabetic retinopathy using SD-OCT images." *Frontiers in Bioscience (Elite Edition)*, vol. 10, pp. 197–207, 2018.

26. M. M. Fraz, P. Remagnino, A. Hoppe, B. Uyyanonvara, A. R. Rudnicka, C. G. Owen, and S. A. Barman, "Blood vessel segmentation methodologies in retinal images a survey," *Computer Methods and Programs in Biomedicine*, vol. 108, no. 1, pp. 407–433, 2012.

27. N. Eladawi, M. Elmogy, M. Ghazal, O. Helmy, A. Aboelfetouh, A. Riad, S. Schaal, and A. El-Baz, "Classification of retinal diseases based on OCT images," *Frontiers In Bioscience, Landmark*, vol. 23, pp. 247–264, 2018.

28. N. Eladawi, M. Elmogy, O. Helmy, A. Aboelfetouh, A. Riad, H. Sandhu, S. Schaal, and A. El-Baz, "Automatic blood vessels segmentation based on different retinal maps from OCTA scans," *Computers in Biology and Medicine*, vol. 89, pp. 150–161, 2017.

29. A. El-Baz and G. Gimel'farb, "Em based approximation of empirical distributions with linear combinations of discrete gaussians," in *2007 IEEE International Conference on Image Processing*, vol. 4, San Antonio, TX, 2007, pp. IV–373–IV–376.

30. A. El-Baz, A. Elnakib, F. Khalifa, M. A. El-Ghar, P. McClure, A. Soliman, and G. Gimelrfarb, "Precise segmentation of 3-D magnetic resonance angiography," *IEEE Transactions on Biomedical Engineering*, vol. 59, no. 7, pp. 2019–2029, 2012.

31. A. E. Tanboly, M. Ismail, A. Switala, M. Mahmoud, A. Soliman, T. Neyer, A. Palacio, A. Hadayer, M. El-Azab, S. Schaal, and A. El-Baz, "A novel automatic segmentation of healthy and diseased retinal layers from OCT scans," in *2016 IEEE International Conference on Image Processing*, Phoenix, AZ, 2016, pp. 116–120.

32. ZEISS. (2017) Angioplex OCT angiography: http://www.zeiss.com/meditec/us/c/oct-angiogr aphy.html.

33. A. H. Mahmoud, "Utilizing radiation for smart robotic applications using visible, thermal, and polarization images." Ph.D. dissertation, University of Louisville, 2014.

34. A. Mahmoud, A. El-Barkouky, J. Graham, and A. Farag, "Pedestrian detection using mixed partial derivative based his togram of oriented gradients," in *2014 IEEE International Conference on Image Processing*, Paris, France, IEEE, 2014, pp. 2334–2337.

35. A. El-Barkouky, A. Mahmoud, J. Graham, and A. Farag, "An interactive educational drawing system using a humanoid robot and light polarization," in *2013 IEEE International Conference on Image Processing*, Melbourne, Australia, IEEE, 2013, pp. 3407–3411.

36. A. H. Mahmoud, M. T. El-Melegy, and A. A. Farag, "Direct method for shape recovery from polarization and shading," in *2012 International Conference on Image Processing*, Orlando, FL, IEEE, 2012, pp. 1769–1772.

37. A. M. Ali, A. A. Farag, and A. El-Baz, "Graph cuts framework for kidney segmentation with prior shape constraints," in *Proceedings of International Conference on Medical Image Computing and Computer-Assisted Intervention, (MICCAI'07)*, vol. 1, Brisbane, Australia, October 29–November 2, 2007, pp. 384–392.

38. A. S. Chowdhury, R. Roy, S. Bose, F. K. A. Elnakib, and A. El-Baz, "Non-rigid biomedical image registration using graph cuts with a novel data term," in *Proceedings of IEEE International Symposium on Biomedical Imaging: From Nano to Macro, (ISBI'12)*, Barcelona, Spain, May 2–5, 2012, pp. 446–449.

39. A. El-Baz, A. A. Farag, S. E. Yuksel, M. E. El-Ghar, T. A. Eldiasty, and M. A. Ghoneim, "Application of deformable models for the detection of acute renal rejection," in *Deformable Models*, Springer, New York, NY, 2007, pp. 293–333.

40. A. El-Baz, A. Farag, R. Fahmi, S. Yuksel, M. A. El-Ghar, and T. Eldiasty, "Image analysis of renal DCE MRI for the detection of acute renal rejection," in *Proceedings of IAPR International Conference on Pattern Recognition (ICPR'06)*, Hong Kong, August 20–24, 2006, pp. 822–825.

41. A. El-Baz, A. Farag, R. Fahmi, S. Yuksel, W. Miller, M. A. El-Ghar, T. El-Diasty, and M. Ghoneim, "A new CAD system for the evaluation of kidney diseases using DCE-MRI," in *Proceedings of International Conference on Medical Image Computing and Computer-Assisted Intervention, (MICCAI'08)*, Copenhagen, Denmark, October 1–6, 2006, pp. 446–453.

42. A. El-Baz, G. Gimel'farb, and M. A. El-Ghar, "A novel image analysis approach for accurate identification of acute renal rejection," in *Proceedings of IEEE International Conference on Image Processing, (ICIP'08)*, San Diego, California, USA, October 12–15, 2008, pp. 1812–1815.

43. A. El-Baz, G. Gimel'farb, and M. A. El-Ghar, "Image analysis approach for identification of renal transplant rejection," in *Proceedings of IAPR International Conference on Pattern Recognition, (ICPR'08)*, Tampa, Florida, USA, December 8–11, 2008, pp. 1–4.

44. A. El-Baz, G. Gimel'farb, and M. A. El-Ghar, "New motion correction models for automatic identification of renal transplant rejection," in *Proceedings of International Conference on Medical Image Computing and Computer-Assisted Intervention, (MICCAI'07)*, Brisbane, Australia, October 29–November 2, 2007, pp. 235–243.

45. A. Farag, A. El-Baz, S. Yuksel, M. A. El-Ghar, and T. Eldiasty, "A framework for the detection of acute rejection with dynamic contrast enhanced magnetic resonance imaging," in *Proceedings of IEEE International Symposium on Biomedical Imaging: From Nano to Macro, (ISBI'06)*, Arlington, Virginia, USA, April 6–9, 2006, pp. 418–421.

46. F. Khalifa, G. M. Beache, M. A. El-Ghar, T. El-Diasty, G. Gimel'farb, M. Kong, and A. El-Baz, "Dynamic contrast-enhanced MRI- based early detection of acute renal transplant rejection," *IEEE Transactions on Medical Imaging*, vol. 32, no. 10, pp. 1910–1927, 2013.

47. F. Khalifa, A. El-Baz, G. Gimel'farb, and M. A. El-Ghar, "Non-invasive image-based approach for early detection of acute renal rejection," in *Proceedings of International Conference Medical Image Computing and Computer-Assisted Intervention, (MICCAI'10)*, Beijing, China, September 20–24, 2010, pp. 10–18.

48. F. Khalifa, A. El-Baz, G. Gimel'farb, R. Ouseph, and M. A. El-Ghar, "Shape-appearance guided level-set deformable model for image segmentation," in *Proceedings of IAPR International Conference on Pattern Recognition, (ICPR'10)*, Istanbul, Turkey, August 23–26, 2010, pp. 4581–4584.

49. F. Khalifa, M. A. El-Ghar, B. Abdollahi, H. Frieboes, T. El-Diasty, and A. El-Baz, "A comprehensive non-invasive framework for automated evaluation of acute renal transplant rejection using DCE-MRI," *NMR in Biomedicine*, vol. 26, no. 11, pp. 1460–1470, 2013.

50. F. Khalifa, M. A. El-Ghar, B. Abdollahi, H. B. Frieboes, T. El-Diasty, and A. El-Baz, "Dynamic contrast-enhanced MRI-based early detection of acute renal transplant rejection," in *2014 Annual Scientific Meeting and Educational Course Brochure of the Society of Abdominal Radiology, (SAR'14)*, Boca Raton, Florida, March 23–28, 2014, p. CID: 1855912.

51. F. Khalifa, A. Elnakib, G. M. Beache, G. Gimel'farb, M. A. El-Ghar, G. Sokhadze, S. Manning, P. McClure, and A. El-Baz, "3D kidney segmentation from CT images using a level set approach guided by a novel stochastic speed function," in *Proceedings of International Conference Medical Image Computing and Computer-Assisted Intervention, (MICCAI'11)*, Toronto, Canada, September 18–22, 2011, pp. 587–594.

52. F. Khalifa, G. Gimel'farb, M. A. El-Ghar, G. Sokhadze, S. Manning, P. McClure, R. Ouseph, and A. El-Baz, "A new deformable model-based segmentation approach for accurate extraction of the kidney from abdominal CT images," in *Proceedings of IEEE International Conference on Image Processing, (ICIP'11)*, Brussels, Belgium, September 11–14, 2011, pp. 3393–3396.

53. M. Mostapha, F. Khalifa, A. Alansary, A. Soliman, J. Suri, and A. El-Baz, "Computer-aided diagnosis systems for acute renal transplant rejection: Challenges and methodologies," in *Abdomen and Thoracic Imaging*, A. El-Baz, L. saba, and J. Suri, Eds. Springer, 2014, pp. 1–35.

54. M. Shehata, F. Khalifa, E. Hollis, A. Soliman, E. Hosseini-Asl, M. A. El-Ghar, M. El-Baz, A. C. Dwyer, A. El-Baz, and R. Keynton, "A new non-invasive approach for early classification of renal rejection types using diffusion-weighted MRI," in *IEEE International Conference on Image Processing (ICIP), 2016*, IEEE, 2016, pp. 136–140.

55. F. Khalifa, A. Soliman, A. Takieldeen, M. Shehata, M. Mostapha, A. Shaffie, R. Ouseph, A. Elmaghraby, and A. El-Baz, "Kidney segmentation from CT images using a 3D NMF-guided active contour model," in *IEEE 13th International Symposium on Biomedical Imaging (ISBI), 2016*, IEEE, 2016, pp. 432–435.

56. M. Shehata, F. Khalifa, A. Soliman, A. Takieldeen, M. A. El-Ghar, A. Shaffie, A. C. Dwyer, R. Ouseph, A. El-Baz, and R. Keynton, "3D diffusion MRI-based CAD system for early diagnosis of acute renal rejection," in *Biomedical Imaging (ISBI), 2016 IEEE 13th International Symposium on IEEE*, 2016, pp. 1177–1180.

57. M. Shehata, F. Khalifa, A. Soliman, R. Alrefai, M. A. El-Ghar, A. C. Dwyer, R. Ouseph, and A. El-Baz, "A level set-based framework for 3D kidney segmentation from diffusion mr images," in *IEEE International Conference on Image Processing (ICIP), 2015*, IEEE, 2015, pp. 4441–4445.

58. M. Shehata, F. Khalifa, A. Soliman, M. A. El-Ghar, A. C. Dwyer, G. Gimelfarb, R. Keynton, and A. El-Baz, "A promising non–invasive CAD system for kidney function assessment," in *International Conference on Medical Image Computing and Computer-Assisted Intervention*, Springer, 2016, pp. 613–621.

59. F. Khalifa, A. Soliman, A. Elmaghraby, G. Gimelfarb, and A. El-Baz, "3d kidney segmentation from abdominal images using spatial-appearance models," *Computational and Mathematical Methods in Medicine*, vol. 2017, pp. 1–10, 2017.

60. E. Hollis, M. Shehata, F. Khalifa, M. A. El-Ghar, T. El-Diasty, and A. El-Baz, "Towards non-invasive diagnostic techniques for early detection of acute renal transplant rejection: A review," *The Egyptian Journal of Radiology and Nuclear Medicine*, vol. 48, no. 1, pp. 257–269, 2016.

61. M. Shehata, F. Khalifa, A. Soliman, M. A. El-Ghar, A. C. Dwyer, and A. El-Baz, "Assessment of renal transplant using image and clinical-based biomarkers," in *Proceedings of 13th Annual Scientific Meeting of American Society for Diagnostics and Interventional Nephrology (ASDIN'17)*, New Orleans, LA, USA, February 10–12, 2017, 2017.

62. M. Shehata, F. Khalifa, A. Soliman, M. A. El-Ghar, A. C. Dwyer, and A. El-Baz, "Early assessment of acute renal rejection," in *Proceedings of 12th Annual Scientific Meeting of American Society for Diagnostics and Interventional Nephrology (ASDIN'16)*, Pheonix, AZ, USA, February 19–21, 2016, 2017.

63. A. Eltanboly, M. Ghazal, H. Hajjdiab, A. Shalaby, A. Switala, A. Mahmoud, P. Sahoo, M. El-Azab, and A. El-Baz, "Level sets-based image segmentation approach using statistical shape priors," *Applied Mathematics and Computation*, vol. 340, pp. 164–179, 2019.

64. M. Shehata, A. Mahmoud, A. Soliman, F. Khalifa, M. Ghazal, M. A. El-Ghar, M. El-Melegy, and A. El-Baz, "3d kidney segmentation from abdominal diffusion MRI using an appearance-guided deformable boundary," *PloS One*, vol. 13, no. 7, p. e0200082, 2018.

65. F. Khalifa, G. Beache, A. El-Baz, and G. Gimel'farb, "Deformable model guided by stochastic speed with application in cine images segmentation," in *Proceedings of IEEE International Conference on Image Processing, (ICIP'10)*, Hong Kong, September 26–29, 2010, pp. 1725–1728.

66. F. Khalifa, G. M. Beache, A. Elnakib, H. Sliman, G. Gimel'farb, K. C. Welch, and A. El-Baz, "A new shape-based framework for the left ventricle wall segmentation from cardiac first-pass perfusion MRI," in *Proceedings of IEEE International Symposium on Biomedical Imaging: From Nano to Macro, (ISBI'13)*, San Francisco, CA, April 7–11, 2013, pp. 41–44.

67. F. Khalifa, G. M. Beache, A. Elnakib, H. Sliman, G. Gimel'farb, K. C. Welch, and A. El-Baz, "A new nonrigid registration framework for improved visualization of transmural perfusion gradients on cardiac first–pass perfusion MRI," in *Proceedings of IEEE International Symposium on Biomedical Imaging: From Nano to Macro, (ISBI'12)*, Barcelona, Spain, May 2–5, 2012, pp. 828–831.

68. F. Khalifa, G. M. Beache, A. Firjani, K. C. Welch, G. Gimel'farb, and A. El-Baz, "A new nonrigid registration approach for motion correction of cardiac first-pass perfusion MRI," in *Proceedings of IEEE International Conference on Image Processing, (ICIP'12)*, Lake Buena Vista, Florida, September 30–October 3, 2012, pp. 1665–1668.

69. F. Khalifa, G. M. Beache, G. Gimel'farb, and A. El-Baz, "A novel CAD system for analyzing cardiac first-pass MR images," in *Proceedings of IAPR International Conference on Pattern Recognition (ICPR'12)*, Tsukuba Science City, Japan, November 11–15, 2012, pp. 77–80.

70. F. Khalifa, G. M. Beache, G. Gimel'farb, and A. El-Baz, "A novel approach for accurate estimation of left ventricle global indexes from short-axis cine MRI," in *Proceedings of IEEE International Conference on Image Processing, (ICIP'11)*, Brussels, Belgium, September 11–14, 2011, pp. 2645–2649.

71. F. Khalifa, G. M. Beache, G. Gimel'farb, G. A. Giridharan, and A. El-Baz, "A new image-based framework for analyzing cine images," in *Handbook of Multi Modality State-of-the-Art Medical Image Segmentation and Registration Methodologies*, A. El-Baz, U. R. Acharya, M. Mirmedhdi, and J. S. Suri, Eds. Springer, New York, 2011, vol. 2, ch. 3, pp. 69–98.

72. F. Khalifa, G. M. Beache, G. Gimel'farb, G. A. Giridharan, and A. El-Baz, "Accurate automatic analysis of cardiac cine images," *IEEE Transactions on Biomedical Engineering*, vol. 59, no. 2, pp. 445–455, 2012.

73. F. Khalifa, G. M. Beache, M. Nitzken, G. Gimel'farb, G. A. Giridharan, and A. El-Baz, "Automatic analysis of left ventricle wall thickness using short-axis cine CMR images," in *Proceedings of IEEE International Symposium on Biomedical Imaging: From Nano to Macro, (ISBI'11)*, Chicago, Illinois, March 30–April 2, 2011, pp. 1306–1309.

74. M. Nitzken, G. Beache, A. Elnakib, F. Khalifa, G. Gimel'farb, and A. El-Baz, "Accurate modeling of tagged cmr 3D image appearance characteristics to improve cardiac cycle strain estimation," in *2012 19th IEEE International Conference on Image Processing (ICIP)*, Orlando, Florida, USA, IEEE, September 2012, pp. 521–524.

75. M. Nitzken, G. Beache, A. Elnakib, F. Khalifa, G. Gimel'farb, and A. El-Baz, "Improving full-cardiac cycle strain estimation from tagged cmr by accurate modeling of 3D image appearance characteristics," in *2012 9th IEEE International Symposium on Biomedical Imaging (ISBI)*, Barcelona, Spain, IEEE, May 2012, pp. 462–465, (Selected for oral presentation).

76. M. J. Nitzken, A. S. El-Baz, and G. M. Beache, "Markov–Gibbs random field model for improved full-cardiac cycle strain estimation from tagged CMR," *Journal of Cardiovascular Magnetic Resonance*, vol. 14, no. 1, pp. 1–2, 2012.

77. H. Sliman, A. Elnakib, G. Beache, A. Elmaghraby, and A. El-Baz, "Assessment of myocardial function from cine cardiac MRI using a novel 4D tracking approach," *J Comput Sci Syst Biol*, vol. 7, pp. 169–173, 2014.

78. H. Sliman, A. Elnakib, G. M. Beache, A. Soliman, F. Khalifa, G. Gimel'farb, A. Elmaghraby, and A. El-Baz, "A novel 4D PDE-based approach for accurate assessment of myocardium function using cine cardiac magnetic resonance images," in *Proceedings of IEEE International Conference on Image Processing (ICIP'14)*, Paris, France, October 27–30, 2014, pp. 3537–3541.

79. H. Sliman, F. Khalifa, A. Elnakib, G. M. Beache, A. Elmaghraby, and A. El-Baz, "A new segmentation-based tracking framework for extracting the left ventricle cavity from cine cardiac MRI," in *Proceedings of IEEE International Conference on Image Processing, (ICIP'13)*, Melbourne, Australia, September 15–18, 2013, pp. 685–689.

80. H. Sliman, F. Khalifa, A. Elnakib, A. Soliman, G. M. Beache, A. Elmaghraby, G. Gimel'farb, and A. El-Baz, "Myocardial borders segmentation from cine MR images using bi-directional coupled parametric deformable models," *Medical Physics*, vol. 40, no. 9, pp. 1–13, 2013.

81. H. Sliman, F. Khalifa, A. Elnakib, A. Soliman, G. M. Beache, G. Gimel'farb, A. Emam, A. Elmaghraby, and A. El-Baz, "Accurate segmentation framework for the left ventricle wall from cardiac cine MRI," in *Proceedings of International Symposium on Computational Models for Life Science, (CMLS'13)*, vol. 1559, Sydney, Australia, November 27–29, 2013, pp. 287–296.

82. I. Reda, M. Ghazal, A. Shalaby, M. Elmogy, A. AbouEl-Fetouh, B. O. Ayinde, M. AbouEl-Ghar, A. Elmaghraby, R. Keynton, and A. El-Baz, "A novel ADCS-based CNN classification system for precise diagnosis of prostate cancer," in *2018 24th International Conference on Pattern Recognition (ICPR)*, IEEE, 2018, pp. 3923–3928.

83. I. Reda, A. Khalil, M. Elmogy, A. Abou El-Fetouh, A. Shalaby, M. Abou El-Ghar, A. Elmaghraby, M. Ghazal, and A. El-Baz, "Deep learning role in early diagnosis of prostate cancer," *Technology in Cancer Research & Treatment*, vol. 17, p. 1533034618775530, 2018.

84. B. Abdollahi, A. C. Civelek, X.-F. Li, J. Suri, and A. El-Baz, "PET/CT nodule segmentation and diagnosis: A survey," in *Multi Detector CT Imaging*, L. Saba and J. S. Suri, Eds. Taylor, Francis, 2014, ch. 30, pp. 639–651.

85. B. Abdollahi, A. El-Baz, and A. A. Amini, "A multi-scale non-linear vessel enhancement technique," in *Engineering in Medicine and Biology Society, EMBC, 2011 Annual International Conference of the IEEE*, IEEE, 2011, pp. 3925–3929.

86. B. Abdollahi, A. Soliman, A. Civelek, X.-F. Li, G. Gimel'farb, and A. El-Baz, "A novel Gaussian scale space-based joint MGRF framework for precise lung segmentation," in *Proceedings of IEEE International Conference on Image Processing, (ICIP'12)*, IEEE, 2012, pp. 2029–2032.

87. B. Abdollahi, A. Soliman, A. Civelek, X.-F. Li, G. Gimelfarb, and A. El-Baz, "A novel 3D joint MGRF framework for precise lung segmentation," in *International Workshop on Machine Learning in Medical Imaging*, Springer, 2012, pp. 86–93.

88. A. M. Ali, A. S. El-Baz, and A. A. Farag, "A novel framework for accurate lung segmentation using graph cuts," in *Proceedings of IEEE International Symposium on Biomedical Imaging: From Nano to Macro, (ISBI'07)*, IEEE, 2007, pp. 908–911.

89. A. El-Baz, G. M. Beache, G. Gimel'farb, K. Suzuki, and K. Okada, "Lung imaging data analysis," *International Journal of Biomedical Imaging*, vol. 2013, pp. 1–2, 2013.

90. A. El-Baz, G. M. Beache, G. Gimel'farb, K. Suzuki, K. Okada, A. Elnakib, A. Soliman, and B. Abdollahi, "Computer-aided diagnosis systems for lung cancer: Challenges and methodologies," *International Journal of Biomedical Imaging*, vol. 2013, pp. 1–46, 2013.

91. A. El-Baz, A. Elnakib, M. Abou El-Ghar, G. Gimel'farb, R. Falk, and A. Farag, "Automatic detection of 2D and 3D lung nodules in chest spiral CT scans," *International Journal of Biomedical Imaging*, vol. 2013, pp. 1–11, 2013.

92. A. El-Baz, A. A. Farag, R. Falk, and R. La Rocca, "A unified approach for detection, visualization, and identification of lung abnormalities in chest spiral CT scans," in *International Congress Series*, vol. 1256, Elsevier, 2003, pp. 998–1004.

93. A. El-Baz, A. A. Farag, R. Falk, and R. La Rocca, "Detection, visualization and identification of lung abnormalities in chest spiral CT scan: Phase-I," in *Proceedings of International conference on Biomedical Engineering*, Cairo, Egypt, vol. 12, no. 1, 2002.

94. A. El-Baz, A. Farag, G. Gimel'farb, R. Falk, M. A. El-Ghar, and T. Eldiasty, "A framework for automatic segmentation of lung nodules from low dose chest CT scans," in *Proceedings of International Conference on Pattern Recognition, (ICPR'06)*, vol. 3, IEEE, 2006, pp. 611–614.

95. A. El-Baz, A. Farag, G. Gimelfarb, R. Falk, and M. A. El-Ghar, "A novel level set-based computer-aided detection system for automatic detection of lung nodules in low dose chest computed tomography scans," *Lung Imaging and Computer Aided Diagnosis*, vol. 10, pp. 221–238, 2011.

96. A. El-Baz, G. Gimel'farb, M. Abou El-Ghar, and R. Falk, "Appearance-based diagnostic system for early assessment of malignant lung nodules," in *Proceedings of IEEE International Conference on Image Processing, (ICIP'12)*, IEEE, 2012, pp. 533–536.

97. A. El-Baz, G. Gimel'farb, and R. Falk, "A novel 3D framework for automatic lung segmentation from low dose CT images," in *Lung Imaging and Computer Aided Diagnosis*, A. El-Baz and J. S. Suri, Eds. CRC Press, 2011, ch. 1, pp. 1–16.

98. A. El-Baz, G. Gimel'farb, R. Falk, and M. El-Ghar, "Appearance analysis for diagnosing malignant lung nodules," in *Proceedings of IEEE International Symposium on Biomedical Imaging: From Nano to Macro (ISBI'10)*, IEEE, 2010, pp. 193–196.

99. A. El-Baz, G. Gimel'farb, R. Falk, and M. A. El-Ghar, "A novel level set-based CAD system for automatic detection of lung nodules in low dose chest CT scans," in *Lung Imaging and Computer Aided Diagnosis*, A. El-Baz and J. S. Suri, Eds. CRC Press, 2011, vol. 1, ch. 10, pp. 221–238.

100. A. El-Baz, G. Gimel'farb, R. Falk, and M. A. El-Ghar, "A new approach for automatic analysis of 3D low dose CT images for accurate monitoring the detected lung nodules," in *Proceedings of International Conference on Pattern Recognition, (ICPR'08)*, IEEE, 2008, pp. 1–4.

101. A. El-Baz, G. Gimel'farb, R. Falk, and M. A. El-Ghar, "A novel approach for automatic follow-up of detected lung nodules," in *Proceedings of IEEE International Conference on Image Processing, (ICIP'07)*, vol. 5, IEEE, 2007, pp. V–501.

102. A. El-Baz, G. Gimel'farb, R. Falk, and M. A. El-Ghar, "A new CAD system for early diagnosis of detected lung nodules," in *Image Processing, 2007. ICIP 2007. IEEE International Conference on*, vol. 2, IEEE, 2007, pp. II–461.

103. A. El-Baz, G. Gimel'farb, R. Falk, M. A. El-Ghar, and H. Refaie, "Promising results for early diagnosis of lung cancer," in *Proceedings of IEEE International Symposium on Biomedical Imaging: From Nano to Macro, (ISBI'08)*, IEEE, 2008, pp. 1151–1154.

104. A. El-Baz, G. L. Gimel'farb, R. Falk, M. Abou El-Ghar, T. Holland, and T. Shaffer, "A new stochastic framework for accurate lung segmentation," in *Proceedings of Medical Image Computing and Computer-Assisted Intervention, (MICCAI'08)*, 2008, pp. 322–330.

105. A. El-Baz, G. L. Gimel'farb, R. Falk, D. Heredis, and M. Abou El-Ghar, "A novel approach for accurate estimation of the growth rate of the detected lung nodules," in *Proceedings of International Workshop on Pulmonary Image Analysis*, 2008, pp. 33–42.

106. A. El-Baz, G. L. Gimel'farb, R. Falk, T. Holland, and T. Shaffer, "A framework for unsupervised segmentation of lung tissues from low dose computed tomography images," in *Proceedings of British Machine Vision, (BMVC'08)*, 2008, pp. 1–10.

107. A. El-Baz, G. Gimelfarb, R. Falk, and M. A. El-Ghar, "3D MGRF-based appearance modeling for robust segmentation of pulmonary nodules in 3D LDCT chest images," in *Lung Imaging and Computer Aided Diagnosis*, A. El-Baz and J. S. Suri, Eds. CRC Press, 2011, ch. 3, pp. 51–63.

108. A. El-Baz, G. Gimelfarb, R. Falk, and M. A. El-Ghar, "Automatic analysis of 3D low dose CT images for early diagnosis of lung cancer," *Pattern Recognition*, vol. 42, no. 6, pp. 1041–1051, 2009.

109. A. El-Baz, G. Gimelfarb, R. Falk, M. A. El-Ghar, S. Rainey, D. Heredia, and T. Shaffer, "Toward early diagnosis of lung cancer," in *Proceedings of Medical Image Computing and Computer-Assisted Intervention, (MICCAI'09)*, Springer, 2009, pp. 682–689.

110. A. El-Baz, G. Gimelfarb, R. Falk, M. A. El-Ghar, and J. Suri, "Appearance analysis for the early assessment of detected lung nodules," in *Lung Imaging and Computer Aided Diagnosis*, A. El-Baz and J. S. Suri, Eds. CRC Press, 2011, ch. 17, pp. 395–404.

111. A. El-Baz, F. Khalifa, A. Elnakib, M. Nitkzen, A. Soliman, P. McClure, G. Gimel'farb, and M. A. El-Ghar, "A novel approach for global lung registration using 3D Markov Gibbs appearance model," in *Proceedings of International Conference Medical Image Computing and Computer-Assisted Intervention, (MICCAI'12)*, Nice, France, October 1–5, 2012, pp. 114–121.

112. A. El-Baz, M. Nitzken, A. Elnakib, F. Khalifa, G. Gimel'farb, R. Falk, and M. A. El-Ghar, "3D shape analysis for early diagnosis of malignant lung nodules," in *Proceedings of International Conference Medical Image Computing and Computer-Assisted Intervention, (MICCAI'11)*, Toronto, Canada, September 18–22, 2011, pp. 175–182.

113. A. El-Baz, M. Nitzken, G. Gimelfarb, E. Van Bogaert, R. Falk, M. A. El-Ghar, and J. Suri, "Three-dimensional shape analysis using spherical harmonics for early assessment of detected lung nodules," in *Lung Imaging and Computer Aided Diagnosis*, A. El-Baz and J. S. Suri, Eds. CRC Press, 2011, ch. 19, pp. 421–438.

114. A. El-Baz, M. Nitzken, F. Khalifa, A. Elnakib, G. Gimel'farb, R. Falk, and M. A. El-Ghar, "3D shape analysis for early diagnosis of malignant lung nodules," in *Proceedings of International Conference on Information Processing in Medical Imaging, (IPMI'11)*, Monastery Irsee, Germany (Bavaria), July 3–8, 2011, pp. 772–783.

115. A. El-Baz, M. Nitzken, E. Vanbogaert, G. Gimel'Farb, R. Falk, and M. Abo El-Ghar, "A novel shape-based diagnostic approach for early diagnosis of lung nodules," in *2011 IEEE International Symposium on Biomedical Imaging: From Nano to Macro*, IEEE, 2011, pp. 137–140.

116. A. El-Baz, P. Sethu, G. Gimel'farb, F. Khalifa, A. Elnakib, R. Falk, and M. A. El-Ghar, "Elastic phantoms generated by microfluidics technology: Validation of an imaged-based approach for accurate measurement of the growth rate of lung nodules," *Biotechnology Journal*, vol. 6, no. 2, pp. 195–203, 2011.

117. A. El-Baz, P. Sethu, G. Gimel'farb, F. Khalifa, A. Elnakib, R. Falk, and M. A. El-Ghar, "A new validation approach for the growth rate measurement using elastic phantoms generated by state-of-the-art microfluidics technology," in *Proceedings of IEEE International Conference on Image Processing, (ICIP'10)*, Hong Kong, September 26–29, 2010, pp. 4381–4383.

118. A. El-Baz, P. Sethu, G. Gimel'farb, F. Khalifa, A. Elnakib, R. Falk, and M. A. E.-G. J. Suri, "Validation of a new imaged-based approach for the accurate estimating of the growth rate of detected lung nodules using real CT images and elastic phantoms generated by state-of-the-art microfluidics technology," in *Handbook of Lung Imaging and Computer Aided Diagnosis*, A. El-Baz and J. S. Suri, Eds. Taylor & Francis, New York, 2011, vol. 1, ch. 18, pp. 405–420.

119. A. El-Baz, A. Soliman, P. McClure, G. Gimel'farb, M. A. El-Ghar, and R. Falk, "Early assessment of malignant lung nodules based on the spatial analysis of detected lung nodules," in *Proceedings of IEEE International Symposium on Biomedical Imaging: From Nano to Macro, (ISBI'12)*, IEEE, 2012, pp. 1463–1466.

120. A. El-Baz, S. E. Yuksel, S. Elshazly, and A. A. Farag, "Non-rigid registration techniques for automatic follow-up of lung nodules," in *Proceedings of Computer Assisted Radiology and Surgery, (CARS'05)*, vol. 1281, Elsevier, 2005, pp. 1115–1120.

121. A. S. El-Baz and J. S. Suri, *Lung Imaging and Computer Aided Diagnosis*. CRC Press, 2011.

122. A. Soliman, F. Khalifa, N. Dunlap, B. Wang, M. El-Ghar, and A. El-Baz, "An iso-surfaces based local deformation handling framework of lung tissues," in *2016 IEEE 13th International Symposium on Biomedical Imaging (ISBI)*, IEEE, 2016, pp. 1253–1259.

123. A. Soliman, F. Khalifa, A. Shaffie, N. Dunlap, B. Wang, A. Elmaghraby, and A. El-Baz, "Detection of lung injury using 4D-CT chest images," in *2016 IEEE 13th International Symposium on Biomedical Imaging (ISBI)*, IEEE, 2016, pp. 1274–1277.

124. A. Soliman, F. Khalifa, A. Shaffie, N. Dunlap, B. Wang, A. Elmaghraby, G. Gimel'farb, M. Ghazal, and A. El-Baz, "A comprehensive framework for early assessment of lung injury," in *2017 IEEE International Conference on Image Processing (ICIP)*, IEEE, 2017, pp. 3275–3279.

125. A. Shaffie, A. Soliman, M. Ghazal, F. Taher, N. Dunlap, B. Wang, A. Elmaghraby, G. Gimel'farb, and A. El-Baz, "A new framework for incorporating appearance and shape features of lung nodules for precise diagnosis of lung cancer," in *2017 IEEE International Conference on Image Processing (ICIP)*, IEEE, 2017, pp. 1372–1376.

126. A. Soliman, F. Khalifa, A. Shaffie, N. Liu, N. Dunlap, B. Wang, A. Elmaghraby, G. Gimel'farb, and A. El-Baz, "Image-based cad system for accurate identification of lung injury," in *2016 IEEE International Conference on Image Processing (ICIP)*, IEEE, 2016, pp. 121–125.

127. B. Dombroski, M. Nitzken, A. Elnakib, F. Khalifa, A. El-Baz, and M. F. Casanova, "Cortical surface complexity in a population-based normative sample," *Translational Neuroscience*, vol. 5, no. 1, pp. 17–24, 2014.

128. A. El-Baz, M. Casanova, G. Gimel'farb, M. Mott, and A. Switala, "An MRI-based diagnostic framework for early diagnosis of dyslexia," *International Journal of Computer Assisted Radiology and Surgery*, vol. 3, no. 3–4, pp. 181–189, 2008.

129. A. El-Baz, M. Casanova, G. Gimel'farb, M. Mott, A. Switala, E. Vanbogaert, and R. McCracken, "A new CAD system for early diagnosis of dyslexic brains," in *Proceedings of International Conference on Image Processing (ICIP'2008)*, IEEE, 2008, pp. 1820–1823.

130. A. El-Baz, M. F. Casanova, G. Gimel'farb, M. Mott, and A. E. Switwala, "A new image analysis approach for automatic classification of autistic brains," in *Proceedings of IEEE International Symposium on Biomedical Imaging: From Nano to Macro (ISBI'2007)*, IEEE, 2007, pp. 352–355.

131. A. El-Baz, A. Elnakib, F. Khalifa, M. A. El-Ghar, P. McClure, A. Soliman, and G. Gimel'farb, "Precise segmentation of 3-D magnetic resonance angiography," *IEEE Transactions on Biomedical Engineering*, vol. 59, no. 7, pp. 2019–2029, 2012.

132. A. El-Baz, A. Farag, G. Gimel'farb, M. A. El-Ghar, and T. Eldiasty, "Probabilistic modeling of blood vessels for segmenting MRA images," in *18th International Conference on Pattern Recognition (ICPR'06)*, vol. 3, IEEE, 2006, pp. 917–920.

133. A. El-Baz, A. A. Farag, G. Gimelfarb, M. A. El-Ghar, and T. Eldiasty, "A new adaptive probabilistic model of blood vessels for segmenting MRA images," in *Medical Image Computing and Computer-Assisted Intervention–MICCAI 2006*, vol. 4191, Springer, 2006, pp. 799–806.

134. A. El-Baz, A. A. Farag, G. Gimelfarb, and S. G. Hushek, "Automatic cerebrovascular segmentation by accurate probabilistic modeling of TOF-MRA images," in *Medical Image Computing and Computer-Assisted Intervention–MICCAI 2005*, Springer, 2005, pp. 34–42.

135. A. El-Baz, A. Farag, A. Elnakib, M. F. Casanova, G. Gimel'farb, A. E. Switala, D. Jordan, and S. Rainey, "Accurate automated detection of autism related corpus callosum abnormalities," *Journal of Medical Systems*, vol. 35, no. 5, pp. 929–939, 2011.

136. A. El-Baz, A. Farag, and G. Gimelfarb, "Cerebrovascular segmentation by accurate probabilistic modeling of TOF-MRA images," in *Image Analysis*, vol. 3540, Springer, 2005, pp. 1128–1137.

137. A. El-Baz, G. Gimelfarb, R. Falk, M. A. El-Ghar, V. Kumar, and D. Heredia, "A novel 3D joint Markov-gibbs model for extracting blood vessels from PC–MRA images," in *Medical Image Computing and Computer-Assisted Intervention–MICCAI 2009*, vol. 5762, Springer, 2009, pp. 943–950.

138. A. Elnakib, A. El-Baz, M. F. Casanova, G. Gimel'farb, and A. E. Switala, "Image-based detection of corpus callosum variability for more accurate discrimination between dyslexic and normal brains," in *Proc. IEEE International Symposium on Biomedical Imaging: From Nano to Macro (ISBI'2010)*, IEEE, 2010, pp. 109–112.

139. A. Elnakib, M. F. Casanova, G. Gimel'farb, A. E. Switala, and A. El-Baz, "Autism diagnostics by centerline-based shape analysis of the corpus callosum," in *Proceedings of IEEE International Symposium on Biomedical Imaging: From Nano to Macro (ISBI'2011)*, IEEE, 2011, pp. 1843–1846.

140. A. Elnakib, M. Nitzken, M. Casanova, H. Park, G. Gimel'farb, and A. El-Baz, "Quantification of age-related brain cortex change using 3D shape analysis," in *2012 21st International Conference on Pattern Recognition (ICPR)*, IEEE, 2012, pp. 41–44.

141. M. Mostapha, A. Soliman, F. Khalifa, A. Elnakib, A. Alansary, M. Nitzken, M. F. Casanova, and A. El-Baz, "A statistical framework for the classification of infant dt images," in *2014 IEEE International Conference on Image Processing (ICIP)*, IEEE, 2014, pp. 2222–2226.

142. M. Nitzken, M. Casanova, G. Gimel'farb, A. Elnakib, F. Khalifa, A. Switala, and A. El-Baz, "3D shape analysis of the brain cortex with application to dyslexia," in *2011 18th IEEE International Conference on Image Processing (ICIP)*, Brussels, Belgium, IEEE, September 2011, pp. 2657–2660, (Selected for oral presentation. Oral acceptance rate is 10 percent and the overall acceptance rate is 35 percent).

143. F. E.-Z. A. El-Gamal, M. M. Elmogy, M. Ghazal, A. Atwan, G. N. Barnes, M. F. Casanova, R. Keynton, and A. S. El-Baz, "A novel cad system for local and global early diagnosis of alzheimer's disease based on pib-pet scans," in *2017 IEEE International Conference on Image Processing (ICIP)*, IEEE, 2017, pp. 3270–3274.

144. M. Ismail, A. Soliman, M. Ghazal, A. E. Switala, G. Gimelfarb, G. N. Barnes, A. Khalil, and A. El-Baz, "A fast stochastic framework for automatic mr brain images segmentation," *PloS one*, vol. 12, no. 11, e0187391, 2017.

145. M. M. Ismail, R. S. Keynton, M. M. Mostapha, A. H. ElTanboly, M. F. Casanova, G. L. Gimel'farb, and A. El-Baz, "Studying autism spectrum disorder with structural and diffusion magnetic resonance imaging: a survey," *Frontiers in Human Neuroscience*, vol. 10, p. 211, 2016.

146. A. Alansary, M. Ismail, A. Soliman, F. Khalifa, M. Nitzken, A. Elnakib, M. Mostapha, A. Black, K. Stinebruner, M. F. Casanova et al., "Infant brain extraction in T1-weighted MR images using BET and refinement using LCDG and MGRF models," *IEEE Journal of Biomedical and Health Informatics*, vol. 20, no. 3, pp. 925–935, 2016.

147. M. Ismail, A. Soliman, A. ElTanboly, A. Switala, M. Mahmoud, F. Khalifa, G. Gimel'farb, M. F. Casanova, R. Keynton, and A. El-Baz, "Detection of white matter abnormalities in MR brain images for diagnosis of autism in children," in *2016 IEEE 13th International Symposium on Biomedical Imaging*, IEEE, 2016, pp. 6–9.

148. M. Ismail, M. Mostapha, A. Soliman, M. Nitzken, F. Khalifa, A. Elnakib, G. Gimel'farb, M. Casanova, and A. El-Baz, "Segmentation of infant brain mr images based on adaptive shape prior and higher-order MGRF," in *2015 IEEE International Conference on Image Processing*, 2015, pp. 4327–4331.

149. E. H. Asl, M. Ghazal, A. Mahmoud, A. Aslantas, A. Shalaby, M. Casanova, G. Barnes, G. Gimelfarb, R. Keynton, and A. El-Baz, "Alzheimers disease diagnostics by a 3D deeply supervised adaptable convolutional network," *Frontiers in Bioscience (Landmark Edition)*, vol. 23, pp. 584–596, 2018.

150. A. Mahmoud, A. El-Barkouky, H. Farag, J. Graham, and A. Farag, "A non-invasive method for measuring blood flow rate in superficial veins from a single thermal image," in *Proceedings of the IEEE Conference on Computer Vision and Pattern Recognition Workshops*, 2013, pp. 354–359.

151. A. El-baz, A. Shalaby, F. Taher, M. El-Baz, M. Ghazal, M. A. El-Ghar, A. Takieldeen, and J. Suri, "Probabilistic modeling of blood vessels for segmenting magnetic resonance angiography images," *Medical Research Archives*, vol. 5, no. 3, 2017.
152. A. S. Chowdhury, A. K. Rudra, M. Sen, A. Elnakib, and A. El-Baz, "Cerebral white matter segmentation from MRI using probabilistic graph cuts and geometric shape priors." in *ICIP*, 2010, pp. 3649–3652.
153. Y. Gebru, G. Giridharan, M. Ghazal, A. Mahmoud, A. Shalaby, and A. El-Baz, "Detection of cerebrovascular changes using magnetic resonance angiography," in *Cardiovascular Imaging and Image Analysis*, A. El-Baz and J. S. Suri, Eds. CRC Press, 2018, pp. 1–22.
154. A. Mahmoud, A. Shalaby, F. Taher, M. El-Baz, J. S. Suri, and A. El-Baz, "Vascular tree segmentation from different image modalities," in *Cardiovascular Imaging and Image Analysis*, A. El-Baz and J. S. Suri, Eds. CRC Press, 2018, pp. 43–70.
155. F. Taher, A. Mahmoud, A. Shalaby, and A. El-Baz, "A review on the cerebrovascular segmentation methods," in *2018 IEEE International Symposium on Signal Processing and Information Technology (ISSPIT)*, IEEE, 2018, pp. 359–364.
156. H. Kandil, A. Soliman, L. Fraiwan, A. Shalaby, A. Mahmoud, A. ElTanboly, A. Elmaghraby, G. Giridharan, and A. El-Baz, "A novel MRA framework based on integrated global and local analysis for accurate segmentation of the cerebral vascular system," in *2018 IEEE 15th International Symposium on Biomedical Imaging (ISBI 2018)*, IEEE, 2018, pp. 1365–1368.

Chapter 13

Computer Aided Diagnosis System for Early Detection of Diabetic Retinopathy Using OCT Images

Ahmed ElTanboly, Ahmed Shalaby, Ali Mahmoud, Mohammed Ghazal, Andrew Switala, Fatma Taher, Jasjit S. Suri, Robert Keynton, and Ayman El-Baz

13.1 Introduction

A majority of ophthalmologists depend on visual interpretation for the identification of disease types. However, inaccurate diagnosis will affect the treatment procedure which may lead to fatal results. Hence, there is a need for a computer automated diagnosing (CAD) system that yields highly accurate results. Optical coherence tomography (OCT) has become a powerful modality for the non-invasive diagnosis of various retinal abnormalities such as glaucoma, diabetic macular edema, and macular degeneration. The problem with diabetic retinopathy (DR) is that the patient is not aware of the disease until the changes in the retina have progressed to a level at which treatment tends to be less effective. Therefore, automated early detection could limit the severity of the disease and assist ophthalmologists in investigating and treating it more efficiently, which is the main goal of this chapter. One of the challenges for OCT is segmentation errors of the retinal layers, which is resolved as a second goal of this chapter.

Mizutani et al. [1] investigated a computerized method for the detection of micro aneurysms on retinal fundus images that are considered to be early signs of DR. His scheme was developed by using the training cases, and when the method was evaluated, the sensitivity for detecting micro aneurysms was 65% at 27 false positives per image. Jaafar et al. [2] suggested an automated method for the detection of hard and soft exudates in fundus images as the earliest signs of diabetic retinopathy; its success is subjective to the existence of those candidates. Pachiyappan et al. [3] describe a system for detecting of the macular abnormalities caused due to DR by applying morphological operations, filters, and thresholds on the fundus images of the patient.

Most CAD systems for early DR detection being introduced in the literature have been proposed from fundus images. Fundus photography uses the same concept of the indirect ophthalmoscope for a wide view of the retina. One of the reasons fundus pictures are more common in CAD systems is that they can give a good presentation of systemic diseases. However, one of its crucial drawbacks is it gives pictures in 2D with no appreciation for depth. To the best of our knowledge, there are no CAD systems in the literature that aim at early detection of DR using OCT scans, and we are the first group proposing such a CAD system.

Since one key component of the CAD system is segmentation of the retinal layers from OCT images, we briefly overview its work in the literature. An automated approach was proposed by Rossant et al. [4] to segment eight layers using active contours, k-means, and random Markov fields and modeled the approximated parallelism between layers based on Kalman filter. The method performed well but failed for blurred images. Yaz et al. [5] presented a semi-automated approach to extract nine layers from OCT images using Chan-Vese's energy-minimizing active contour without edges model along with shape priors. Their method, however, required user initialization and was never tested on human retinas nor on diseased cases. Kafieh et al. [6] used graph-based diffusion maps to segment the intraretinal layers in OCT scans from normal controls and glaucoma patients. Ehnes et al. [7] developed a graph-based algorithm which could segment up to 11 layers yet worked only with high-contrast images.

In addition to the limitations of DR CAD systems discussed above, OCT layer segmentation in the literature also suffers from some limitations that can be summarized as follows: (i) most of the proposed approaches achieve good accuracy with OCT images that have a high signal to noise ratio (SNR), whereas they fail with those of low SNR; (ii) most of the proposed works in the literature have been tested on normal OCT images yet not on diseased cases; and finally, (iii) the majority of the previous work was able to segment up to eight layers, while methods that segmented more [7] worked on high-contrast normal images.

To overcome the aforementioned limitations, we propose an innovative CAD system for early detection of DR. The algorithm starts with segmenting 12 distinct retinal layers, followed by extracting three discriminant features that are able to distinguish normal from DR images.

13.2   Methods

In this chapter, a new automated non-invasive framework for early diagnosis of DR from OCT images is developed. Figure 13.1 summarizes the main steps of the proposed framework. It performs sequentially three steps. First, a novel segmentation method of 12 distinct retinal layers is adopted using a novel joint model that combines shape, intensity, and spatial information. Second, three global features are extracted based on curvature, reflectivity, and thickness of layers exhibiting statistical significance. Finally, a Deep Fusion Classification Network (DFCN) is used to classify the test subject as normal or DR. *To our knowledge, and in addition to being the first group segmenting 12 distinct layers of the retina, we are the first group to propose a CAD system for early detection of DR from OCT scans.* Details of the CAD system are discussed in the following sections.

13.2.1   OCT Layers Segmentation

Let $g = \{g(x) : x \in R; g(x) \in Q\}$ and $m = \{l(x) : x \in R; l(x) \in L\}$ be a grayscale image taking values from Q, i.e., g: R → Q, with the associated region map taking values from L, i.e., m: R → L, respectively. R denotes a finite arithmetic lattice, Q is a finite set of integer gray values, and L is a set of region labels. An input OCT image, g, co-aligned to the training database, and its map, m, are described with a joint probability model:

$$P(g,m) = P(g \mid m)P(m) \tag{13.1}$$

which combines a conditional distribution of the images given the map $P(g|m)$, and an unconditional probability distribution of maps $P(m) = P_{sp}(m)P_V(m)$. Here, $P_{sp}(m)$ denotes a weighted shape prior, and $P_V(m)$ is a Gibbs probability distribution with potentials V, that specifies a MGRF model.

1. *Adaptive shape model* $P_{sp}(m)$: The shape model is constructed using 12 OCT scans (six men, six women), selected to capture the biological variability of the whole data set. Their "ground truth" segmentations were delineated under the supervision of retina specialists. Figure 13.2 represents a manually segmented retina with 12 distinct layers. Using one optimal scan as a reference (no tilt, centrally

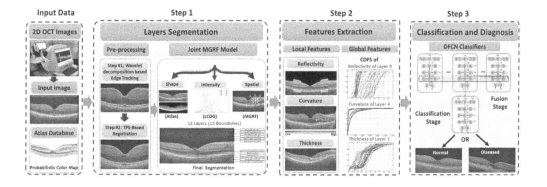

FIGURE 13.1:   Illustration of the basic steps of the proposed CAD system framework.

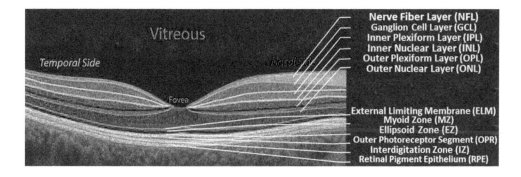

FIGURE 13.2:   A typical OCT scan of a normal subject showing the 12 distinct layers.

located fovea), the others were co-registered using a thin plate spline (TPS) [8]. The same deformations were applied to their respective ground truth segmentations, then averaged to produce a probabilistic shape prior of the typical retina. The image to be segmented must be first aligned to the shape database. In this chapter, a new image alignment approach is proposed that integrates TPS with multi-resolution edge tracking method that identifies control points for initializing the alignment process. First, the "à trous" algorithm [9] was used to construct an undecimated wavelet decomposition of each scan. The retina takes on a three-banded appearance, with two hyperreflective bands separated by a hyporeflective band corresponding roughly to layers from ONL to MZ. Contours following the gradient maxima of this wavelet component provided initial estimates of the vitreous/NFL, MZ/EZ, and RPE/choroid boundaries. The fourth gradient maximum could estimate the OPL/ONL boundary. The foveal peak was identified as the point of closest approach of the vitreous/NFL and MZ/EZ contours. Control points were then located on these boundaries at the foveal peak and at uniform intervals therefrom. Finally, the optimized TPS was employed in order to align the input image to the shape database using the control points identified.

2. *First-order intensity model $P(g|m)$*: In order to make the segmentation process adaptive and not biased to only the shape information, we model the empirical gray level distribution of the OCT images. The first-order visual appearance of each label of the image is modeled by separating a mixed distribution of pixel intensities into individual components associated with the dominant modes of the mixture. The latter is approximated using the LCDG approach in [10], which employs positive and negative Gaussian components and is based on a modified version of the classical expectation maximization (EM) algorithm. For further details on the modified EM algorithm, please refer to [10].

3. *Second-order MGRF model $P_V(m)$*: In order to improve the spatial homogeneity of the segmentation, the MGRF Potts model that accounts for spatial information was incorporated with the shape and intensity information [11]. This model is identified using the nearest pixels' 8-neighborhood v_s and analytical bi-valued Gibbs potentials as:

$$P(m) \propto \exp\left(\sum_{(x,y)\in R} \sum_{(\xi,\zeta)\in v_s} V(l_{x,y}, l_{x+\xi, y+\zeta}) \right) \qquad (13.2)$$

where V is the bi-value Gibbs potential, that depends on the equality of the nearest pair of labels. Complete segmentation steps of the proposed framework are summarized in Algorithm 1 in the supplementary materials.

13.2.2 Feature Extraction

Three distinct retinal features are extracted from the segmented OCT scans. The first feature is the "reflectivity" of the retinal layers, that was obtained from two regions per scan, comprising the thickest portions of the retina on the nasal and temporal sides of the foveal peak. Mean reflectivity is expressed on a normalized scale, calibrated such that the formed vitreous has a mean value of 0 normalized reflectivity scale (NRS), and the retinal pigment epithelium has a mean value of 1000 NRS. The average gray level within a segment was calculated using Huber's M-estimate, which is resistant to outlying values that may be present, such as very bright pixels in the innermost segment that properly belong to the internal limiting membrane and not the NFL. The second feature is "curvature" of the retinal layers, which calculates the curvature values for each point across the layer. First, a locally weighted polynomial is applied for smoothing the surface, then Menger curvature is calculated for each point. The third and last feature is the "thickness" of the retinal layers, which uses Laplace's equation [12] to calculate the streamlines between the two surfaces for each retinal layer.

For each subject, these three features are described as a whole with a cumulative probability distribution function (CDF) of the extracted retina layers. The CDFs are considered global discriminatory characteristics, being able to distinguish between normal and DR cases. In our system, the CDFs for a training set of the OCT images are used for deep learning of a multistage classifier with a stack of nonnegatively constrained autoencoders (SNCAE). Examples for CDFs of significant layers are given in the supplementary materials.

It is worth noting that more conventional classification techniques, employing directly the pixelwise values of curvature, reflectivity, and thickness as discriminative features, encounter at least two serious difficulties. Various input data sizes for each subject require unification by either data truncation for large volumes or zero padding for small ones. Both ways may decrease the accuracy of the classification. In addition, large data sizes lead to considerable time expenditures for training and classification. Contrastingly, our classifier exploits only the 100-component CDFs to describe the entire raw data estimated at retinal layer. This fixed data size helps overcome the above challenges.

13.2.3   Deep Fusion Classification Network

In this step, after segmenting the 12 retinal layers and extracting their global discriminatory features (the CDFs of the curvature, reflectivity, and thickness), our CAD system classifies normal and DR subjects. Since these data are considered to be huge, we suggest using a deep learning network that has the ability to learn these features and fuse them together. To learn characteristics of both normal and DR subjects, CDFs were calculated for each feature and fed into the proposed network. To build the classification model, a deep neural network with a two-stage structure of stacked autoencoders (SAE) was employed. *The first stage* is composed of several deep networks built with an SAE and output layer of softmax regression for each input feature (one SAE for each feature per significant layer, as shown in Figure 13.3). Each SAE compresses its input feature (u_f) to capture most prominent variations and is built separately by greedy unsupervised pre-training [13]. The softmax output layer facilitates the subsequent supervised back-propagation-based fine-tuning of the entire classifier by minimizing the total loss (negative log-likelihood) for given training labeled data. Using the autoencoder with a non-negativity constraint NCAE [14] yields both more reasonable data codes (features) during its unsupervised pre-training and better classification performance after the supervised refinement. The SNCAE of f-th input feature converts an n-dimensional input column vector $u_f = [u_{1:f}, ..., u_{n:f}]^T$, where T denotes transposition, into an s-dimensional column vector $h_f = [h_{1:f}, ..., h_{s:f}]^T$ of hidden codes. More details about the SNCAE are available in supplementary materials.

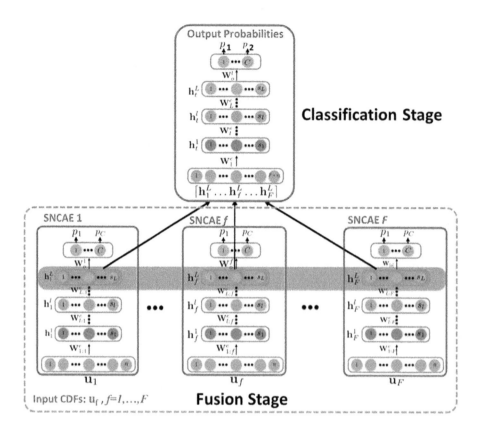

FIGURE 13.3:   Structure of DFCN.

After separately pre-training and supervised fine-tuning to each SNCAE of each u_f, in *the second stage*, the top-most hidden activators $h_f^{[L]}$ of each SNCAE; $f=1, \ldots$, are extracted and concatenated, as feature fusion, to capture the discriminative information within each u_f using class labels. Finally, a SNCAE-based classifier is trained using the fused feature $u_t = \left[h_f^{[1]}, \ldots, h_F^{[L]} \right]$ with layer-wise pre-training and fine-tuning for accurate classification. Since the final decision is taken based on $h_t^{[L]}$ extracted from u_t, and subsequently from lowest input features u_f, the developed network is called Deep Fusion Classification Network.

13.3   Experimental Results and Conclusions

In order to test and validate the proposed segmentation method, spectral domain OCT scans (Zeiss Cirrus HD-OCT 5000) were prospectively collected from 52 subjects (26 normal, 26 diseased) aged 40–79 years. Subjects with high myopia ($\leq$ −6.0 diopters) and tilted OCT were excluded. The proposed segmentation approach was validated using ground truth for subjects, which was created by manual delineations of retina layers with the aid of retina specialists.

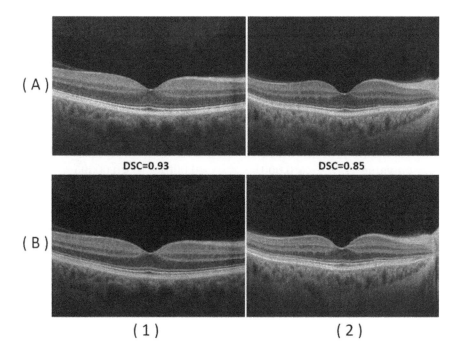

FIGURE 13.4: Segmentation results for different OCT images in row (A) for normal (1) and DR (2) cases. Results of the proposed approach are displayed in row (B). The DSC score is displayed above each result.

Figure 13.4 shows the segmentation of 12 distinct retinal layers for two different examples. More segmentation results are given in the supplementary materials. In addition to the visual results in this figure, the robustness and accuracy of our approach are evaluated using both agreement coefficient (AC) and Dice similarity coefficient (DSC) metrics and the average deviation (AD) distance metric comparing our segmentation with the ground truth. Mean boundary error was 6.87 micrometers from ground truth, averaged across all 13 boundaries. The inner (vitreous) boundary of the retina was placed most accurately, with 2.78 μm mean error. The worst performance was on the outer (choroid) boundary, with 11.6 μm, whereas only the RPE/choroid boundary was reliably detected by the other approach [15]. Table 13.1 summarizes the quantitative comparison of our segmentation method and the other method versus the ground truth, based on the three evaluation metrics for all subjects. Statistical analysis using paired t-test demonstrates a significant difference in terms of all three metrics of our method over [15], as confirmed by p-values <0.05. This analysis clearly demonstrates the promise of the developed approach for the segmentation of the OCT scans.

After segmenting the 12 retinal layers and extracting their features, statistical analysis was conducted on those features (curvature, reflectivity, and thickness) for all 12 layers extracted from available subjects. The purpose of this analysis was to find out whether these features are significant to discriminate between normal and diabetic subjects. The statistical results are shown in Table 13.2. According to the unpaired t-test results, the curvature of INL, the reflectivity of MZ, and the thickness of NFL show statistically significant differences between normal and diseased cases (p-values

TABLE 13.1: Comparative Segmentation Accuracy of the Proposed Segmentation and [15] Using (DSC), (AC), and (AD) Metrics. Values Are Represented as Mean $\pm$ Standard Deviation

	Evaluation Metric		
	DSC	AC,%	AD, μm
Our	0.76 ± 0.16	73.2 ± 4.5	6.87 ± 2.8
Other [15]	0.41 ± 0.263	2.25 ± 9.7	15.1 ± 8.6
p-value	<0.0001	<0.0001	<0.00395

<0.05). For now, these results encouraged us to explore the classification potential of those features on three layers only.

Finally, a deep fusion classification network (DFCN) is used to classify the test subject as normal or DR based on the extracted CDFs of the most significant features (the curvature of layer INL, the reflectivity of layer MZ, and thickness of layer NFL). We trained our classifier by 52 OCT datasets (26 normal and 26 DR).

A quantitative comparison between the proposed classifier and other known classifiers (K-Star, K-nearest neighbor, and Random Forest classifiers implemented in Weka toolbox) [26] is summarized in Table 13.3. We perform a leave-one-out cross-validation test with the whole datasets. Our DFCN classifier achieves an overall accuracy of 100% (with AUC of 0.976) for all testing datasets.

In summary, this book chapter has proposed a novel computer aided diagnosis (CAD) system for early detection of diabetic retinopathy (DR) using OCT images. The framework includes a new approach for the segmentation of the 12 distinct retinal layers. A multistage deep fusion classification network (DFCN), trained by a stacked non-negativity constraint autoencoder (SNCAE), is used to classify the subject as normal or diabetic based on three discriminant features (curvature, reflectivity, and thickness) of the statistically significant layers of the retina. Applications of the proposed approach yield promising results that could, in the near future, replace the use of current technologies for early detection of DR.

This work could also be applied to various other applications in medical imaging, such as the kidney, the heart, the prostate, and the lung, as well as several non-medical applications [16–19].

One application is renal transplant functional assessment, especially with developing noninvasive CAD systems for renal transplant function assessment, utilizing different image modalities (e.g., ultrasound, computed tomography (CT), MRI, etc.). Accurate assessment of renal transplant function is critically important for graft survival. Although transplantation can improve a patient's wellbeing, there is a potential post-transplantation risk of kidney dysfunction that, if not treated in a timely manner, can lead to the loss of the entire graft and even patient death. In particular, dynamic and diffusion MRI-based systems have been clinically used to assess transplanted kidneys with the advantage of providing information on each kidney separately. For more details about renal transplant functional assessment, please read [20–37, 37–47].

The heart is also an important application for this work. The clinical assessment of myocardial perfusion plays a major role in the diagnosis, management, and prognosis of ischemic heart disease patients. Thus, there have been ongoing efforts to develop

TABLE 13.2: The Statistical Analysis Results. All Values Are Represented as Mean±StD

Retinal Layers	Reflectivity		Curvature, mm^{-1}		Thickness, μm	
	Normal	Diabetic	Normal	Diabetic	Normal	Diabetic
(NFL)	667.1±147	665.1±165	0.704±1.72	0.738±2.03	19.13±12.98	21.03±12.17
(GCL)	584.1±103	589.6±111	0.891±3.68	1.02±4.63	34.73±18.17	31.51±17.93
(IPL)	637.5±96.8	630.7±117	0.861±3.71	1.15±4.74	33.72±15.29	33.62±16.19
(INL)	479.1±95.8	463.8±101	1.19±4.85	2.17±6.18	32.01±15.52	27.91±15.09
(OPL)	546.3±93.7	513.8±108	1.17±4.86	1.75±5.62	27.20±14.31	30.63±13.94
(ONL)	302.9±74.3	292.0±108	0.842±3.13	1.72±4.95	67.07±23.62	61.60±22.62
(ELM)	396.3±135	465.4±236	0.134±0.247	0.102±0.164	13.78±3.71	14.71±2.99
(MZ)	414.0±160	512.4±309	0.113±0.196	0.101±0.168	15.51±4.34	14.30±3.96
(EZ)	996.1±189	1038±281	0.115±0.202	0.101±0.168	14.55±3.66	14.97±3.05
(OPR)	822.9±209	883.2±271	0.126±0.249	0.111±0.198	14.85±5.54	15.10±5.51
(IZ)	1121±180	1164±191	0.083±0.138	0.061±0.093	14.31±3.81	15.56±3.10
(RPE)	1046±97.9	1041±135	0.077±0.165	0.057±0.106	23.48±6.39	23.34±6.75

TABLE 13.3: Classification Accuracy, Sensitivity, Specificity, and the Area under the ROC Curve (AUC) for our CAD System and Different Classifiers from Weka Tool

Classifier	Accuracy	Sensitivity	Specificity	AUC
SNCAE (Proposed)	100%	100%	100%	**0.976**
K* (K-Star)	95.2%	95.2%	95.3%	0.936
KNN-Classifier (IBK)	89.67%	89.6%	89.7%	0.897
Random Forest	84.7%	84.8%	84.6%	0.871

automated systems for accurate analysis of myocardial perfusion using first-pass images [12, 48–63].

Moreover, the work could be applied for prostate cancer which is the most common cancer in American men and its related mortality rate is the second after lung cancer. Fortunately, the mortality rate can be reduced if prostate cancer is detected in its early stages. Early detection enables physicians to treat prostate cancer before it develops to clinically significant disease [64–67].

Abnormalities of the lung could also be another promising area of research and a related application to this work. Radiation-induced lung injury is the main side effect of radiation therapy for lung cancer patients. Although higher radiation doses increase the radiation therapy effectiveness for tumor control, this can lead to lung injury as a greater quantity of normal lung tissues is included in the treated area. Almost one-third of patients who undergo radiation therapy develop lung injury following radiation treatment. The severity of radiation-induced lung injury ranges from ground-glass opacities and consolidation at the early phase to fibrosis and traction bronchiectasis in the late phase. Early detection of lung injury will thus help to improve management of the treatment [68–110].

This work can also be applied to other brain abnormalities, such as dyslexia, in addition to autism. Dyslexia is one of the most complicated developmental brain disorders that affect children's learning abilities. Dyslexia leads to the failure to develop age-appropriate reading skills in spite of normal intelligence level and adequate reading instructions. Neuropathological studies have revealed an abnormal anatomy of some structures, such as the corpus callosum, in dyslexic brains. There has been a lot of work in the literature that aims at developing CAD systems for diagnosing this disorder, along with other brain disorders [111–132].

For the vascular system [133], this work could also be applied for the extraction of blood vessels, e.g., from phase contrast (PC) magnetic resonance angiography (MRA). Accurate cerebrovascular segmentation using non-invasive MRA is crucial for the early diagnosis and timely treatment of intracranial vascular diseases [115, 116, 134–139].

References

1. A. Mizutani, C. Muramatsu, Y. Hatanaka, S. Suemori, T. Hara, and H. Fujita, "Automated microaneurysm detection method based on double ring filter in retinal fundus images," in *SPIE Medical Imaging*, 2009.

2. H. F. Jaafar, A. K. Nandi, and W. Al-Nuaimy, "Automated detection of exudates in retinal images using a split-and-merge algorithm," in *Signal Processing Conference*. 2010, pp. 1622–1626.

3. A. Pachiyappan, U. N. Das, T. V. Murthy, and R. Tatavarti, "Automated diagnosis of diabetic retinopathy and glaucoma using fundus and OCT images," *Lipids in Health and Disease*, *11*(1), p. 73, 2012.

4. F. Rossant, I. Ghorbel, I. Bloch, M. Paques, and S. Tick, "Automated segmentation of retinal layers in OCT imaging and derived ophthalmic measures," in *ISBI'09*, 2009, pp. 1370–1373.

5. A. Yazdanpanah, G. Hamarneh, B. R. Smith, and M. V. Sarunic, "Segmentation of intra-retinal layers from optical coherence tomography images using active contour approach," *IEEE Transactions on Medical Imaging*, *30*(2), pp. 484–496, 2011.

6. R. Kafieh, H. Rabbani, M. D. Abramoff, and M. Sonka, "Intra-retinal layer segmentation of 3D optical coherence tomography using coarse grained diffusion map," Medical Image Analysis, *17*(8), pp. 907–928, 2013.

7. A. Ehnes, Y. Wenner, C. Friedburg, M. N. Preising, W. Bowl, W. Sekundo, E. M. zu Bexten, K. Stieger, and B. Lorenz, "Optical coherence tomography (OCT) device independent intra-retinal layer segmentation," *Translational Vision Science & Technology*, *3*(1), 1, 2014.

8. J. Lim, and M. H. Yang, "A direct method for modeling non-rigid motion with thin plate spline," in *CVPR 2005*, 2005, pp. 1196–1202.

9. E. Lega, H. Scholl, J. M. Alimi, A. Bijaoui, and P. Bury, "A parallel algorithm for structure detection based on wavelet and segmentation analysis," *Parallel Computing*, *21*(2), pp. 265–285, 1995.

10. A. El-Baz, A. Elnakib, F. Khalifa, M. A. El-Ghar, P. McClure, A. Soliman, and G. Gimelrfarb, "Precise segmentation of 3-D magnetic resonance angiography," *IEEE Transactions on Biomedical Engineering*, *59*(7), pp. 2019–2029, 2012.

11. A. Alansary, M. Ismail, A. Soliman, F. Khalifa, M. Nitzken, A. Elnakib, M. Mostapha, A. Black, K. Stinebruner, M. F. Casanova, and J. M. Zurada, "Infant brain extraction in T1-weighted MR images using BET and refinement using LCDG and MGRF models," *IEEE Journal of Biomedical and Health Informatics*, *20*(3), pp. 925–935, 2015.

12. F. Khalifa, G. M. Beache, G. Gimel'farb, G. A. Giridharan, and A. El-Baz, "Accurate automatic analysis of cardiac cine images," *IEEE Transactions on Biomedical Engineering*, *59*(2), pp. 445–455, 2012.

13. Y. Bengio, P. Lamblin, D. Popovici, and H. Larochelle, "Greedy layer-wise training of deep networks," *Advances in Neural Information Processing Systems*, *19*, pp. 153–160, 2007.

14. E. Hosseini-Asl, J. M. Zurada, and O. Nasraoui, "Deep learning of part-based representation of data using sparse autoencoders with nonnegativity constraints," *IEEE Transactions on Neural Networks and Learning Systems*, *27*(12), pp. 2486–2498, 2015.

15. S. J. Chiu, X. T. Li, P. Nicholas, C. A. Toth, J. A. Izatt, and S. Farsiu, "Automatic segmentation of seven retinal layers in SDOCT images congruent with expert manual segmentation," *Optics Express*, *18*(18), pp. 19413–19428, 2010.

16. A. H. Mahmoud, "Utilizing radiation for smart robotic applications using visible, thermal, and polarization images," Ph.D. dissertation, University of Louisville, 2014.

17. A. Mahmoud, A. El-Barkouky, J. Graham, and A. Farag, "Pedestrian detection using mixed partial derivative based histogram of oriented gradients," in *2014 IEEE International Conference on Image Processing (ICIP)*. IEEE, 2014, pp. 2334–2337.

18. A. El-Barkouky, A. Mahmoud, J. Graham, and A. Farag, "An interactive educational drawing system using a humanoid robot and light polarization," in *2013 IEEE International Conference on Image Processing*. IEEE, 2013, pp. 3407–3411.

19. A. H. Mahmoud, M. T. El-Melegy, and A. A. Farag, "Direct method for shape recovery from polarization and shading," in *2012 19th IEEE International Conference on Image Processing*. IEEE, 2012, pp. 1769–1772.

20. A. M. Ali, A. A. Farag, and A. El-Baz, "Graph cuts framework for kidney segmentation with prior shape constraints," in *Proceedings of International Conference on Medical Image Computing and Computer-Assisted Intervention (MICCAI'07)*, Volume *1*. Brisbane, Australia, October 29–November 2, 2007, pp. 384–392.

21. A. S. Chowdhury, R. Roy, S. Bose, F. K. A. Elnakib, and A. El-Baz, "Non-rigid bio-medical image registration using graph cuts with a novel data term," in *Proceedings of IEEE International Symposium on Biomedical Imaging: From Nano to Macro (ISBI'12)*, Barcelona, Spain, May 2–5, 2012, pp. 446–449.

22. A. El-Baz, A. A. Farag, S. E. Yuksel, M. E. El-Ghar, T. A. Eldiasty, and M. A. Ghoneim, "Application of deformable models for the detection of acute renal rejection," in *Deformable Models*. Springer, New York, NY, 2007, pp. 293–333.

23. A. El-Baz, A. Farag, R. Fahmi, S. Yuksel, M. A. El-Ghar, and T. Eldiasty, "Image analysis of renal DCE MRI for the detection of acute renal rejection," in *Proceedings of IAPR International Conference on Pattern Recognition (ICPR'06)*. Hong Kong, August 20–24, 2006, pp. 822–825.

24. A. El-Baz, A. Farag, R. Fahmi, S. Yuksel, W. Miller, M. A. El-Ghar, T. El-Diasty, and M. Ghoneim, "A new CAD system for the evaluation of kidney diseases using DCE-MRI," in *Proceedings of International Conference on Medical Image Computing and Computer-Assisted Intervention (MICCAI'08)*. Copenhagen, Denmark, October 1–6, 2006, pp. 446–453.

25. A. El-Baz, G. Gimel'farb, and M. A. El-Ghar, "A novel image analysis approach for accurate identification of acute renal rejection," in *Proceedings of IEEE International Conference on Image Processing (ICIP'08)*. San Diego, California, USA, October 12–15, 2008, pp. 1812–1815.

26. A. El-Baz, G. Gimel'farb, and M. A. El-Ghar, "Image analysis approach for identification of renal transplant rejection," in *Proceedings of IAPR International Conference on Pattern Recognition (ICPR'08)*. Tampa, Florida, USA, December 8–11, 2008, pp. 1–4.

27. A. El-Baz, G. Gimel'farb, and M. A. El-Ghar, "New motion correction models for automatic identification of renal transplant rejection," in *Proceedings of International Conference on Medical Image Computing and Computer-Assisted Intervention (MICCAI'07)*. Brisbane, Australia, October 29–November 2, 2007, pp. 235–243.

28. A. Farag, A. El-Baz, S. Yuksel, M. A. El-Ghar, and T. Eldiasty, "A framework for the detection of acute rejection with dynamic contrast enhanced magnetic resonance imaging," in *Proceedings of IEEE International Symposium on Biomedical Imaging: From Nano to Macro (ISBI'06)*, Arlington, Virginia, USA, April 6–9, 2006, pp. 418–421.

29. F. Khalifa, G. M. Beache, M. A. El-Ghar, T. El-Diasty, G. Gimel'farb, M. Kong, and A. El-Baz, "Dynamic contrast-enhanced MRI-based early detection of acute renal transplant rejection," *IEEE Transactions on Medical Imaging*, 32(10), pp. 1910–1927, 2013.

30. F. Khalifa, A. El-Baz, G. Gimel'farb, and M. A. El-Ghar, "Non-invasive image-based approach for early detection of acute renal rejection," in *Proceedings of International Conference Medical Image Computing and Computer-Assisted Intervention (MICCAI'10)*. Beijing, China, September 20–24, 2010, pp. 10–18.

31. F. Khalifa, A. El-Baz, G. Gimel'farb, R. Ouseph, and M. A. El-Ghar, "Shape-appearance guided level-set deformable model for image segmentation," in *Proceedings of IAPR International Conference on Pattern Recognition (ICPR'10)*. Istanbul, Turkey, August 23–26, 2010, pp. 4581–4584.

32. F. Khalifa, M. A. El-Ghar, B. Abdollahi, H. Frieboes, T. El-Diasty, and A. El-Baz, "A comprehensive non-invasive framework for automated evaluation of acute renal transplant rejection using DCE-MRI," *NMR in Biomedicine*, 26(11), pp. 1460–1470, 2013.

33. F. Khalifa, M. A. El-Ghar, B. Abdollahi, H. B. Frieboes, T. El-Diasty, and A. El-Baz, "Dynamic contrast-enhanced MRI-based early detection of acute renal transplant rejection," in *2014 Annual Scientific Meeting and Educational Course Brochure of the Society of Abdominal Radiology (SAR'14)*, Boca Raton, Florida, March 23–28, 2014, p. CID: 1855912.

34. F. Khalifa, A. Elnakib, G. M. Beache, G. Gimel'farb, M. A. El-Ghar, G. Sokhadze, S. Manning, P. McClure, and A. El-Baz, "3D kidney segmentation from CT images using a level set approach guided by a novel stochastic speed function," in *Proceedings of International Conference Medical Image Computing and Computer-Assisted Intervention (MICCAI'11)*. Toronto, Canada, September 18–22, 2011, pp. 587–594.

35. F. Khalifa, G. Gimel'farb, M. A. El-Ghar, G. Sokhadze, S. Manning, P. McClure, R. Ouseph, and A. El-Baz, "A new deformable model-based segmentation approach for accurate extraction of the kidney from abdominal CT images," in *Proceedings of IEEE International Conference on Image Processing (ICIP'11)*. Brussels, Belgium, September 11–14, 2011, pp. 3393–3396.

36. M. Mostapha, F. Khalifa, A. Alansary, A. Soliman, J. Suri, and A. El-Baz, "Computer-aided diagnosis systems for acute renal transplant rejection: Challenges and methodologies," in A. El-Baz, L. Saba, and J. Suri, eds. *Abdomen and Thoracic Imaging*. Springer, 2014, pp. 1–35.

37. M. Shehata, F. Khalifa, E. Hollis, A. Soliman, E. Hosseini-Asl, M. A. El-Ghar, M. El-Baz, A. C. Dwyer, A. El-Baz, and R. Keynton, "A new non-invasive approach for early classification of renal rejection types using diffusion-weighted MRI," in *2016 IEEE International Conference on Image Processing (ICIP)*. IEEE, 2016, pp. 136–140.

38. F. Khalifa, A. Soliman, A. Takieldeen, M. Shehata, M. Mostapha, A. Shaffie, R. Ouseph, A. Elmaghraby, and A. El-Baz, "Kidney segmentation from CT images using a 3D NMF-guided active contour model," in *2016 IEEE 13th International Symposium on Biomedical Imaging (ISBI)*. IEEE, 2016, pp. 432–435.

39. M. Shehata, F. Khalifa, A. Soliman, A. Takieldeen, M. A. El-Ghar, A. Shaffie, A. C. Dwyer, R. Ouseph, A. El-Baz, and R. Keynton, "3D diffusion MRI-based CAD system for early diagnosis of acute renal rejection," in *2016 IEEE 13th International Symposium on Biomedical Imaging (ISBI)*. IEEE, 2016, pp. 1177–1180.

40. M. Shehata, F. Khalifa, A. Soliman, R. Alrefai, M. A. El-Ghar, A. C. Dwyer, R. Ouseph, and A. El-Baz, "A level set-based framework for 3D kidney segmentation from diffusion MR images," in *2015 IEEE International Conference on Image Processing (ICIP)*. IEEE, 2015, pp. 4441–4445.

41. M. Shehata, F. Khalifa, A. Soliman, M. A. El-Ghar, A. C. Dwyer, G. Gimelfarb, R. Keynton, and A. El-Baz, "A promising non-invasive CAD system for kidney function assessment," in *International Conference on Medical Image Computing and Computer-Assisted Intervention*. Springer, 2016, pp. 613–621.

42. F. Khalifa, A. Soliman, A. Elmaghraby, G. Gimelfarb, and A. El-Baz, "3D kidney segmentation from abdominal images using spatial-appearance models," *Computational and Mathematical Methods in Medicine, 2017*, pp. 1–10, 2017.

43. E. Hollis, M. Shehata, F. Khalifa, M. A. El-Ghar, T. El-Diasty, and A. El-Baz, "Towards non-invasive diagnostic techniques for early detection of acute renal transplant rejection: A review," *The Egyptian Journal of Radiology and Nuclear Medicine, 48*(1), pp. 257–269, 2016.

44. M. Shehata, F. Khalifa, A. Soliman, M. A. El-Ghar, A. C. Dwyer, and A. El-Baz, "Assessment of renal transplant using image and clinical-based biomarkers," in *Proceedings of 13th Annual Scientific Meeting of American Society for Diagnostics and Interventional Nephrology (ASDIN'17)*, New Orleans, LA, USA, February 10–12, 2017, 2017.

45. M. Shehata, F. Khalifa, A. Soliman, M. A. El-Ghar, A. C. Dwyer, and A. El-Baz, "Early assessment of acute renal rejection," in *Proceedings of 12th Annual Scientific Meeting of American Society for Diagnostics and Interventional Nephrology (ASDIN'16)*, Pheonix, AZ, USA, February 19–21, 2016, 2017.

46. A. Eltanboly, M. Ghazal, H. Hajjdiab, A. Shalaby, A. Switala, A. Mahmoud, P. Sahoo, M. El-Azab, and A. El-Baz, "Level sets-based image segmentation approach using statistical shape priors," *Applied Mathematics and Computation, 340*, pp. 164–179, 2019.

47. M. Shehata, A. Mahmoud, A. Soliman, F. Khalifa, M. Ghazal, M. A. El-Ghar, M. El-Melegy, and A. El-Baz, "3D kidney segmentation from abdominal diffusion MRI using an appearance-guided deformable boundary," *PLoS One, 13*(7), p. e0200082, 2018.

48. F. Khalifa, G. Beache, A. El-Baz, and G. Gimel'farb, "Deformable model guided by stochastic speed with application in cine images segmentation," in *Proceedings of IEEE International Conference on Image Processing (ICIP'10)*. Hong Kong, September 26–29, 2010, pp. 1725–1728.

49. F. Khalifa, G. M. Beache, A. Elnakib, H. Sliman, G. Gimel'farb, K. C. Welch, and A. El-Baz, "A new shape-based framework for the left ventricle wall segmentation from cardiac first-pass perfusion MRI," in *Proceedings of IEEE International Symposium on Biomedical Imaging: From Nano to Macro (ISBI'13)*, San Francisco, CA, April 7–11, 2013, pp. 41–44.

50. F. Khalifa, G. M. Beache, A. Elnakib, H. Sliman, G. Gimel'farb, K. C. Welch, and A. El-Baz, "A new nonrigid registration framework for improved visualization of transmural perfusion gradients on cardiac first-pass perfusion MRI," in *Proceedings of IEEE International Symposium on Biomedical Imaging: From Nano to Macro (ISBI'12)*, Barcelona, Spain, May 2–5, 2012, pp. 828–831.

51. F. Khalifa, G. M. Beache, A. Firjani, K. C. Welch, G. Gimel'farb, and A. El-Baz, "A new non-rigid registration approach for motion correction of cardiac first-pass perfusion MRI," in *Proceedings of IEEE International Conference on Image Processing (ICIP'12)*. Lake Buena Vista, Florida, September 30–October 3, 2012, pp. 1665–1668.

52. F. Khalifa, G. M. Beache, G. Gimel'farb, and A. El-Baz, "A novel CAD system for analyzing cardiac first-pass MR images," in *Proceedings of IAPR International Conference on Pattern Recognition (ICPR'12)*. Tsukuba Science City, Japan, November 11–15, 2012, pp. 77–80.

53. F. Khalifa, G. M. Beache, G. Gimel'farb, and A. El-Baz, "A novel approach for accurate estimation of left ventricle global indexes from short-axis cine MRI," in *Proceedings of IEEE International Conference on Image Processing (ICIP'11)*. Brussels, Belgium, September 11–14, 2011, pp. 2645–2649.

54. F. Khalifa, G. M. Beache, G. Gimel'farb, G. A. Giridharan, and A. El-Baz, "A new image-based framework for analyzing cine images," in A. El-Baz, U. R. Acharya, M. Mirmedhdi, and J. S. Suri, eds. *Handbook of Multi Modality State-of-the-Art Medical Image Segmentation and Registration Methodologies*, Volume 2, Chapter 3. Springer, New York, 2011, pp. 69–98.

55. F. Khalifa, G. M. Beache, M. Nitzken, G. Gimel'farb, G. A. Giridharan, and A. El-Baz, "Automatic analysis of left ventricle wall thickness using short-axis cine CMR images," in *Proceedings of IEEE International Symposium on Biomedical Imaging: From Nano to Macro (ISBI'11)*, Chicago, Illinois, March 30–April 2, 2011, pp. 1306–1309.

56. M. Nitzken, G. Beache, A. Elnakib, F. Khalifa, G. Gimel'farb, and A. El-Baz, "Accurate modeling of tagged CMR 3D image appearance characteristics to improve cardiac cycle strain estimation," in *2012 19th IEEE International Conference on Image Processing (ICIP)*. Orlando, Florida, USA: IEEE, September, 2012, pp. 521–524.

57. M. Nitzken, G. Beache, A. Elnakib, F. Khalifa, G. Gimel'farb, and A. El-Baz, "Improving full-cardiac cycle strain estimation from tagged CMR by accurate modeling of 3D image appearance characteristics," in 2012 9th IEEE International Symposium on *Biomedical Imaging (ISBI)*. Barcelona, Spain: IEEE, May, 2012, pp. 462–465 (Selected for oral presentation).

58. M. J. Nitzken, A. S. El-Baz, and G. M. Beache, "Markov-Gibbs random field model for improved full-cardiac cycle strain estimation from tagged CMR," *Journal of Cardiovascular Magnetic Resonance, 14*(1), pp. 1–2, 2012.

59. H. Sliman, A. Elnakib, G. Beache, A. Elmaghraby, and A. El-Baz, "Assessment of myocardial function from cine cardiac MRI using a novel 4D tracking approach," *Journal of Computer Science and Systems Biology, 7*, pp. 169–173, 2014.

60. H. Sliman, A. Elnakib, G. M. Beache, A. Soliman, F. Khalifa, G. Gimel'farb, A. Elmaghraby, and A. El-Baz, "A novel 4D PDE-based approach for accurate assessment of myocardium function using cine cardiac magnetic resonance images," in *Proceedings of IEEE International Conference on Image Processing (ICIP'14)*. Paris, France, October 27–30, 2014, pp. 3537–3541.

61. H. Sliman, F. Khalifa, A. Elnakib, G. M. Beache, A. Elmaghraby, and A. El-Baz, "A new segmentation-based tracking framework for extracting the left ventricle cavity from cine cardiac MRI," in *Proceedings of IEEE International Conference on Image Processing (ICIP'13)*. Melbourne, Australia, September 15–18, 2013, pp. 685–689.

62. H. Sliman, F. Khalifa, A. Elnakib, A. Soliman, G. M. Beache, A. Elmaghraby, G. Gimel'farb, and A. El-Baz, "Myocardial borders segmentation from cine MR images using bi-directional coupled parametric deformable models," *Medical Physics, 40*(9), pp. 1–13, 2013.

63. H. Sliman, F. Khalifa, A. Elnakib, A. Soliman, G. M. Beache, G. Gimel'farb, A. Emam, A. Elmaghraby, and A. El-Baz, "Accurate segmentation framework for the left ventricle wall from cardiac cine MRI," in *Proceedings of International Symposium on Computational Models for Life Science (CMLS'13)*, Volume *1559*, Sydney, Australia, November 27–29, 2013, pp. 287–296.

64. I. Reda, M. Ghazal, A. Shalaby, M. Elmogy, A. AbouEl-Fetouh, B. O. Ayinde, M. AbouEl-Ghar, A. Elmaghraby, R. Keynton, and A. El-Baz, "A novel ADCS-based CNN classification system for precise diagnosis of prostate cancer," in *2018 24th International Conference on Pattern Recognition (ICPR)*. IEEE, 2018, pp. 3923–3928.

65. I. Reda, A. Khalil, M. Elmogy, A. Abou El-Fetouh, A. Shalaby, M. Abou El-Ghar, A. Elmaghraby, M. Ghazal, and A. El-Baz, "Deep learning role in early diagnosis of prostate cancer," *Technology in Cancer Research and Treatment*, *17*, p. 1533034618775530, 2018.

66. I. Reda, B. O. Ayinde, M. Elmogy, A. Shalaby, M. El-Melegy, M. A. El-Ghar, A. A. El-fetouh, M. Ghazal, and A. El-Baz, "A new CNN-based system for early diagnosis of prostate cancer," in *2018 IEEE 15th International Symposium on Biomedical Imaging (ISBI 2018)*. IEEE, 2018, pp. 207–210.

67. I. Reda, A. Shalaby, M. Elmogy, A. A. Elfotouh, F. Khalifa, M. A. El-Ghar, E. Hosseini-Asl, G. Gimel'farb, N. Werghi, and A. El-Baz, "A comprehensive non-invasive framework for diagnosing prostate cancer," *Computers in Biology and Medicine*, *81*, pp. 148–158, 2017.

68. B. Abdollahi, A. C. Civelek, X.-F. Li, J. Suri, and A. El-Baz, "PET/CT nodule segmentation and diagnosis: A survey," in L. Saba and J. S. Suri, eds. *Multi Detector CT Imaging*, Chapter 30. Taylor & Francis, 2014, pp. 639–651.

69. B. Abdollahi, A. El-Baz, and A. A. Amini, "A multi-scale non-linear vessel enhancement technique," in *2011 Annual International Conference of the IEEE Engineering in Medicine and Biology Society (EMBC)*. IEEE, 2011, pp. 3925–3929.

70. B. Abdollahi, A. Soliman, A. Civelek, X.-F. Li, G. Gimel'farb, and A. El-Baz, "A novel Gaussian scale space-based joint MGRF framework for precise lung segmentation," in *Proceedings of IEEE International Conference on Image Processing (ICIP'12)*. IEEE, 2012, pp. 2029–2032.

71. B. Abdollahi, A. Soliman, A. Civelek, X.-F. Li, G. Gimelfarb, and A. El-Baz, "A novel 3D joint MGRF framework for precise lung segmentation," in *Machine Learning in Medical Imaging*. Springer, 2012, pp. 86–93.

72. A. M. Ali, A. S. El-Baz, and A. A. Farag, "A novel framework for accurate lung segmentation using graph cuts," in *Proceedings of IEEE International Symposium on Biomedical Imaging: From Nano to Macro (ISBI'07)*. IEEE, 2007, pp. 908–911.

73. A. El-Baz, G. M. Beache, G. Gimel'farb, K. Suzuki, and K. Okada, "Lung imaging data analysis," *International Journal of Biomedical Imaging*, *2013*, pp. 1–2, 2013.

74. A. El-Baz, G. M. Beache, G. Gimel'farb, K. Suzuki, K. Okada, A. Elnakib, A. Soliman, and B. Abdollahi, "Computer-aided diagnosis systems for lung cancer: Challenges and methodologies," *International Journal of Biomedical Imaging*, *2013*, pp. 1–46, 2013.

75. A. El-Baz, A. Elnakib, M. Abou El-Ghar, G. Gimel'farb, R. Falk, and A. Farag, "Automatic detection of 2D and 3D lung nodules in chest spiral CT scans," *International Journal of Biomedical Imaging*, *2013*, pp. 1–11, 2013.

76. A. El-Baz, A. A. Farag, R. Falk, and R. La Rocca, "A unified approach for detection, visualization, and identification of lung abnormalities in chest spiral CT scans," in *International Congress Series*, Volume *1256*. Elsevier, 2003, pp. 998–1004.

77. A. El-Baz, A. A. Farag, R. Falk, and R. La Rocca, "Detection, visualization and identification of lung abnormalities in chest spiral CT scan: Phase-I," in *Proceedings of International Conference on Biomedical Engineering*, Volume *12*, No. *1*. Cairo, Egypt, 2002.

78. A. El-Baz, A. Farag, G. Gimel'farb, R. Falk, M. A. El-Ghar, and T. Eldiasty, "A framework for automatic segmentation of lung nodules from low dose chest CT scans," in *Proceedings of International Conference on Pattern Recognition (ICPR'06)*, Volume *3*. IEEE, 2006, pp. 611–614.

79. A. El-Baz, A. Farag, G. Gimel'farb, R. Falk, and M. A. El-Ghar, "A novel level set-based computer-aided detection system for automatic detection of lung nodules in low dose chest computed tomography scans," in A. El-Baz and J. Suri, eds. *Lung Imaging and Computer Aided Diagnosis*, Chapter 10. CRC Press, 2011, pp. 221–238, 2011.

80. A. El-Baz, G. Gimel'farb, M. Abou El-Ghar, and R. Falk, "Appearance-based diagnostic system for early assessment of malignant lung nodules," in *Proceedings of IEEE International Conference on Image Processing (ICIP'12)*. IEEE, 2012, pp. 533–536.

81. A. El-Baz, G. Gimel'farb, and R. Falk, "A novel 3D framework for automatic lung segmentation from low dose CT images," in A. El-Baz and J. S. Suri, eds. *Lung Imaging and Computer Aided Diagnosis*, Chapter 1. CRC Press, 2011, pp. 1–16.

82. A. El-Baz, G. Gimel'farb, R. Falk, and M. El-Ghar, "Appearance analysis for diagnosing malignant lung nodules," in A. El-Baz and J. S. Suri, eds. *Proceedings of IEEE International Symposium on Biomedical Imaging: From Nano to Macro (ISBI'10)*. IEEE, 2010, pp. 193–196.

83. A. El-Baz, G. Gimel'farb, R. Falk, and M. A. El-Ghar, "A novel level set-based CAD system for automatic detection of lung nodules in low dose chest CT scans," in A. El-Baz and J. Suri, eds. *Lung Imaging and Computer Aided Diagnosis*, Volume *1*, Chapter 10. CRC Press, 2011, pp. 221–238.

84. A. El-Baz, G. Gimel'farb, R. Falk, and M. A. El-Ghar, "A new approach for automatic analysis of 3D low dose CT images for accurate monitoring the detected lung nodules," in *Proceedings of International Conference on Pattern Recognition (ICPR'08)*. IEEE, 2008, pp. 1–4.

85. A. El-Baz, G. Gimel'farb, R. Falk, and M. A. El-Ghar, "A novel approach for automatic follow-up of detected lung nodules," in *Proceedings of IEEE International Conference on Image Processing (ICIP'07)*, Volume *5*. IEEE, 2007, pp. V-501.

86. A. El-Baz, G. Gimel'farb, R. Falk, and M. A. El-Ghar, "A new CAD system for early diagnosis of detected lung nodules," in *2007 IEEE International Conference on Image Processing, 2007 (ICIP 2007)*, Volume *2*. IEEE, 2007, pp. II-461.

87. A. El-Baz, G. L. Gimel'farb, R. Falk, M. Abou El-Ghar, and H. Refaie, "Promising results for early diagnosis of lung cancer," in *Proceedings of IEEE International Symposium on Biomedical Imaging: From Nano to Macro (ISBI'08)*. IEEE, 2008, pp. 1151–1154.

88. A. El-Baz, G. L. Gimel'farb, R. Falk, M. Abou El-Ghar, T. Holland, and T. Shaffer, "A new stochastic framework for accurate lung segmentation," in *Proceedings of Medical Image Computing and Computer-Assisted Intervention (MICCAI'08)*, 2008, pp. 322–330.

89. A. El-Baz, G. L. Gimel'farb, R. Falk, D. Heredis, and M. Abou El-Ghar, "A novel approach for accurate estimation of the growth rate of the detected lung nodules," in *Proceedings of International Workshop on Pulmonary Image Analysis*, 2008, pp. 33–42.

90. A. El-Baz, G. L. Gimel'farb, R. Falk, T. Holland, and T. Shaffer, "A framework for unsupervised segmentation of lung tissues from low dose computed tomography images," in *Proceedings of British Machine Vision (BMVC'08)*, 2008, pp. 1–10.

91. A. El-Baz, G. Gimelfarb, R. Falk, and M. A. El-Ghar, "3D MGRF-based appearance modeling for robust segmentation of pulmonary nodules in 3D LDCT chest images," in A. El-Baz and J. Suri, eds. *Lung Imaging and Computer Aided Diagnosis*, Chapter 3, CRC Press, 2011, pp. 51–63.

92. A. El-Baz, G. Gimelfarb, R. Falk, and M. A. El-Ghar, "Automatic analysis of 3D low dose CT images for early diagnosis of lung cancer," *Pattern Recognition*, *42*(6), pp. 1041–1051, 2009.

93. A. El-Baz, G. L. Gimel'farb, R. Falk, M. Abou El-Ghar, S. Rainey, D. Heredia, and T. Shaffer, "Toward early diagnosis of lung cancer," in *Proceedings of Medical Image Computing and Computer-Assisted Intervention (MICCAI'09)*. Springer, 2009, pp. 682–689.

94. A. El-Baz, G. Gimelfarb, R. Falk, M. A. El-Ghar, and J. Suri, "Appearance analysis for the early assessment of detected lung nodules," in A. El-Baz and J. Suri, eds. *Lung Imaging and Computer Aided Diagnosis*, Chapter 17, CRC Press, 2011, pp. 395–404.

95. A. El-Baz, F. Khalifa, A. Elnakib, M. Nitkzen, A. Soliman, P. McClure, G. Gimel'farb, and M. A. El-Ghar, "A novel approach for global lung registration using 3D Markov Gibbs appearance model," in *Proceedings of International Conference Medical Image Computing and Computer-Assisted Intervention (MICCAI'12)*. Nice, France, October 1–5, 2012, pp. 114–121.

96. A. El-Baz, M. Nitzken, A. Elnakib, F. Khalifa, G. Gimel'farb, R. Falk, and M. A. El-Ghar, "3D shape analysis for early diagnosis of malignant lung nodules," in *Proceedings of International Conference Medical Image Computing and Computer-Assisted Intervention (MICCAI'11)*. Toronto, Canada, September 18–22, 2011, pp. 175–182.

97. A. El-baz, M. Nitzken, G. Gimelfarb, E. Van Bogaert, R. Falk, M. A. El-Ghar, and J. Suri, "Three-dimensional shape analysis using spherical harmonics for early assessment of detected lung nodules," in A. El-Baz and J. Suri, eds. *Lung Imaging and Computer Aided Diagnosis*, Chapter 19, CRC Press, 2011, pp. 421–438.

98. A. El-baz, M. Nitzken, F. Khalifa, A. Elnakib, G. Gimel'farb, R. Falk, and M. A. El-Ghar, "3D shape analysis for early diagnosis of malignant lung nodules," in *Proceedings of International Conference on Information Processing in Medical Imaging (IPMI'11)*. Monastery Irsee, Germany (Bavaria), July 3–8, 2011, pp. 772–783.

99. A. El-baz, M. Nitzken, E. Vanbogaert, G. Gimel'Farb, R. Falk, and M. Abo El-Ghar, "A novel shape-based diagnostic approach for early diagnosis of lung nodules," in *2011 IEEE International Symposium on Biomedical Imaging: From Nano to Macro*. IEEE, 2011, pp. 137–140.

100. A. El-baz, P. Sethu, G. Gimel'farb, F. Khalifa, A. Elnakib, R. Falk, and M. A. El-Ghar, "Elastic phantoms generated by microfluidics technology: Validation of an imaged-based approach for accurate measurement of the growth rate of lung nodules," *Biotechnology Journal*, 6(2), pp. 195–203, 2011.

101. A. El-baz, P. Sethu, G. Gimel'farb, F. Khalifa, A. Elnakib, R. Falk, and M. A. El-Ghar, 'A new validation approach for the growth rate measurement using elastic phantoms generated by state-of-the-art microfluidics technology," in *Proceedings of IEEE International Conference on Image Processing (ICIP'10)*. Hong Kong, September 26–29, 2010, pp. 4381–4383.

102. A. El-baz, P. Sethu, G. Gimel'farb, F. Khalifa, A. Elnakib, R. Falk, and M. A. E.-G. J. Suri, "Validation of a new imaged-based approach for the accurate estimating of the growth rate of detected lung nodules using real CT images and elastic phantoms generated by state-of-the-art microfluidics technology," in A. El-Baz and J. S. Suri, eds. *Handbook of Lung Imaging and Computer Aided Diagnosis*, Volume 1, Chapter 18. Taylor & Francis, New York, 2011, pp. 405–420.

103. A. El-baz, A. Soliman, P. McClure, G. Gimel'farb, M. A. El-Ghar, and R. Falk, "Early assessment of malignant lung nodules based on the spatial analysis of detected lung nodules," in *Proceedings of IEEE International Symposium on Biomedical Imaging: From Nano to Macro (ISBI'12)*. IEEE, 2012, pp. 1463–1466.

104. A. El-baz, S. E. Yuksel, S. Elshazly, and A. A. Farag, "Non-rigid registration techniques for automatic follow-up of lung nodules," in *Proceedings of Computer Assisted Radiology and Surgery (CARS'05)*, Volume 1281. Elsevier, 2005, pp. 1115–1120.

105. A. S. El-Baz and J. S. Suri, *Lung Imaging and Computer Aided Diagnosis*. CRC Press, 2011.

106. A. Soliman, F. Khalifa, N. Dunlap, B. Wang, M. El-Ghar, and A. El-Baz, 'An iso-surfaces based local deformation handling framework of lung tissues," in *2016 IEEE 13th International Symposium on Biomedical Imaging (ISBI)*. IEEE, 2016, pp. 1253–1259.

107. A. Soliman, F. Khalifa, A. Shaffie, N. Dunlap, B. Wang, A. Elmaghraby, and A. El-Baz, "Detection of lung injury using 4D-CT chest images," in *2016 IEEE 13th International Symposium on Biomedical Imaging (ISBI)*. IEEE, 2016, pp. 1274–1277.

108. A. Soliman, F. Khalifa, A. Shaffie, N. Dunlap, B. Wang, A. Elmaghraby, G. Gimel'farb, M. Ghazal, and A. El-Baz, "A comprehensive framework for early assessment of lung injury," in *2017 IEEE International Conference on Image Processing (ICIP)*. IEEE, 2017, pp. 3275–3279.

109. A. Shaffie, A. Soliman, M. Ghazal, F. Taher, N. Dunlap, B. Wang, A. Elmaghraby, G. Gimel'farb, and A. El-Baz, "A new framework for incorporating appearance and shape features of lung nodules for precise diagnosis of lung cancer," in *2017 IEEE International Conference on Image Processing (ICIP)*. IEEE, 2017, pp. 1372–1376.

110. A. Soliman, F. Khalifa, A. Shaffie, N. Liu, N. Dunlap, B. Wang, A. Elmaghraby, G. Gimel'farb, and A. El-Baz, "Image-based CAD system for accurate identification of lung injury," in *2016 IEEE International Conference on Image Processing (ICIP)*. IEEE, 2016, pp. 121–125.

111. B. A. Dombroski, M. Nitzken, A. Elnakib, F. Khalifa, A. El-Baz, and M. F. Casanova, "Cortical surface complexity in a population-based normative sample," *Translational Neuroscience*, 5(1), pp. 17–24, 2014.

112. A. El-baz, M. Casanova, G. Gimel'farb, M. Mott, and A. Switala, 'An MRI-based diagnostic framework for early diagnosis of dyslexia," *International Journal of Computer Assisted Radiology and Surgery*, 3(3–4), pp. 181–189, 2008.

113. A. El-Baz, M. Casanova, G. Gimel'farb, M. Mott, A. Switala, E. Vanbogaert, and R. McCracken, "A new CAD system for early diagnosis of dyslexic brains," in *Proceedings of International Conference on Image Processing (ICIP'2008)*. IEEE, 2008, pp. 1820–1823.

114. A. El-Baz, M. F. Casanova, G. Gimel'farb, M. Mott, and A. E. Switwala, "A new image analysis approach for automatic classification of autistic brains," in *Proceedings of IEEE International Symposium on Biomedical Imaging: From Nano to Macro (ISBI'2007)*. IEEE, 2007, pp. 352–355.

115. A. El-baz, A. Farag, G. Gimel'farb, M. A. El-Ghar, and T. Eldiasty, "Probabilistic modeling of blood vessels for segmenting MRA images," in *18th International Conference on Pattern Recognition (ICPR'06)*, Volume *3*. IEEE, 2006, pp. 917–920.

116. A. El-Baz, A. A. Farag, G. Gimelfarb, M. A. El-Ghar, and T. Eldiasty, "A new adaptive probabilistic model of blood vessels for segmenting MRA images," in *Medical Image Computing and Computer-Assisted Intervention-MICCAI 2006*, Volume *4191*. Springer, 2006, pp. 799–806.

117. A. El-baz, A. A. Farag, G. Gimelfarb, and S. G. Hushek, "Automatic cerebrovascular segmentation by accurate probabilistic modeling of TOF-MRA images," in *Medical Image Computing and Computer-Assisted Intervention-MICCAI 2005*. Springer, 2005, pp. 34–42.

118. A. El-Baz, A. Farag, A. Elnakib, M. F. Casanova, G. Gimel'farb, A. E. Switala, D. Jordan, and S. Rainey, "Accurate automated detection of autism related corpus callosum abnormalities," *Journal of Medical Systems*, 35(5), pp. 929–939, 2011.

119. A. El-Baz, A. Farag, and G. Gimelfarb, "Cerebrovascular segmentation by accurate probabilistic modeling of TOF-MRA images," in *Image Analysis*, Volume *3540*. Springer, 2005, pp. 1128–1137.

120. A. El-baz, G. Gimelfarb, R. Falk, M. A. El-Ghar, V. Kumar, and D. Heredia, "A novel 3D joint Markov-gibbs model for extracting blood vessels from PC-MRA images," in *Medical Image Computing and Computer-Assisted Intervention-MICCAI 2009*, Volume *5762*. Springer, 2009, pp. 943–950.

121. A. Elnakib, A. El-Baz, M. F. Casanova, G. Gimel'farb, and A. E. Switala, "Image-based detection of corpus callosum variability for more accurate discrimination between dyslexic and normal brains," in *Proceedings of IEEE International Symposium on Biomedical Imaging: From Nano to Macro (ISBI'2010)*. IEEE, 2010, pp. 109–112.

122. A. Elnakib, M. F. Casanova, G. Gimel'farb, A. E. Switala, and A. El-Baz, "Autism diagnostics by centerline-based shape analysis of the corpus callosum," in *Proceedings of IEEE International Symposium on Biomedical Imaging: From Nano to Macro (ISBI'2011)*. IEEE, 2011, pp. 1843–1846.

123. A. Elnakib, M. Nitzken, M. Casanova, H. Park, G. Gimel'farb, and A. El-Baz, "Quantification of age-related brain cortex change using 3D shape analysis," in *2012 21s International Conference on Pattern Recognition (ICPR)*. IEEE, 2012, pp. 41–44.

124. M. Mostapha, A. Soliman, F. Khalifa, A. Elnakib, A. Alansary, M. Nitzken, M. F. Casanova, and A. El-Baz, "A statistical framework for the classification of infant dt images," in *2014 IEEE International Conference on Image Processing (ICIP)*. IEEE, 2014, pp. 2222–2226.

125. M. Nitzken, M. Casanova, G. Gimel'farb, A. Elnakib, F. Khalifa, A. Switala, and A. El-Baz, "3D shape analysis of the brain cortex with application to dyslexia," in *2011 18th IEEE International Conference on Image Processing (ICIP)*. Brussels, Belgium, IEEE, September 2011, pp. 2657–2660 (Selected for oral presentation. Oral acceptance rate is 10 percent and the overall acceptance rate is 35 percent).

126. F. E.-Z. A. El-Gamal, M. M. Elmogy, M. Ghazal, A. Atwan, G. N. Barnes, M. F. Casanova, R. Keynton, and A. S. El-Baz, "A novel CAD system for local and global early diagnosis of alzheimer's disease based on pib-pet scans," in *2017 IEEE International Conference on Image Processing (ICIP)*. IEEE, 2017, pp. 3270–3274.

127. M. Ismail, A. Soliman, M. Ghazal, A. E. Switala, G. Gimelfarb, G. N. Barnes, A. Khalil, and A. El-Baz, "A fast stochastic framework for automatic MR brain images segmentation," *PLoS one, 12*(11), e0187391, 2017.

128. M. M. Ismail, R. S. Keynton, M. M. Mostapha, A. H. ElTanboly, M. F. Casanova, G. L. Gimel'farb, and A. El-Baz, "Studying autism spectrum disorder with structural and diffusion magnetic resonance imaging: A survey," *Frontiers in Human Neuroscience, 10*, p. 211, 2016.

129. A. Alansary, M. Ismail, A. Soliman, F. Khalifa, M. Nitzken, A. Elnakib, M. Mostapha, A. Black, K. Stinebruner, M. F. Casanova, and J. Zurada, "Infant brain extraction in t1-weighted MR images using bet and refinement using lcdg and mgrf models," *IEEE journal of biomedical and health informatics, 20*(3), pp. 925–935, 2016.

130. M. Ismail, A. Soliman, A. ElTanboly, A. Switala, M. Mahmoud, F. Khalifa, G. Gimel'farb, M. F. Casanova, R. Keynton, and A. El-Baz, "Detection of white matter abnormalities in MR brain images for diagnosis of autism in children," in *2016 IEEE 13th International Symposium on Biomedical Imaging (ISBI).* IEEE, 2016, pp. 6–9.

131. M. Ismail, M. Mostapha, A. Soliman, M. Nitzken, F. Khalifa, A. Elnakib, G. Gimel'farb, M. Casanova, and A. El-Baz, "Segmentation of infant brain MR images based on adaptive shape prior and higher-order mgrf," in *2015 IEEE International Conference on Image Processing (ICIP).* IEEE, 2015, pp. 4327–4331.

132. E. H. Asl, M. Ghazal, A. Mahmoud, A. Aslantas, A. Shalaby, M. Casanova, G. Barnes, G. Gimelfarb, R. Keynton, and A. El-Baz, "Alzheimers disease diagnostics by a 3D deeply supervised adaptable convolutional network," *Frontiers in Bioscience (Landmark Edition), 23*, pp. 584–596, 2018.

133. A. Mahmoud, A. El-Barkouky, H. Farag, J. Graham, and A. Farag, "A non-invasive method for measuring blood flow rate in superficial veins from a single thermal image," in *Proceedings of the IEEE Conference on Computer Vision and Pattern Recognition Workshops.* 2013, pp. 354–359.

134. A. El-baz, A. Shalaby, F. Taher, M. El-Baz, M. Ghazal, M. A. El-Ghar, A. Takieldeen, and J. Suri, "Probabilistic modeling of blood vessels for segmenting magnetic resonance angiography images," *Medical Research Archives, 5*(3), 2017.

135. A. S. Chowdhury, A. K. Rudra, M. Sen, A. Elnakib, and A. El-Baz, "Cerebral white matter segmentation from MRI using probabilistic graph cuts and geometric shape priors." in *ICIP,* 2010, pp. 3649–3652.

136. Y. Gebru, G. Giridharan, M. Ghazal, A. Mahmoud, A. Shalaby, and A. El-Baz, "Detection of cerebrovascular changes using magnetic resonance angiography," in *Cardiovascular Imaging and Image Analysis,* A. El-Baz and J. S. Suri, eds. CRC Press, 2018, pp. 1–22.

137. A. Mahmoud, A. Shalaby, F. Taher, M. El-Baz, J. S. Suri, and A. El-Baz, "Vascular tree segmentation from different image modalities," in *Cardiovascular Imaging and Image Analysis,* A. El-Baz and J. S. Suri, eds. CRC Press, 2018, pp. 43–70.

138. F. Taher, A. Mahmoud, A. Shalaby, and A. El-Baz, "A review on the cerebrovascular segmentation methods," in *2018 IEEE International Symposium on Signal Processing and Information Technology (ISSPIT).* IEEE, 2018, pp. 359–364.

139. H. Kandil, A. Soliman, L. Fraiwan, A. Shalaby, A. Mahmoud, A. ElTanboly, A. Elmaghraby, G. Giridharan, and A. El-Baz, "A novel MRA framework based on integrated global and local analysis for accurate segmentation of the cerebral vascular system," in *2018 IEEE 15th International Symposium on Biomedical Imaging (ISBI 2018).* IEEE, 2018, pp. 1365–1368.

Index

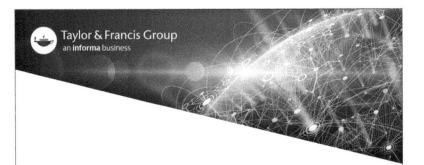